Element	Symbol	Atomic Number	Atomic Weight
Neon	Ne	10	20.18
Nickel	Ni	28	58.7
Nitrogen	N	7	14.01
Oxygen	O	8	16.00
Phosphorous	P	15	30.97
Plutonium	Pu	94	244
Potassium	K	19	39.10
Radium	Ra	88	226.0
Radon	Rn	86	222
Selenium	Se	34	79.0
Silicon	Si	14	28.09
Silver	Ag	47	107.9
Sodium	Na	11	22.99
Strontium	Sr	38	87.62
Sulfur	S	16	32.06
Tin	Sn	50	118.7
Titanium	Ti	22	47.9
Tungsten	W	74	183.8
Uranium	U	92	238.0
Vanadium	V	23	50.94
Xenon	Xe	54	131.3

Physical Constants:

Avogadro's number, N_A = 6.02 x 10^{23} (atoms, molecules, or ions)/mol

Faraday's constant = 96,500 coulomb/eq = 23,100 cal/volt-eq

Gas constant, 1.99 cal/K-mole = 0.0821 L-atm/K-mol = 8.31 J/K-mol

Gravity constant = 981 cm/s^2

Atmosphere = 760 torr

Speed of light = 3.00 x 10^8 m/sec

Environmental Engineering

PWS Series in Engineering

Anderson, *Thermodynamics*
Askeland, *The Science and Engineering of Materials, Third Edition*
Borse, *FORTRAN 77 and Numerical Methods for Engineers, Second Edition*
Bolluyt/Stewart/Oladipupo, *Modeling for Design Using SilverScreen*
Clements, *68000 Family Assembly Language*
Clements, *Microprocessor Systems Design, Second Edition*
Clements, *Principles of Computer Hardware, Second Edition*
Das, *Principles of Foundation Engineering, Third Edition*
Das, *Principles of Geotechnical Engineering, Third Edition*
Das, *Principles of Soil Dynamics*
Duff/Ross, *Freehand Sketching for Engineering Design*
El-Wakil/Askeland, *Materials Science and Engineering Lab Manual*
Fleischer, *Introduction to Engineering Economy*
Gere/Timoshenko, *Mechanics of Materials, Third Edition*
Glover/Sarma, *Power System Analysis and Design, Second Edition*
Janna, *Design of Fluid Thermal Systems*
Janna, *Introduction to Fluid Mechanics, Third Edition*
Kassimali, *Structural Analysis*
Keedy, *An Introduction to CAD Using CADKEY 5 and 6, Third Edition*
Keedy/Teske, *Engineering Design Using CADKEY 5 and 6*
Knight, *The Finite Element Method in Mechanical Design*
Knight, *A Finite Element Method Primer for Mechanical Design*
Logan, *A First Course in the Finite Element Method, Second Edition*
McDonald, *Continuum Mechanics*
McGill/King, *Engineering Mechanics: Statics, Third Edition*
McGill/King, *Engineering Mechanics: An Introduction to Dynamics, Third Edition*
McGill/King, *Engineering Mechanics: Statics and An Introduction to Dynamics, Third Edition*
Meissner, *Fortran 90*
Raines, *Software for Mechanics of Materials*
Ray, *Environmental Engineering*
Reed-Hil/Abbaschian, *Physical Metallurgy Principles, Third Edition*
Reynolds, *Unit Operations and Processes in Environmental Engineering*
Russ, *CD-ROM for Materials Science*
Schmidt/Wong, *Fundamentals of Surveying, Third Edition*
Segui, *Fundamentals of Structural Steel Design*
Segui, *LRFD Steel Design*
Shen/Kong, *Applied Electromagnetism, Second Edition*
Sule, *Manufacturing Facilities, Second Edition*
Vardeman, *Statistics for Engineering Problem Solving*
Weinman, *VAX FORTRAN, Second Edition*
Weinman, *FORTRAN for Scientists and Engineers*
Wempner, *Mechanics of Solids*
Wolff, *Spreadsheet Applications in Geotechnical Engineering*
Zirkel/Berlinger, *Understanding FORTRAN 77 and 90*

Environmental Engineering

Bill T. Ray

Southern Illinois University, Carbondale

PWS Publishing Company

I(T)P   An International Thomson Publishing Company

New York • London • Bonn • Boston • Detroit • Madrid • Melbourne • Mexico City • Paris
Singapore • Tokyo • Toronto • Washington • Albany NY • Belmont CA • Cincinnati OH

PWS PUBLISHING COMPANY
20 Park Plaza, Boston, MA 02116-4324

 This book is printed on recycled, acid-free paper

I ⓣ Pᵀᴹ   International Thomson Publishing
         The trademark ITP is used under license

For more information contact:
PWS Publishing Co.
20 Park Plaza
Boston, MA 02116

International Thomson Publishing Europe
Berkshire House 168-173
High Holborn
London WC1V7AA
England

Thomas Nelson Australia
102 Dodds Street
South Melbourne, 3205
Victoria, Australia

Nelson Canada
1120 Birchmount Road
Scarborough, Ontario
Canada M1K5G4

International Thomson Editores
Campos Eliseos 385, Piso 7
Col. Polanco
11560 Mexico D.F., Mexico

International Thomson Publishing GmbH
Königswinterer Strasse 418
53227 Bonn, Germany

International Thomson Publishing Asia
221 Henderson Road
#05-10 Henderson Building
Singapore 0315

International Thomson Publishing Japan
Hirakawacho Kyowa Building, 31
2-2-1 Hirakawacho
Chiyoda-ku, Tokyo 102
Japan

Library of Congress Cataloging-in-Publication Data
Ray, Bill T.
    Environmental engineering / Bill T. Ray.
        p.   cm. -- (PWS series in engineering)
    Includes bibliographical references and index.
    ISBN 0-534-20652-2
    1. Environmental engineering.   2. Environmental sciences.
I. Title.   II. Series.
TD146.R38   1995   94-26681
628--dc20   CIP

Printed and bound in the United States of America
95 96 97 98 99—10 9 8 7 6 5 4 3 2 1

Sponsoring Editor: Jonathan Plant
Editorial Assistant: Cynthia Harris
Developmental Editor: Mary Thomas
Production Editor: Kirby Lozyniak
Marketing Manager: Nathan Wilbur
Manufacturing Coordinator: Marcia Locke
Interior Designer: Carol H. Rose

Cover Designer: Kathleen A. Wilson
Interior Illustrator: ST Associates
Compositor: Doyle Graphics, Ltd.
Cover Photo: © by Peter Turner, from the Image Bank.
Cover Printer: John Pow Company
Text Printer: Quebecor Printing, Martinsburg

Contents

Chapter **5**

Organic Chemistry   118

Chapter **6**   **Microbiology and Microbial Growth**

Chapter *7*

Analysis of Treatment Processes   184

Chapter **8** **Water Quality** 221

Chapter **9**   # Water Treatment   262

Chapter 11 **Solid Waste Disposal** 347

Chapter **12** **Hazardous Waste Treatment and Disposal** 388

Chapter **13 Air Pollution and Control**   427

Chapter **14 Summing Up, The Global Picture**   460

Preface

Students in all engineering disciplines can benefit from a course emphasizing the natural environment, the impact of human activity on the environment, the technology intended to reduce that impact, and the changes in human activity that can also reduce the adverse impact on the environment. This book is intended to accompany such an introductory sophomore or junior level course in environmental engineering.

In the not so distant past, engineering students interested in the environment studied "sanitary engineering." Their course of study most likely included only water and wastewater treatment. In contrast, today's coursework usually includes air pollution control, solid and hazardous waste management, hazardous waste remediation (cleaning up past mistakes), and modeling of natural systems in addition to the traditional water and wastewater treatment. This text is intended to present an overview of these areas of study as well as the science and engineering fundamentals necessary to grasp the environmental topics.

The text is divided into two parts. The first part deals with the basic science and engineering principles required to understand both natural and engineered systems. Because many engineering students have a poor background in chemistry, two chapters are dedicated to improving these skills. Chapter 4 reviews general and physical chemistry principles and Chapter 5 provides a brief overview of organic chemistry. Students may wish to refer to this material as a refresher. Both chapters are intended to present chemistry in a manner that relates it to environmental engineering.

The second part of the text presents an engineering approach to understanding the natural environment and specific treatment and mitigation methods. The text is written in a simple style and uses many figures, example problems, and photos to clarify the explanatory text. Case studies document the use of particular treatment techniques to solve real-life problems.

The units used in this textbook include both the SI system and the conventional English system. Both systems of measurement are in common use in the field today. Although there is a trend toward the metric system, it will be many years before the existing English system is replaced completely. Practicing engineers must be familiar with both systems of measurement. Students should be exposed to both.

The student will need a background in general chemistry and mathematics through at least differential calculus. Physics, biology, geology, fluid mechanics, soil mechanics, or later mathematics courses would be supportive, but are not required.

ACKNOWLEDGEMENTS

There are many people that have assisted me in the preparation of this text. I would like to thank Dr. Bruce A. DeVantier, of Southern Illinois University at Carbondale, for providing suggestions and using the manuscript numerous times in his introductory environmental engineering course, CE314. Special thanks goes to the CE314 students, whose feedback was of great value in accuracy checking and identifying those sections of the first draft that needed clarification.

I would also like to express my appreciation to U.S. EPA Region 8, Denver, Colorado, and William R. Thomas, Department of the Army, Commerce City, Colorado, for their assistance in providing information about groundwater contamination at the Rocky Mountain Arsenal, Denver, Colorado. Thanks is also due to Lisa Zinner of the Washington State Department of Ecology, Nancy Kelley Beaton of Camp Dresser & McKee, Inc., and Doug Hermann of Gallatin National, for their assistance with the chapter on solid waste. Thanks also to Jim Roth of Crawford, Murphy, and Tilly for information used in the chapter on water treatment and to Bradley Paul for providing references and data on mine refuse. I am also grateful to Susan Fahnestock of Sverdrup Corporation for providing information on the Bissell Point Wastewater Treatment Plant in Chapter 10.

Thanks also to Luther Haliday, Administrative Assistant, City of Carbondale, for assistance with solid waste collection information and photographs of solid waste collection equipment; Richard Erickson of Gundle Lining Systems for supplying photographs of solid waste liner systems; Patricia Eggert, Kim Clark, and Lori Fenimore of DuPont Chemical Corp., John Fralich of the Leach Co., Steven A. Jaasund, Geoenergy International Corp., Gene Winkler of Amerex, Inc., and Doris Crawford at Envirex Inc. for supplying numerous photographs.

A special thanks is due to Jonathan Plant, Engineering Editor, and Mary Thomas, Assistant Engineering Editor, for their assistance and support throughout this project. I am especially indebted to Kirby Lozyniak,

Production Editor, for her support and assistance in the final stages of the project.

I would like to thank Braja M. Das for encouraging me to write the book and for his assistance in arranging my first meeting with PWS Publishing. I would also like to thank the following reviewers for their assistance in the development of this text:

Dr. Paul R. Anderson
Illinois Institute of Technology

Dr. Robert P. Carnahan
University of South Florida

Dr. Brian Dempsey
Pennsylvania State University

Dr. Robert H. Easton
Rochester Institute of Technology

Dr. D. M. Griffin, Jr.
Louisiana Technical University

Dr. John W. Klock
Arizona State University

Dr. Ross E. McKinney
University of Kansas

Dr. James R. Mihelcic
Michigan Technological University

Dr. Larry Moore
Memphis State University

Dr. Robert M. Sykes
Ohio State University

Last, but certainly not least, I am indebted to my wife, Rita Gautney, for reading, editing, and suggesting changes in various chapters, especially during the early stages of this project.

Bill T. Ray

To my uncle, Donald L. Marshall.
He provided the inspiration and encouragement that led to my becoming
an engineer, for which I will be forever grateful.

1  Introduction

"ALONG THE ROADS, LAUREL, VIBURNUM AND ALDER, GREAT FERNS AND WILDFLOWERS DELIGHTED THE TRAVELER'S EYE THROUGH MUCH OF THE YEAR. EVEN IN WINTER THE ROADSIDES WERE PLACES OF BEAUTY, WHERE COUNTLESS BIRDS CAME TO FEED ON THE BERRIES AND ON THE SEED HEADS OF THE DRIED WEEDS RISING ABOVE THE SNOW.... THEN A STRANGE BLIGHT CREPT OVER THE AREA AND EVERYTHING BEGAN TO CHANGE. SOME EVIL SPELL HAD SETTLED ON THE COMMUNITY."

RACHEL CARSON
Silent Spring
Houghton Mifflin Company, 1962

Hiking trail in the Great Smokey Mountain National Forest.

INTRODUCTION

Environmental engineering is the application of science and engineering principles to minimize the adverse effects of human activity on the environment. Although it is impossible to eliminate negative impacts, human effects can be diminished and brought under control through public education, conservation, regulation, and the application of good engineering practice.

Current environmental problems have two major causes. One is the increasing number of humans on earth—the world population has swelled from less than 100 million people only 2500 years ago to almost 5.5 billion today. One of the first applications of "environmental engineering" was the removal of sewage from cities, a practice that became increasingly impor-tant as populations grew. At first there was no treatment; wastes were simply conveyed to the nearest stream or other water body. At the same time, some municipalities began to treat drinking water before distributing it. Because sewage disposal eventually caused obvious damage to natural waters, civil engineers and others began in the early 1900s to develop methods of treating wastewater prior to discharge. This has evolved into a large industry.

The second major factor affecting the environment is the rising standard of living, not just in this nation, but in many places around the world. A higher living standard generates greater consumption of natural resources and more pollution.

The lofty standard of living enjoyed by people of North America and Western Europe is due in part to development of the synthetic organic chemical industry over the past 50 years and to the successful exploitation of fossil fuels for energy production. These industries produced toxic and hazardous chemicals in great quantities long before they were known to be harmful. Unlike sewage, even small amounts of synthetic chemicals can be harmful, but the technology to detect these chemicals at low levels did not exist when the new industries appeared. Therefore, it was impossible to identify them as the factors in human health or environmental problems. There were no attempts to control chemical production, use, or disposal.

Then, in the late 1960s and early 1970s, synthetic organic chemicals came under increasing scrutiny. Rachael Carson's now famous *Silent Spring* connected the use of chemical pesticides to a dramatic decrease in wildlife, particularly birds [1]. As awareness of the health and environmental threats inherent in the use of chemical increased, environmental regulations were passed, and environmental engineering began to come of age.

Technological advances have spurred continual growth of environmental engineering. For example, the development of analytical chemistry techniques have made it possible to determine the presence of specific organic compounds in a matter of minutes. And while they were measured in parts per million 25 years ago, detection levels today are in the parts per billion to parts per trillion range. As advanced testing identifies new problems, engineering solutions are mandated.

INFORMATION SOURCES

Environmental engineers are constantly in need of up-to-date information on topics ranging from current federal or state regulatory requirements (which are constantly changing) to innovative treatment methods. This section introduces you to commonly available sources of information.

Current information about environmental issues can be found in a variety of publications ranging from technical journals published by professional organizations such as the American Society of Civil Engineers, *(Journal of the Environmental Engineering Division)*, the *Journal of the Air Pollution Control Association*, and *Environmental Science and Technology* published by the American Chemical Society to reference books to federal and state publications. Many of these are contained in your college or university library. Bibliographies, various types of indexes, and article summaries, or abstracts, are available to help identify sources of information on specific subjects.

The federal government publishes many different documents dealing with environmental pollution and its control. Most of these come from the U.S. Environmental Protection Agency, although the Department of Energy, the Department of Defense, and other federal agencies also publish documents dealing with the environment. State governments and universities publish environmental documents as well.

Use of Technical Journals

Technical journals contain a wealth of information on almost any environmental topic imaginable. Several commonly available technical journals are listed in Table 1.1.

Many articles published in technical journals are referenced in abstracts such as the *Science Citation Index* (available in both printed form and as a CD-ROM for microcomputers), *Chemical Abstracts*, *WILSONDISC*, *Applied Science and Technology Index* (a CD-ROM data base from H. W. Wilson Company) and *Pollution Abstracts*. These publications list articles by topic and author, and they often contain a brief summary of the article's contents. Following is a demonstration of their use.

EXAMPLE 1.1	**LOCATING AND USING INFORMATION FROM A TECHNICAL JOURNAL**

Use your library's technical journal index to find information about toxic waste incineration.

SOLUTION: You could use one of your library's indexes to locate articles about toxic waste incineration. As an example, several articles are listed in the computer data base, *WILSONDISC*, which stores information on a CD-ROM. *WILSONDISC* has several databases; you should use the *Applied Science and Technology Index*. Using the *WILSONDISC* multiple subject search (Wilsearch) for "toxic waste; incineration," you will find several entries. This data base is updated frequently, so if you repeat this procedure later, you may find a slightly different set of references.

An alternate procedure is to use *Chemical Abstracts*, which will yield similar articles. *Chemical Abstracts* is available in both a printed and a computer version.

TABLE 1.1 Selected Technical Journals Related to Environmental Issues

FIELD	JOURNAL	PUBLISHER
Environmental quality	*Environmental Science and Technology*	American Chemical Society
	Water Resources Research	American Geophysical Union
	Water, Air and Soil Pollution	Kluwer Academic Publications
Water treatment	*Journal of the American Water Works Association*	American Water Works Association
	Journal of the Environmental Engineering Division	American Society of Civil Engineers
Wastewater treatment	*Water Environment Research*	Water Environment Federation
	Journal of the Environmental Engineering Division	American Society of Civil Engineers
Solid waste	*BioCycle*	J. G. Press, Inc.
Hazardous waste	*Hazardous Waste and Hazardous Materials*	Mary Ann Liebert, Inc.
	Ground Water	Ground Water Publications, Inc.
Air pollution and control	*Journal of the Air and Waste Management Association*	Air and Waste Management Association
General	*Chemical and Engineering News*	American Chemical Society
	Civil Engineering	American Society of Civil Engineers

Use of Environmental Engineering Reference Books

Technical organizations, consulting engineering firms, practicing professionals, and university professors all publish reference books. You can find references for a given topic by using the library's computer data base or card catalog. An abbreviated list of environmental reference books appears in Table 1.2.

TABLE 1.2   Selected Environmental Reference Books

FIELD	REFERENCE/AUTHORS	PUBLISHER
Water treatment	*Water Treatment Plant Design,* 2nd ed., American Water Works Association and American Society of Civil Engineers, 1990	McGraw-Hill, Inc.
	Water Quality and Treatment, 4th ed., American Water Works Association, 1990	McGraw-Hill, Inc.
	Water Treatment Principles and Design, James M. Montgomery Consulting Engineers, Inc., 1985	John Wiley & Sons
Wastewater treatment	*Design of Municipal Wastewater Treatment Plants,* Vols. I and II, Water Environment Federation, American Society of Civil Engineers, 1992	Water Environment Federation and American Society of Civil Engineers
	Wastewater Engineering, Treatment, Disposal and Reuse, Metcalf and Eddy, Inc., 1991	McGraw-Hill, Inc.
Solid waste disposal	*Integrated Solid Waste Management, Engineering Principles and Management Issues,* G. Tchobanogious, E. H. Thelsen, and S. Vigil, 1993	McGraw-Hill, Inc.
Hazardous waste disposal	*Handbook of Hazardous Waste Disposal,* Freeman	McGraw-Hill, Inc.
	Hazardous Waste Management, C. A. Wentz, 1989	McGraw-Hill, Inc.
Air pollution and control	*Air Pollution and Control; Traditional and Hazardous Pollutants,* H. E. Hesketh	Technomic Publishing Co., Inc.

EXAMPLE 1.2	LOCATING AND USING ENVIRONMENTAL ENGINEERING REFERENCE BOOKS

Use the reference *Water Quality and Treatment,* 4th ed., to determine the health effects of chemicals found in drinking water.

SOLUTION: Look in the index for the topic "health effects, of pathogens." The index indicates pages 66 to 68. The reference describes the different effects — toxic, neurotoxic, carcinogenic, mutagenic, and teratogenic — of chemicals on living organisms.

Use of Government Documents

Each year the Environmental Protection Agency (EPA) publishes several hundred documents dealing with a wide range of environmental subjects, including hazardous waste, solid waste, water quality, wastewater treatment, air quality, air pollution control, and drinking water treatment. Topics range from setting up a community recycling program to the fate of dioxins in soils to effects of air pollution on forests. These publications are initially available to the public free of charge. After the initial printing has been distributed, the documents are available from the National Technical Information Service (NTIS) for a reasonable fee. Listed in Table 1.3 are several information sources with public access. The people at these locations are courteous, knowledgeable, and helpful.

EXAMPLE 1.3	OBTAINING AND USING THE FEDERAL REGISTER

Landfills represent a hazard to low-flying aircraft because scavenging birds often congregate there searching for food. There is a significant risk that an aircraft could fly into one or more birds during takeoff or landing, resulting in aircraft damage or possibly a crash. As a result, the EPA requires landfills to keep a minimum distance from aircraft runways or show that they can operate without danger to such aircraft. What is that distance for both turbojet and piston type aircraft?

TABLE 1.3   Selected Government Sources
for Environmental Publications

SOURCE	TELEPHONE NUMBER	ADDRESS
Center for Environmental Research Information (CERI)	(513) 569-7562	ORD Publications P.O. Box 19962 Cincinnati, OH 45219-0962
Superintendent of Documents	(202) 783-3238	Superintendent of Documents Government Printing Office Washington, DC 20402
RCRA Docket Information Center (RIC)	(800) 424-9346	RCRA Docket Information Center (RIC) Office of Solid Waste (OS-305) U.S. Environmental Protection Agency 401 M Street, S.W. Washington, DC 20460
National Technical Information Service (NTIS)	(703) 487-4650	National Technical Information Service U.S. Department of Commerce Springfield, VA 22161

SOLUTION: This information is contained in the *Federal Register*, a collection of documents that contains all federal regulations. The register is likely to be in the federal documents section of your university library, although in this setting it may be on microfiche or microfilm rather than available in paper copy. You can also contact the RCRA Docket Information Center (RIC) at 1-800-424-9346 or write to the address below for a copy of the appropriate sections:

> RCRA Docket Information Center (RIC)
> Office of Solid Waste (OS-305)
> U.S. Environmental Protection Agency
> 401 M Street, S.W.
> Washington, DC 20460

The particular issue of the *Federal Register* you will need is Vol. 56, No. 196, Wednesday, October 9, 1991, Rules and Regulations. This contains pages 50978 to 51119. After obtaining the document, you can find the regulations dealing with landfill location restrictions. It is in Subpart B—Location Restrictions, page 51018, §258.10. Here the *Federal Register* states:

(a) Owners or operators of new MSWLF* units, existing MSWLF units, and lateral expansions that are located within 10,000 feet (3048 meters) of any airport runway and used by turbojet aircraft or within 5,000 feet (1524 meters) of any airport runway end used by only piston-type aircraft must demonstrate that the units are designated and operated so that the MSWLF unit does not pose a bird hazard to aircraft.

Review Questions

1. Define (in your own words, not the textbook's):
   a. Pollution
   b. Environment
   c. Environmental engineering

2. List several adverse environmental effects that you cause.

3. Inventory your room or apartment and list chemicals you use.

4. Obtain a list of the most recent publications available from the Center for Environmental Research Information, CERI.

5. Obtain a copy of the EPA document *Recycling Works! State and Local Solutions to Solid Waste Management* by calling or writing to the RCRA Docket Information Center (RIC). Request the document designated EPA/530-SW-89-014. You may also obtain a copy from your school library or through interlibrary loan.

   a. Determine the average amount of solid waste each American throws away every year (Hint: page 3).
   b. Explain how Hamburg, New York, encourages people to separate their trash for recycling. (Hint: first column of page 16).

   If you order this document (as opposed to locating it in the library), it will probably take five to seven weeks to arrive.

6. Obtain a copy of your state's solid waste regulations. Some states have more or stricter regulations than the federal standards. Do your state's restrictions concerning landfills and airport runways differ from the federal standards noted in Example 1.3?

7. As directed by your instructor, obtain a copy of the most recent EPA regulations in one of the following areas: hazardous waste, solid waste, air quality, or water quality.

---

* MSWLF is a municipal solid waste landfill.

8. Use *Chemical Abstracts* to locate information about the National Acid Precipitation Assessment Program.

9. Use the *Science Citation Index* to locate an article by Perry McCarty on the transport of hazardous chemicals by microscopic clay (colloidal) particles.

10. Use the *Pollution Abstracts* to locate information on hazardous waste landfill procedures.

11. Use the *Applied Science and Technology Index* from *WILSONDISC* to find the ten references on toxic waste incineration mentioned in Example 1.1.

References

1. Carson, Rachael, *Silent Spring*, Houghton Mifflin Company, Boston, Massachusetts, 1982.

2 ⚖ Laws and Regulations

``. . . DEMOCRACY IS THE WORST FORM OF
GOVERNMENT EXCEPT ALL THOSE
OTHER FORMS THAT HAVE BEEN TRIED
FROM TIME TO TIME.''

WINSTON CHURCHILL
House of Commons
November 11, 1947

The United States Congress

INTRODUCTION

Environmental regulations have existed for centuries. Because of poor air quality near his palace in about A.D. 1300, King Edward II of England reportedly ordered any person burning coal to be hanged. However, no major environmental legislation in any country existed until the second half of the twentieth century. The first significant laws in this area were federal statutes passed in the United States in the 1970s dealing with air and surface water quality and hazardous waste.

Those laws are still in effect in much of their original form. However, they are still in a period of transition. What is legal or accepted practice today may be illegal in a few years. When today's average college student was born, the first significant environmental laws had been passed, but their requirements were just beginning to become effective. You have been witness to the most sweeping environmental regulations in history.

This chapter introduces the process that brings about changes in environmental law, and briefly describes major federal environmental laws. More detailed discussions of each area of legislation will appear where appropriate in later chapters.

DEVELOPMENT OF ENVIRONMENTAL REGULATIONS

The U.S. Congress writes environmental laws. For such legislation to be enacted, lawmakers must perceive that environmental regulation benefits society. Only after legislators see the public interest in and the public's desire for such laws will they be passed. In the United States a law can be passed by a simple majority of the Houe of Representatives and the Senate if the bill is signed by the president. However, if the president vetoes the bill, a two-thirds majority of both houses is required to override. Several important pieces of environmental legislation have been passed over presidential vetoes during the past twenty-five years.*

When Congress passes environmental legislation, it directs the appropriate federal agency to develop and publish regulations to implement it. Before 1970 the U.S. Public Health Service was the agency most concerned about environmental matters. In 1970 Congress created the U.S. Environmental Protection Agency. Since then the EPA has been responsible for enforcing applicable federal laws. In many cases the laws allow the states to adopt and enforce the federal laws.

There are laws, for example, to protect people from the toxic effects of copper and lead in drinking water. These require the EPA to determine

---

*The Clean Water Act of 1972 (PL 92-500) was passed over President Nixon's veto. The Resource Conservation and Recovery Act and the reauthorization of the Clean Water Act in 1987 were passed over President Reagan's vetoes.

acceptable levels of the contaminants and what must be done to bring excessively high levels into compliance. In this case the EPA set maximum contaminant levels for metals, and public utilities are now required to test their drinking water for actual amounts. If a public water supply has levels in excess of the limits, responsible officials must initiate a treatment plan to reduce the contamination. Individual states may adopt the federal regulations and obtain EPA permission to enforce them.

Why Are Environmental Laws Passed?

Hindsight tells us that the United States should have passed hazardous waste laws in about 1940. Regulations would then have been in place as the petrochemical industry developed. However, in the 1940s the world did not envision a petrochemical industry—or the hazardous wastes it was later to generate.

Hazardous wastes became more and more of an issue during the 1970s. Although many toxic compounds existed before the 1940s, during that decade synthetic organic chemistry was born, and with it came catalytic synthesis of gasolines from heavier crude oils. That infant enterprise bloomed into the massive petrochemical industry of today. Many synthetic organic chemicals are carcinogenic, but carcinogens often have latency periods of 10 to 20 years. Thus, it took decades for scientists, engineers, and physicians to recognize the link, which is tenuous even today, between particular chemicals in the environment and adverse health effects. One initial missing element was the ability to detect chemicals at extremely low levels. One cannot regulate something that cannot be detected. Where typical laboratory detection limits were in the mg/L range in the 1950s, they are in the μg/L or ng/L range today. In other words, detection limits are three to six orders of magnitude lower.

The point is, people in industry did not even imagine the hazards they were creating in the 1940s and 1950s. These dangers were not realized until the 1960s and 1970s. Even today we do not understand what effects extended exposure to low levels of many chemicals may cause. However, technical people have gradually become aware of environmental problems related to a wide array of synthetic chemicals, many of them herbicides and insecticides. After these problems were understood and made public, citizens and Congress had to be convinced of the seriousness of the risks, so laws could be passed to protect human health and the environment. Unfortunately, this is often a slow process.

The U.S. Environmental Protection Agency

Congress created the U.S. Environmental Protection Agency in December of 1970, giving it several missions: to establish standards to protect the environment consistent with U.S. goals; to conduct research on the adverse effects of pollution and methods and equipment for controlling it; to gather information on pollution and its effects; to use this information to strengthen environmental protection; to help others protect the environment through grants and technical assistance; and to assist the Council on Environmental quality in developing and recommending to the president new policies on protecting the environment.

The EPA is the primary agency responsible for protecting the environment, although several other agencies are also involved in particular areas. One exception is the control of nuclear wastes, where primary responsibility lies with the U.S. Department of Energy. The EPA's duties include enforcement of air quality standards, drinking water quality standards, stream discharge standards, solid and hazardous waste disposal standards, and the cleanup of abandoned hazardous waste sites. In many cases the agency encourages or allows individual states to take over the primary enforcement of these standards, but federal officials maintain overall responsibility. The EPA has divided the nation into ten regions, with an office and administrator for each. The regions are shown in Figure 2.1, the office addresses in Table 2.1.

State agencies enforce state environmental laws and regulations—and in many cases the federal regulations as well. In general, for a state to enforce the federal regulations, it must first adopt regulations equivalent to

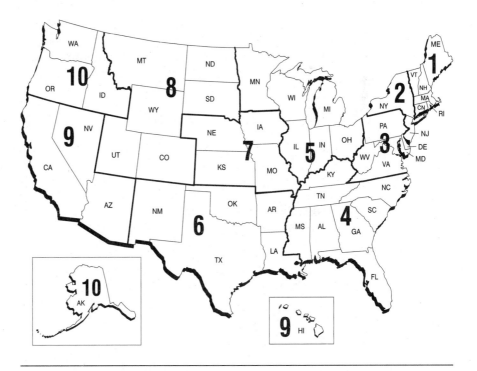

FIGURE 2.1 U.S. EPA Regions.

TABLE 2.1   U.S. EPA Regional Office Addresses

U.S. EPA Region 1 JFK Building Boston, MA 02203	U.S. EPA Region 2 26 Federal Plaza New York, NY 10278
U.S. EPA Region 3 841 Chestnut Street Philadelphia, PA 19107	U.S. EPA Region 4 345 Courtland Street, NE Atlanta, GA 30365
U.S. EPA Region 5 77 West Jackson Blvd. Chicago, IL 60604	U.S. EPA Region 6 First Interstate Bank Tower 1445 Ross Avenue, Suite 1200 Dallas, TX 75202
U.S. EPA Region 7 726 Minnesota Avenue Kansas City, KS 66101	U.S. EPA Region 8 999 18th Street, Suite 1500 Denver, CO 80202
U.S. EPA Region 9 75 Hawthorne Street San Francisco, CA 94105	U.S. EPA Region 10 1200 Sixth Avenue Seattle, WA 98101

or stricter than federal requirements. The names of state agencies vary. There is, for example, the New Jersey Department of Environmental Protection, the Arkansas State Department of Pollution Control and Ecology, and the Illinois Environmental Protection Agency.

Methods of Regulation

Speed laws in this country are regulated by posting the legal speed limit on each highway, and spot checking motorists to ensure compliance. There is no way to prevent motorists from speeding if they desire to do so. However, they run a risk of being apprehended by police officers using radar or other speed detection instruments. The penalty for speeding is usually a fine. Environmental laws often work in a similar manner. The city or industry (discharger) must obtain a permit that states the types of pollutants that may be discharged and the allowable amounts. The permit is equivalent to highway speed limit signs. The discharger must monitor the levels of the various pollutants, an action analogous to a driver watching a speedometer, and keep these below the permitted levels. One difference, however, is that the discharger, as a part of the permit, must not only check levels of various pollutants, but also submit reports on these levels every 3 months. This would be equivalent to the government requiring you, to turn in quarterly summaries of the maximum speeds at which you drive each day.

When dischargers violate their permit conditions, they must notify the

EPA and usually the corresponding state agency. The normal time for doing this is 5 business days. In addition, the discharger must describe why the violation occurred and what measures are being taken to prevent future violations.

A Short History of Environmental Legislation

Environmental law, as well as water pollution law, started with little fanfare in 1899. The event was passage of the Rivers and Harbors Act, the purpose of which was to prevent the discharge of materials that would impede river traffic into navigable waterways. There was a 49-year silence before the passage, in 1948, of the Federal Water Pollution Control Act, PL 80-845.* This act, which began limited funding to state and local governments to solve their water problems, also charged the U.S. Surgeon General with development of a comprehensive plan to protect interstate waterways. However, there were no goals, objectives, or standards for any measurement or enforcement of water quality [1].

Solid Waste Regulations.   The first significant step in the federal regulation of solid wastes was the Solid Waste Disposal Act of 1965, PL 89-272. This promoted better management of solid wastes and supported resource recovery. The next significant step was the Resource Recovery Act of 1970, PL 95-512. The Resource Recovery Act redirected the emphasis in solid waste management from disposal to recycling and energy recovery. This act also required that the U.S. Public Health Service investigate and report to Congress on the disposal of hazardous waste in the nation [2].

Hazardous Waste Regulations.   The Resource Conservation and Recovery Act of 1976 (RCRA) increased controls on conventional (nonhazardous) solid wastes and began the regulation of hazardous wastes. In accordance with this act and its subsequent amendments, the EPA has issued regulations for solid and hazardous waste disposal [3]. However, RCRA did not address the problem of what to do about the many hazardous wastes already disposed of improperly. So Congress passed the Comprehensive Environmental Response, Compensation and Liability Act of 1980 (CERCLA or Superfund). This places the responsibility for cleaning up abandoned or improper disposal sites on the companies that produced the wastes. Where several companies placed hazardous substances at a site, all are liable for the full extent of cleanup. Where companies no longer are operating, or are bankrupt and do not have sufficient funds, the government pays for the cleanup with money generated from taxing chemical and petrochemical companies and through general revenues such as those from federal income taxes.

---

*The PL is for Public Law. The first two numbers indicate the session of Congress (e.g. the 80th Congress). The last set of numbers indicates the sequential number of acts passed in that session. So this reference is to the 845th act passed by the 80th Congress.

Drinking Water Quality Regulations.   The first water quality standards in the United States were adopted by the U.S. Public Health Service in 1914. These specified a maximum level of bacterial contamination. There were no requirements for chemical quality [4]. Drinking (or potable) water regulations have gradually been made more stringent over the years. Additional changes occurred in the drinking water regulations in 1925, 1942, 1946, and 1962.

In 1974 Congress passed the Safe Drinking Water Act (SDWA), giving enforcement of drinking water quality to the U.S. Environmental Protection Agency. This act greatly expanded the coverage of the federal regulations. Acting at the direction of Congress in 1975, the EPA promulgated the National Interim Primary Drinking Water Standards, a more encompassing set of drinking water standards than were previously in effect. And these regulations have been modified as more information about hazardous substances has become available. In 1980 standards to prevent corrosion that could cause lead or other metals to be dissolved from distribution or consumer piping were established, and in 1991 other corrosion standards relating to lead were added. Thus, drinking water quality standards are an ever-changing document, being revised as additional information on health hazards is collected and as better control methods are developed.

Wastewater Treatment Regulations.   Congress passed the Water Pollution Control Act in 1948. Although the WPCA was the first federal legislation to address water pollution and water quality in the nation, it did little to correct rapidly arising problems. During the 1950s and 60s federal funding and involvement continued to increase, but Congress was careful not to infringe upon states' rights. Federal legislation was passed in 1956, 1961, and 1966. With each measure there was an increase in federal funding overall and in the percentage of money available for municipalities to use for the construction of wastewater treatment facilities. Yet, although treatment was encouraged in this way, there was no all-encompassing federal plan. Nor, essentially, was there any intervention to require pollution control. That "hands off" attitude ended with the Clean Water Act in 1972.

The Federal Water Pollution Control Act Amendments of 1972 (PL 92-500) established national water quality goals. PL 92-500 is often called the Clean Water Act (CWA). Its objective was to "restore and maintain the chemical, physical, and biological integrity of the Nation's waters." PL 92-500 was the first legislation requiring the EPA to establish criteria for water quality and discharge limits. The Clean Water Act also established a procedure for the permitting of all dischargers into navigable waters [1], and it provided several billion dollars per year for grants to municipalities for construction of wastewater treatment facilities. In 1977, however, the CWA was changed. The amendments, officially titled the Clean Water Act of 1977, reduced the amount of money available for construction grants.

In 1987 Congress passed the Water Quality Act (PL 100-4) over President Reagan's veto. PL 100-4 converted the construction grants to a revolving state loan program over a period of 4 years. [1]. In addition, it required permits for storm water discharge from city streets, parking lots, and industrial areas.

Air Quality Regulations. Since air pollution has two major sources — industry, including power plants, and highway vehicles — the approach to regulating and improving air quality is somewhat different from that applied to other types of pollution. The federal government has approached this by requiring limits on emissions from industries and power plants and by requiring auto manufacturers to produce vehicles with improved emissions. The first federal legislation in this area was the Air Pollution Control Act of 1955, PL 84-159. It did not really control air pollution, but it did provide for research on its effects and training of pollution control personnel. The law was a small first step. Other legislation followed in 1962 and 1963, and in 1965 the Motor Vehicle Air Pollution Control Act, PL 89-272, was passed. This began the auto emissions controls that remain in effect. Other laws were passed in 1967, 1970, and 1977, and controls on industries and power plants have gradually increased. The most recent major legislation is the Clean Air Act Amendments of 1990, which requires reductions in air pollutants in major cities not in compliance with air quality standards. The 1990 measure tightens emission requirements for automobiles and trucks and places additional controls on almost 200 toxic air pollutants [5].

REGULATORY METHODS

Water quality standards can be based on either the quality of the effluent being introduced into the environment or on the quality of the surrounding environment, or both. Each method has advantages. Standards maintaining a set environmental quality are probably best for the environment. But they are difficult to regulate. Where multiple dischargers exist, it is often difficult to prove which is responsible, and in some cases, several dischargers may be responsible for a single adverse condition. An extreme example is metropolitan smog. Do you blame the industries present? If so, which ones? Do you blame the automobiles emitting pollutants? The trucks and busses? In reality, all are partially responsible. So how do you reduce smog? Possible options are shown in Table 2.2.

Environmental Quality–Based Standards

Environmental quality–based standards* focus on the quality of the receiving water or local air. A discharger may release pollutants in any

---

*Environmental quality–based standards are often called **stream-based standards** when applied to water, although the concept applies to lake or ocean water as well. When applied to air quality, these requirements are often termed ambient air quality standards.

TABLE 2.2   Methods of Controlling Air Pollution in Cities

SOURCE	POLLUTANTS	METHODS OF CONTROL
Industries	Volatile organics	Require reduced emissions
	Volatile chlorofluorocarbons	Require reduced emissions
	Particulate inorganics	Require reduced emissions
Automobiles	Hydrocarbons	Improve discharge nozzles at filling stations and ventilation in gasoline tanks
	Products of incomplete combustion	Require better combustion efficiency of auto makers and emission testing and regular engine maintenance of drivers. Limit gasoline suppliers to oxygenated fuels
	Chlorofluorocarbons from air conditioners	Require the redesign of automobile air conditioners so vehicles made in the future can use other refrigerants.

quantity that does not cause the receiving water or local air quality to drop below established minimums. **Receiving water quality standards** have advantages. They maintain the water or air quality above a preset minimum, and dischargers can get rid of larger quantities of water pollutants during high-flow/low-temperature periods or more air pollutants during windy/low-temperature periods. However, there are significant disadvantages as well. One is the difficulty of enforcement, particularly where there are multiple dischargers within a given stream reach or local area. Also, water conditions are hard to monitor. A discharger's required effluent quality varies with the stream flow, wind currents, and temperature, so monitoring must be continuous. In many cases dischargers need to take measurements elsewhere because upstream contaminant levels affect the amounts of pollutants that can be discharged, on site. And to benefit from such standards, industries must have highly trained personnel and real-time monitoring equipment. Thus, it is difficult for most dischargers to maintain compliance.

Effluent-Based Standards

Effluent-based standards concentrate on the quality of the discharger — either water or air. With this type of requirement, a discharger has definite parameters to meet, and workers do not have to concern themselves with variations in stream flow, weather conditions, temperature or other receiving water or air conditions. Complying with effluent-based standards is also easier from the standpoint of consistency. There is a specific allowable level for each pollutant based on discharge concentration, discharge mass, or both. The discharger must keep contaminants below that limit.

For the same reasons it is also easier for regulatory agencies to monitor effluent-based standards. One apparent disadvantage of them, however, is that they do not allow flexibility in protecting the ambient environmental quality. But since streams can better assimilate wastes during cooler weather and high water flows, different effluent standards can be set for different seasons. Such an approach combines advantages of the two regulatory methods, and similar flexibility can be built into air pollution control as well.

ENVIRONMENTAL ETHICS

A corporation is in business to make a profit for its shareholders. Its primary purpose is not to protect the environment. However, businesses are required to comply with environmental regulations — a process that normally requires a significant investment in both capital expenditure and operating costs. When environmental regulations are applied fairly overall, other factors usually play a dominant role in determining the relative profitability of competing companies. However, where one facility operates in a location where environmental regulations are more stringent than those experienced by competitors, it may be required to operate at lower profit, or at a loss. Or it may have to raise its prices. Few companies are successful at selling equivalent products at prices higher than those of the competition. And companies cannot operate at a loss for extended periods. They must either close or move to areas where environmental regulations are less restrictive. Applied uniformly, federal standards are the fairest method of providing environmental protection while allowing businesses to compete.

There are exceptions to the free-market rule that companies cannot sell products for long at uncompetitive prices. Particularly in recent years, some consumers have been willing to spend more for "environmentally friendly" products. A number of such products — including everything from recycled paper made into bathroom tissue to automobiles to fast food to many small consumer goods packaged in recyclable paper — are doing well. So progress is being made.

Review Questions

1. Distinguish between effluent-based standards and water quality–based standards.

2. Distinguish between ambient air quality standards and air emission standards.

3. Using the government documents or science/engineering section of your library, write one- to two-page summaries of

   a. PL 92-500

   b. Superfund

   c. CERCLA

   d. Current wastewater treatment regulations

4. What session of Congress enacted PL 92-500? How many other acts were passed in that session before this legislation?

5. Determine what EPA region you are in. Where is the regional headquarters?

References

1. J. M. Kovalic, *The Clean Water Act of 1987*, Water Pollution Control Federation, Alexandria, Virginia, 1987.

2. U.S. Environmental Protection Agency, "Report to Congress: Disposal of Hazardous Wastes," Office of Solid Waste Management, SW-115, Washington, DC, 1973.

3. U.S. Environmental Protection Agency, "Solid Waste Disposal Facility Criteria; Final Rule," Federal Register, 40 CFR Parts 257 and 258, October 9, 1991.

4. *Water Quality and Treatment*, 3rd ed., The American Water Works Association, McGraw-Hill Book Co., 1971.

5. A. L. Alm, "The Clean Air Act," *Environmental Science and Technology*, vol. 25, no. 3 (March 1991), p. 383.

3 ☀ Ecology and the Environment

"THERE ARE THREE KEY FACTORS . . . [THE FIRST TWO ARE] THE LEVEL OF CONSUMPTION . . . [AND] THE TECHNOLOGY NEEDED TO SATISFY THAT CONSUMPTION, AND DISPOSE OF THE WASTE GENERATED. THESE TWO FACTORS DECIDE HOW MUCH ENVIRONMENTAL DAMAGE IS DONE PER PERSON. MULTIPLY BY THE THIRD FACTOR, POPULATION, AND YOU ARRIVE AT THE TOTAL LEVEL OF DAMAGE.''

PAUL HARRISON
New Scientist
May 19, 1990

A captive mountain lion in coastal South Carolina. Now restricted to sparsely populated areas of the American west, the mountain lion once roamed from coast to coast.

ECOLOGY

Ecology is the study of living organisms and their environments or habitats. When pollutants are discharged into a stream or the air, or placed in the soil, they affect the natural environment. Their effects may be subtle, as with low concentrations of toxic materials, or they may be drastic as in large oil spills or highly contaminated groundwater. The study of ecology is important because it helps us understand how pollution impacts our environment.

Ecosystems

An **ecosystem** is a basic study area for ecologists. It is an organism or a group of organisms and their surroundings. Ecosystems may be as small as an aquarium or as large as one's backyard or the earth itself. Those who study life systems set ecosystem boundaries arbitrarily to correspond with their spheres of interest.

Trophic Levels within an Ecosystem

Ecosystems have four basic components: the **abiotic environment**, producers, consumers, and decomposers. **Producers** utilize energy from the sun and nutrients from the abiotic environment (carbon dioxide from the atmosphere or water, other nutrients from soil or water) to produce protoplasm by means of photosynthesis. Major categories of producers are green plants and phytoplankton. These organisms obtain their carbon for synthesis from carbon dioxide rather than from organic carbon. Thus, they produce organic carbon or biomass—hence, the name producer. Another term for them is **autotrophic organisms** or autotrophs.

   Consumers feed on protoplasm produced from photosynthesis or on organisms from higher levels that indirectly consume protoplasm from photosynthesis. So consumers depend on producers for their energy and synthesis needs.

   Finally, **decomposers** utilize energy from wastes or dead organisms, completing the cycle by returning nutrients to the soil or water and carbon dioxide to the air or water.

Energy and Trophic Levels

As solar energy flows into the earth's biosystem, part of it is absorbed by the earth and a portion is reflected back into space. The absorbed energy is responsible for the production of biomass on earth. Thus, indirectly the sun is the energy source for all the needs of life forms on earth, including human beings.

   Primary production is generation of biomass through photosynthesis. Tropical rain forests, swamps and marshes, algal beds, and estuaries are the highest producers of biomass. Deserts and frozen areas are among the lowest. Table 3.1 [1] compares primary production in the earth's major ecosystems.

TABLE 3.1   Net Primary Production of Biomass

ECOSYSTEM	NET PRIMARY PRODUCTION g/m²/yr	AREA 10⁶ km²
Tropical rain forests	2000	17
Tropical seasonal forests	1500	7.5
Temperate evergreen forests	1300	5
Temperate deciduous forests	1200	7
Cultivated lands	644	14
Temperate grasslands	500	9
Tundra and alpine meadows	144	8
Desert shrubs	71	18
Lakes and streams	500	2.5
Swamps and marshes	2500	2
Algal beds and reefs	2000	0.6
Estuaries	1800	1.4
Total continental	720	149
Total marine	153	361
Total world	320	510

In the ocean (see Figure 3.1) phytoplankton is the primary producer. It is thus the first level in the food chain or the first **trophic level**. Phytoplankton converts inorganic carbon (soluble carbonates in the water) into protoplasm. In turn, phytoplankton is consumed by microscopic animals called zooplankton. These represent the second trophic level in the food chain. Crustaceans feeding on the zooplankton are at the third trophic level. Fish eating the crustaceans would be the fourth trophic level. And seals consuming the fish would be the fifth trophic level.

Similarly on land, the first trophic level, or the producers, are plants. Insects (see Figure 3.2) that eat grass are first-level consumers, so they represent the second trophic level. A rodent eating insects represents the

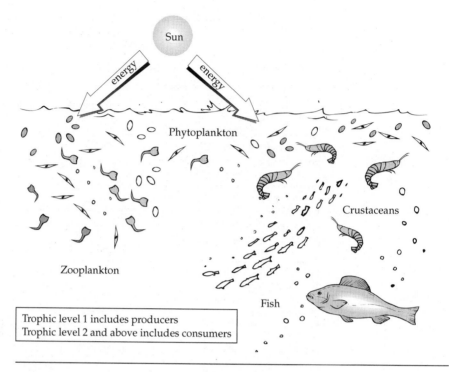

FIGURE 3.1 Trophic levels in an ocean ecosystem.

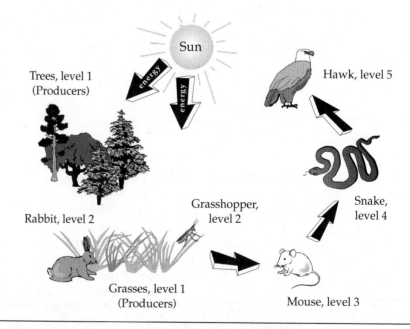

FIGURE 3.2 Trophic levels in a grassland ecosystem.

second level of consumers or the third trophic level. A snake consuming the rodent would represent the third level of consumers and the fourth trophic level. A hawk, at the top of the pyramid, would be living at the fourth consumer level and the fifth trophic level. If the hawk were to eat the rabbit rather than the snake, however, it would be a secondary consumer at the third trophic level.

The amount of biomass produced for a given amount of incoming solar energy is highest at the producer, or first trophic level. Much less biomass is produced at the first consumer level since energy is lost as the biomass is converted. At higher trophic levels even less biomass is produced for a given amount of solar energy input.

Figure 3.3 illustrates the concept of energy loss at each successive trophic level for that same grassland ecosystem. The energy available for synthesis and reproduction is shown for each trophic level. A large amount of energy is directly available from the sun for the producers in this ecosystem, so the most biomass can be produced at this level. At the second trophic level, less energy is available to grasshoppers because energy is lost to respiration, to decomposers, and to other competing primary consumers. In turn, less energy is available to mice because of losses to respiration and decomposition. Also, some grasshoppers are taken by other competing consumers. The same reasoning can be applied on up the food chain to the

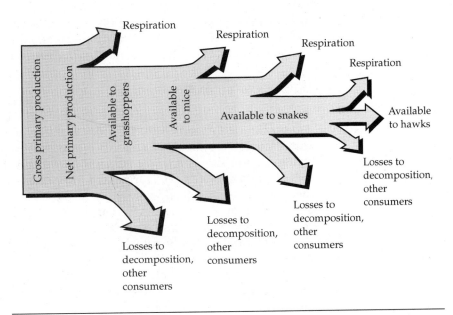

FIGURE 3.3 Biomass production per unit of energy for different trophic levels.

hawk. Very little of the sun's energy (per unit area) is transferred to the hawk. Thus, the hawk requires a large production area for the energy it needs to live. A similar graph could be constructed for biomass. Because of the large losses at each trophic level, a lot of grassland is required to supply enough biomass for the hawk.

Humans are generally primary and secondary consumers and thus operate at the second and third trophic levels. Most humans are omnivores—that is, they consume both plants and animals. Less energy and land area are required to support vegetarian humans. There is a significant energy loss in using grain to produce beef, pork, and poultry. The earth will support a larger number of primary consumers than secondary consumers. As our population continues to rise we will be forced to consume more vegetables and less meat.

Interspecies Relationships

Each species within an ecosystem is affected by other species in the system, and it is important to understand several major interspecies relationships. Predation is the feeding of one organism on another. In classical predator–prey relationships, the predator decreases the prey population until it is so low that the predator begins to die out due to starvation. At that point the prey begins to recover and increase in population. At some point the cycle begins to repeat. However, in nature there are few if any single predator–prey relationships. Most predators have more than one prey, and most prey are consumed by more than one predator. In addition, many other environmental factors— including competition from other organisms and the stresses brought on by drought, flooding, and loss of habitat—impinge on these relationships.

Bioconcentration

Bioconcentration is the increasing accumulation of a compound as it progresses up the food chain. DDT, for example, bioconcentrates or bioaccumulates in ecosystems in which it is used as an insecticide. It was first synthesized in 1874, but it was not until 1939 that its properties as an insect killer were discovered [2]. During and after World War II the use of DDT increased. It was applied extensively in the United States in a variety of situations until its ban by the EPA in 1973. But this poison is still used in many developing countries for mosquito eradication and control.

DDT is only slightly soluble in water, so it was not found in high concentrations in aquatic environments. However, DDT is readily adsorbed to organic matter at the bottom of lakes and rivers. It is also soluble in the fatty tissues of animals. Thus, the substance was found in the aquatic environment—not in the water, but in aquatic sediments and aquatic life. In 1967 Woodwell and others documented the bioconcentration of DDT in an aquatic food chain [3]. They found that DDT amounts would be multiplied more than 500,000 times from water to birds (see Table 3.2). Thus, what appears to be a relatively harmless concentration of DDT in the water, only 0.05 ppb, resulted in extremely high levels in predatory fish and birds.

TABLE 3.2   Bioconcentration of DDT in an Aquatic Environment

SOURCE	CONCENTRATION, ppm
Water	0.00005
Plankton	0.04
Hard clam	0.42
Sheephead minnow	0.94
Chain pickerel (predatory fish)	1.33
Needlefish (predatory fish)	2.07
Heron (feeds on small animals)	3.57
Tern (feeds on small animals)	3.91
Herring gull (scavenger)	6.00
Osprey egg	13.8
Merganser (fish eating duck)	22.8
Cormorant (feeds on larger fish)	26.4
Ring billed gull	75.5

SOURCE: G. M. Woodwell, C. F. Warster, Jr., and P. A. Isaacson, "DDT Residues in an East Coast Estuary: A Case of Biological Concentration of a Persistent Insecticide," *Science*, vol. 156 (May 1967), pp. 812–824.

Biodiversity

Biodiversity, the variety among living organisms and the ecological communities they inhabit, gives ecosystems stability. When consumers have many different producers available for food, a decrease in the population of a single producer does not have a severe effect on the consumer. However, when a consumer has only a few or a single producer as a food source, a decrease in the producer population will impact that particular consumer significantly. And similarly, a predator with many different food sources is not likely to be affected by a reduction in the population of a single prey.

Humans upset the balance of biodiversity in many ways. Reduction of habitat, hunting and fishing of certain species to extinction or near extinction, eradication of "pests,"* and the spread of pollution are a few examples. And there is evidence that we have caused major changes in many marine ecosystems as well [4]. Although the exact causes are often difficult to determine, human activity is probably responsible for the decline in and/or extinction of many aquatic species, including the bottlenose dolphin and striped dolphin, various seals, and coral reefs. It has been estimated that up to 80% of the PCBs contained in a female dolphin can be passed to the first born young through lactation. DDT averages 432 ppm in Mediterranean dolphins [4].

There is evidence that our continued destruction of habitat for animals, fish, and birds could eliminate 15% of the earth's species within the next 25 years [5]. Within the lower 48 states, 99% of all prairies, 90 to 95% of all forests, and 79% of all wetlands have been lost. On a global basis, 42% of all tropical rain forests have been destroyed, most during the average college student's lifetime. Each year an area of tropical forest larger than Washington State is being destroyed. And, over half the world's species inhabit that region alone.

LIMNOLOGY

Limnology is the study of freshwater ecosystems. Our discussion will be limited to freshwater lakes.

Lakes more than a few feet deep will stratify because the temperature in the upper region will approximate the average air temperature, and deeper waters will change temperature much more slowly. And since the density of water changes with temperature, there will be water density gradations in a lake—the most dense at the greatest depth. The water is most dense at 4°C. Figure 3.4 shows the density of water from 0°C to 30°C, Figure 3.5 shows lake stratification.

In summer the upper strata, or **epilimnion**, will have a temperature above 4°C. It will be less dense than the cool lower region, or **hypolimnion**. (The mid-region in lake water is called the **metalimnion**.) Figure 3.4 (page 000) shows lake stratification. As the temperature cools in the fall, the water temperature in the upper region of a lake begins to drop, increasing the water density. As the temperature continues to drop, the density in the upper layer will become greater than that of the lower region. At this point,

---

*A few of the "pests" we have killed to near extinction are the cougar, wolf, grizzly bear, black bear, and coyote.

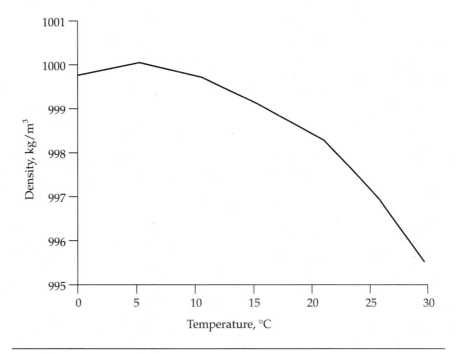

FIGURE 3.4 Density of water versus temperature.

the upper-region water will begin to displace the water in the lower region. When this occurs, the lake will "turn over." This same phenomenon occurs in the late spring as the upper region water warms to 4°C, becoming more dense when it reaches that temperature than the lower region water.

The period between turnovers is characterized by stratification. The epilimnion is separated from the hypolimnion by the metalimnion. Interlayer mixing does not occur. Thus, oxygen that dissolves in the epilimnion does not peetrate to the hypolimnion. This results in decreased oxygen

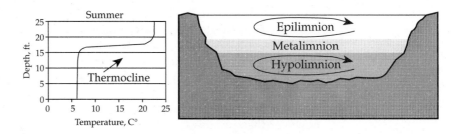

FIGURE 3.5 Lake stratification caused by differences in water temperature and density.

levels in the lower regions of a lake during summer. In some cases, the dissolved oxygen (DO) level will drop to near zero.

The oxygen level in the lower region of a lake is often so low that fish will not survive there. In such cases, fish remain in the epilimnion venturing into the hypolimnion only for brief periods. However, during the spring or fall lake turnovers the two regions mix. This mixing can result in very low DO levels over large portions of a lake. The diminished oxygen levels can result in large fish kills. Such occurrences are natural — they occurred prior to the large numbers of humans on earth. However, humans have greatly increased the occurrence by adding organic matter to lakes, resulting in still lower DO levels in the hypolimnion of many lakes today.

POPULATION

Population is important to environmental engineers for several reasons. An understanding of population history helps us understand the difficulties with future human population growth, and it gives us a better understanding of future energy and natural resource demands. Also, as a matter of practicality, population estimates (or predictions) are often required for completion of engineering projects. If we design a water treatment facility today, it should supply enough water not only when construction is complete in 2 to 5 years, but also have the capacity to serve the anticipated population for 20 or more years after it is built. To do this, we must make reasonable forecasts of population growth.

A History of Population Growth

The U.S. population, as well as the world's, has increased at an alarming rate during the past 200 years. The implications of this from an environmental standpoint are immense. About 100,000 years ago the human population of the *world* was likely less than that of any major U.S. city today. It has been estimated that the humanoid population was only 2 million in 100,000 B.C., increasing to possibly 4 to 10 million by 10,000 B.C. [6,7]. Where one family existed 10,000 years ago, a city of more than 100,000 people will exist today. In addition to the increase in population, there has been a related increase in both the amount of pollution produced per person and the amount of natural resources consumed per person.

Before A.D. 1500 the human population grew slowly. As humans became more adept at making tools and developed industrial and agricultural skills, however, their numbers began to soar. This is shown in Figure 3.6 [6,8]. At the same time, humans began to consume energy and resources at an ever-increasing rate. And per person pollution has also increased with increasing technology. This has degraded water, air, and land quality throughout the world.

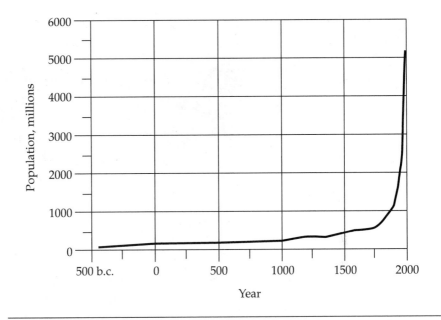

FIGURE 3.6   World population growth since 500 B.C.

Effects of Increased Pollution

The effects of our increased population are evident everywhere on earth. They range from pollution of the air, land, and water to diminished natural habitat.

Energy Consumption.   Energy and natural resource consumption make possible the higher standard of living enjoyed in the more developed countries. This standard of living has been achieved only in the past one or two centuries, and as a result of it, humans are now using energy at a phenomenal rate. Current estimates are that world energy reserves of petroleum will last for another 40 to 50 years, natural gas for possibly 60 years, and coal for another 225 to 250 years [9]. In the short span of a few centuries, humankind will have consumed all of the fossil-fuel energy stored on earth—a resource that took literally millions of years to produce. World and U.S. energy consumption is shown in Figure 3.7.

In 1930 per capita energy consumption in the United States was approximately 70 gigajoules per year. This steadily increased until around 1970. At that point, and until the present, consumption stabilized at around 350 gigajoules per person per year [9]. In 1930 the U.S. population was 123 million. In 1990 it was 249 million [10]. Thus, our nation's energy consumption has increased by a factor of 10 in the past 40 years.

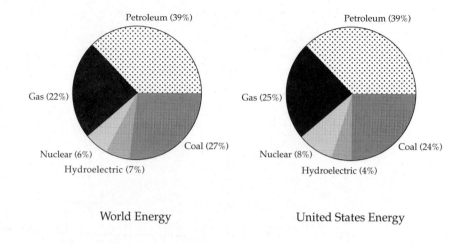

FIGURE 3.7 Energy sources in 1990.

There are many ways to reduce the amount of energy we consume. For example, some refrigerators now available consume 80 to 90% less electricity than conventional models. The average new refrigerator available in 1990 cost $64 per year less to operate than comparable models produced in 1970 [11]. And efficient fluorescent fixtures can reduce the lighting electricity demand by as much as 75 to 85%. Fluorescent lights cost more than incandescent lights, but they use significantly less power and last four to five times as long. They will often pay for themselves in 1 or 2 years of operation [11].

Estimating Population Growth

Population estimation is important for many reasons. A clear idea of the world population at some point in the future facilitates reasonable calculations of world resource consumption and pollution loads. Estimation of smaller population bases, such as those of a city or town, enables engineers to calculate the need for additional water supply or wastewater treatment capacity in the future. Most engineering systems, from water supply to wastewater treatment, are designed for a future capacity, not just the need at the time of design or construction completion. Thus, valid projections of population growth are important on a global scale and the regional or local level.

Many population growth factors cannot be reliably predicted. Yet, reasonable estimates can be made—with the caveat that predictions assume no catastrophic changes. A common method of forecasting global or other large-scale population changes is the exponential model. The form of the equation is:

$$P = P_0 e^{r\Delta t}$$

where

P = population at time t

P_0 = population at time zero

r = population growth rate, year

Δt = time in years.

EXAMPLE 3.1	WORLD POPULATION GROWTH

In 1950 the world population was estimated to be 2.5 billion. In 1990 it was estimated to be 5.5 billion. Assuming this growth rate will be sustained,

a. calculate the world population in A.D. 2000.
b. determine the year the global population will reach 10 billion.

SOLUTION: We must first calculate the population growth rate, r. To do this, we convert Equation 3.1 into log form:

$$\ln \left(\frac{P}{P_0} \right) = r \Delta t$$

Now if we know the population at two different times in the past, we can calculate the population growth rate, r. We know that the population in 1950 was 2.5 billion and that it was 5.5 billion in 1990. The time change, Δt, is thus 40 years. So the growth rate is

$$r = \frac{\ln \left(\dfrac{P}{P_0} \right)}{\Delta t} = \frac{\ln \left(\dfrac{5.5 \times 10^9}{2.5 \times 10^9} \right)}{40 \ \text{yr}}$$

$$= 0.020/\text{yr}$$

a. Estimate the population in the year 2000.

$$P = P_0 e^{r\Delta t} = 5.5 \times e^{0.020(2000 - 1990)}$$

$$= 6.7 \times 10^9 \ \text{in A.D. 2000.}$$

b. Calculate the year the population will reach 10 billion. Use 1990 as the base year.

$$\Delta t = \frac{\ln\left(\dfrac{P}{P_0}\right)}{r} = \frac{\ln\left(\dfrac{10 \times 10^9}{5.5 \times 10^9}\right)}{0.020/\text{yr}}$$

$$= 30 \text{ yr}$$

Since the base year is 1990, the year in which the population will reach 10 billion is 1990 + 30 or A.D. 2020.

HABITAT

Wildlife habitat has been disappearing since the world population began its rapid rise. In the United States, plains, wetlands, and hardwood and softwood forests have vanished. In Europe, many wetlands and forests have been destroyed. In recent years the tropics have begun to lose rain forests at an alarming rate. These losses are due primarily to human activities: consumption of wood, urban and agricultural expansion, and pollution.

Wetlands

Wetlands are much more than swamps to be filled in for development. They are teeming with life. They provide fish and wildlife habitat, they improve water quality, protect surrounding lands from floods and erosion, and they make wonderful recreational areas. What is a wetland? A **wetland** is a semi-aquatic area that is either inundated or saturated by water for varying periods during each year and that supports aquatic vegetation specifically adapted for saturated soil conditions.

Rain Forests

Treed areas with a closed canopy and more than 25 inches of rainfall per year are considered rain forests. These forests are estimated to hold 50% or more of all species, yet they now account for only 7% of the earth's surface [12, 13]. Rain forests are being destroyed at a rate equivalent to the area of the state of Washington each year. Within the life of the typical college student, only portions of three large rain forests may remain intact: the central Zaire forest, the Brazilian Amazon, and the Guyanas [12]. All other large, contiguous forests will probably be destroyed unless there is unexpected intervention and significant change in human attitude and activity.

An estimate of tropical forest depletion is shown in Figure 3.8 [14]. It is estimated that by the year 2050 only about half of the world's tropical forests will remain. Humankind will have planted less than 10% of that amount as replacement. And, replacement, or new growth forests, do not sustain the levels of biodiversity that occur in old growth or natural forests. Yet it is difficult to slow or stop forest destruction when land is needed to feed and house a starving, expanding human population.

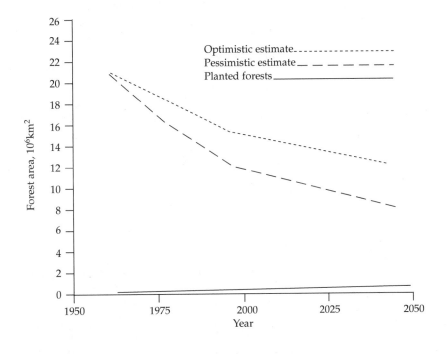

FIGURE 3.8 Estimated depletion of tropical forests through 2050, from Bruenig.

BIOGEOCHEMICAL CYCLES

A **biogeochemical cycle** is a summary of the different chemical repositories where a particular element resides, coupled with the pathways that convert and transport the element from one repository or form to another. The chemical elements, including all the essential elements for life, follow some form of biogeochemical cycle. There are many such cycles. Some of the more important ones are the carbon, oxygen, nitrogen, phosphorus, and sulfur cycles.

Global biogeochemical cycles distribute nutrients throughout the earth's atmosphere and upper geologic layers. Most cycles of interest to environmental engineers—including carbon, nitrogen, oxygen, and water—have at least one gaseous or atmospheric phase. The exception is phosphorus, which has no significant atmospheric presence.

All biogeochemical cycles are composed of a reservoir pool and a cycling pool. The reservoir contains the majority of the element. This nonbiological portion of the element is relatively slow to transform. The reservoir pool can be either solid, contained in sedimentary geologic forms, or gaseous, contained in the atmosphere. The cycling pool is more chemically or biochemically active. Elements in it move rapidly between organisms and their immediate environment.

Many natural systems are cybernetic. A **cybernetic system** is one in which feedback causes self-regulation and stability to some degree. Many natural processes, including the carbon cycle, are somewhat cybernetic. As the carbon dioxide level increases, photosynthetic organisms increase their rate of conversion of carbon dioxide. This results in the consumption of additional amounts of atmospheric and aqueous carbon dioxide. So the carbon increase is dampened by the increased consumption. The problem for humans now is that we are destroying forests that carry out most of the world's photosynthesis—thus negating positive effects of the cybernetic system.

Carbon Cycle

Carbon is the building block of life on earth. The carbon–carbon bond is relatively stable, and it facilitates the construction of complex macromolecules. Compounds containing carbon, excluding the carbonates and cyanides, are considered organic. Organic chemistry is the chemistry of life.

Carbon is present in the air (primarily as CO_2), in both freshwater and seawater (as dissolved carbon dioxide, $[CO_2(aq)]$, carbonic acid $[H_2CO_3]$, bicarbonate $[HCO_3^-]$, and carbonate $[CO_3^{2-}]$), in minerals (as carbonates such as limestone $[MgCO_3]$ and $[CaCO_3]$ or as petroleum or coal), and in all life forms (as protein, fats, and carbohydrates). The major forms of carbon are listed in Table 3.3 with approximate total mass on earth. Note that the major repository of carbon is inorganic minerals. Eighty-three percent of all carbon exists in this form. Seventeen percent, the majority of the remainder, is contained in organic sedimentary minerals [15]. This segment is primarily **kerogen**, the remains of tissues of ancient plants and animals. Only a tiny fraction exists as petroleum and coal. Also note that very little of the carbon on earth is contained in the atmosphere, in surfacewater or groundwater, or in living things. More than 100,000 times as much carbon is contained in sedimentary rocks as in all living things.

Carbon can be converted between these different living and nonliving forms by photosynthesis and degradation. A key to carbon's diversity and distribution is its gaseous phase, which enables it to be transported

TABLE 3.3 Major Forms of Carbon on Earth

SOURCE	MASS, 10^{15} kg	PERCENT
Geologic inorganic minerals	60,000	83
Geologic organic minerals*	12,000	17
Oceanic inorganics	40	0.056
Atmosphere	0.7	0.00097
All life on earth	0.6	0.00083

*Only about 4×10^{15} Kg of this is available as an energy supply for humans.

throughout the earth. A simplified schematic of the carbon cycle is shown in Figure 3.9. Photosynthetic organisms convert inorganic carbon into plant biomass. On land, plants perform the photosynthetic process using atmospheric carbon dioxide. In the oceans and bodies of freshwater photosyn-

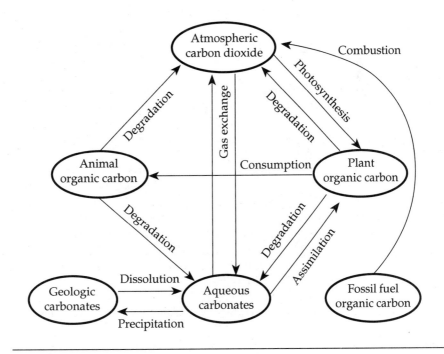

FIGURE 3.9 The carbon cycle.

theticorganisms such as planktonic algae perform synonymous conversion processes using dissolved inorganic carbon. Much of this organic carbon is either converted into animal biomass by consumers or back to inorganic carbon by decomposers. However, a fraction remains as organic decay products on the bottom of oceans or lake floors. These photosynthetic processes are the primary means by which carbon dioxide is removed from the earth's atmosphere. In addition, aquatic organisms utilize carbon in building shell and skeletal structures. This inorganic carbon, primarily carbonate, is eventually deposited on the floor of oceans and lakes. This process also removes a substantial amount of inorganic carbon.

Animals, which exist at the second trophic level, consume plant matter for energy production and biomass synthesis. These life processes transform the plant organic carbon into other forms. The result is the production of carbon dioxide and organic carbon. At death, organic matter is converted into simpler organic compounds and into inorganic forms by decomposers. The decomposition process returns the carbon dioxide to the atmosphere and to water.

The different inorganic forms of carbon can be interconverted from one reservoir to another by chemical processes such as precipitation, dissolution, and gas transfer. The mineral carbonates can be dissolved by undersaturated or acidic waters seeping through the formations. Conversely, additional carbonates can be deposited through chemical precipitation. However, most geologic carbonates have been formed by aquatic invertebrates using dissolved inorganic carbon to form carbonate shells. These shells are eventually deposited on the ocean or lake floors.

Carbon dioxide in the atmosphere dissolves into both ocean and freshwater and is released back into the atmosphere in accordance with Henry's law (see Chapter 4, Physical Chemistry). Where carbonate minerals reach great depths, the heat within the earth causes them to revert to carbon dioxide. This results in the release of carbon dioxide from volcanoes and soda springs [15].

During the past 200 years human activity has been changing the global carbon balance. The largest changes are occurring as a result of the introduction of carbon dioxide into the atmosphere from fossil fuel combustion in developed nations. In addition, large amounts of forest cover have been destroyed in the past century, both in developed and developing nations. This combination has resulted in a marked increase in atmospheric carbon dioxide. Although atmospheric carbon dioxide levels have been recorded for only the past 50 or so years, coring in the polar ice caps has established atmospheric carbon dioxide levels for the past several hundred years (see Figure 3.10) [16]. The world has known reserves of oil, natural gas, and coal sufficient to last for the next 200 to 300 years at current consumption rates [17], but there are currently no practical methods of removing carbon dioxide from combustion processes. So we can expect

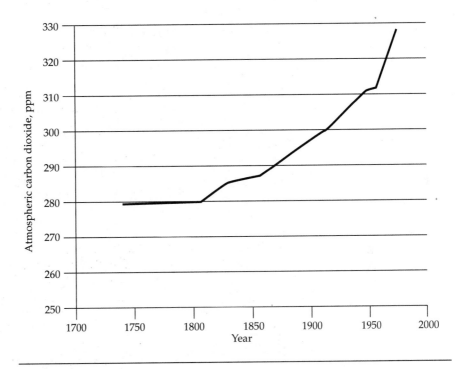

FIGURE 3.10 Global atmospheric carbon dioxide from preindustrialization to present.

global carbon dioxide levels to continue to rise over that same period. There are strong indications that the increased global carbon dioxide is causing a warming of the earth. This "greenhouse effect" will be discussed further in Chapter 14, Air Pollution.

Nitrogen Cycle

Nitrogen is the major constituent of the atmosphere, a major dissolved gas in water, and a component of amino acids, which are the building blocks for proteins comprising all life forms. Amino acids form bonds with other amino acids, resulting in protein structures (see Chapter 5, Organic Chemistry). Thus, like carbon, nitrogen is a chemical of life. It is a major nutrient for both plant and animal forms. In addition, nitrogenous compounds are a major component of all municipal wastewater and of some industrial wastewater. Because they play such a critical role in the environment, nitrogenous compounds are important to environmental engineers.

Figure 3.11 illustrates the nitrogen cycle. As with carbon, nitrogen is converted into different forms by several processes, both biotic and abiotic. Nitrogen gas in the atmosphere is oxidized and deposited on earth by

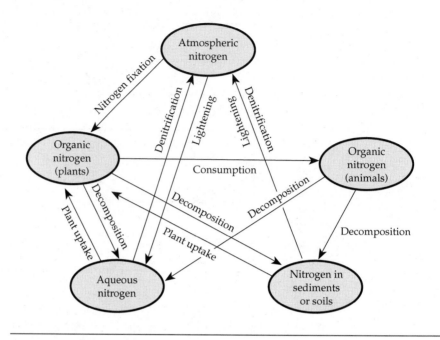

FIGURE 3.11 The nitrogen cycle.

lightning. Some organisms can convert atmospheric nitrogen to nitrate by a process known as **nitrogen fixation**. Nitrogen in the soil or water can be used by plants to produce organic nitrogen, amino acids, and proteins. Animals obtain nitrogen by consuming plants or other animals. Plant and animal life is eventually decomposed and returned to the soil or water as inorganic forms of nitrogen.* Animals expel excess nitrogen as urea, $(NH_2)_2CO$. This is then broken down into ammonia and carbon dioxide. The biological processes of nitrification and denitrification can convert the different inorganic forms of nitrogen to nitrogen gas and back again. **Nitrification** is the conversion of ammonia (or ammonium) to nitrate:

$$NH_3 + 1\tfrac{1}{2}O_2 \rightarrow H^+ + NO_2^- + H_2O \qquad\qquad \textbf{3.2}$$

and then

$$NO_2^- + \tfrac{1}{2}O_2 \rightarrow NO_3^- \qquad\qquad \textbf{3.3}$$

And in the presence of organic matter, nitrate can be converted to atmos-

---

*Organic nitrogen is nitrogen contained in carbon compounds such as amino acids and urea. Inorganic nitrogen is primarily nitrogen gas [N_2], ammonia [NH_3 or NH_4^+], nitrite [NO_2^-], or nitrate [NO_3^-].

pheric nitrogen by the denitrification process:

$$6H^+ + 6NO_3^- + 5CH_3OH \rightarrow 3N_2\uparrow + 5CO_2 + 13H_2O \qquad \textbf{3.4}$$

Human activities produce changes in the nitrogen cycle. Nitrates and ammonia are frequently added to agricultural crops to increase yields. Combustion processes, particularly those in motor vehicles, oxidize atmospheric nitrogen into nitrate, producing nitric acid, a component of acid precipitation. Domestic wastewater, treated and then discharged into lakes, streams, or oceans, often contain large amounts of nitrogen, usually as nitrate or ammonia. The excess nitrogen that is contained in surface-waters often causes increased photosynthetic activity.

Hydrologic Cycle

The hydrologic cycle is the movement of water in the earth's atmosphere, on the surface, and below the surface — a process powered by the sun's energy. Water is of great importance because it is required for all life processes on earth. Also, it receives much of the pollution that humans generate.

A simplified schematic of the hydrologic cycle appears in Figure 3.12.

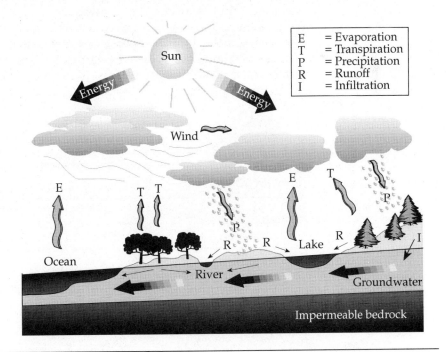

FIGURE 3.12 The hydrologic cycle.

In this cycle the sun's radiant energy supplies the power to evaporate water from lakes, rivers, and plants. Since the sun also supplies the energy to drive the winds, it is responsible for the transport of moisture in the atmosphere. Much water vapor is returned to the ocean by precipitation before ever reaching land. However, winds move part of the water vapor to land, where it is deposited as **precipitation**. Of the water that falls over land, a portion **infiltrates** through the ground to the water table, recharging the groundwater. Part is intercepted by vegetation and directly returned to the atmosphere as **evaporation**. Part of the water absorbed by vegetation is returned to the atmosphere through **transpiration**. Part of the precipitation flows overland to lakes and rivers. And part falls directly onto lakes and rivers. Eventually, when the freshwater is returned to the sea, the cycle is completed.

We discharge our wastes into water. In some cases this water is then treated to remove part of the waste material before being discharged into streams, rivers, lakes, or oceans. In many cases untreated water is discharged into the environment, and nature is left with the task of dealing with the pollution.

Water covers two-thirds of the earth's surface. This makes it seem to be an inexhaustible resource, but in fact the amount of water available for human use is a very small fraction of the total. As shown in Table 3.4, most water on earth is seawater [18]. Seawater is not a feasible large-scale water source because the energy required to purify and desalinate it is too great. The second most abundant source of water is that contained in the pores of rocks and minerals below the earth's surface. Only a fraction of this groundwater is usable because most is either too tightly bound within the pores, too deep to retrieve, or too contaminated by dissolved gases, salinity,

TABLE 3.4   The Hydrosphere

WATER SOURCE	MASS, kg
Oceans	$13{,}700 \times 10^{17}$
Groundwater	$3{,}200 \times 10^{17}$
Water locked in ice	165×10^{17}
Water in lakes, rivers	0.34×10^{17}
Water in atmosphere	0.105×10^{17}
Total yearly stream discharge	0.32×10^{17}

SOURCE: W. Stumm and J. J. Morgan, *Aquatic Chemistry*, 2nd ed.

or other natural contaminants. The third largest water source is ice in the polar caps and on mountaintops, which is not available to most humans. The smallest source is surfacewater contained in lakes and rivers. This is easily used. So the usable sources are a small fraction of pore or groundwater and the water contained in rivers and lakes. (See Problem 9 to calculate usable water as an approximate percentage of all sources.)

Water Budget

The water budget is the sum of the water inputs and outputs for a particular system. Over an extended period of time, precipitation should equal the sum of the **runoff**, **evapotranspiration**, and **infiltration**. An expression for the water budget over an extended period is

$$\sum \text{Inputs} = \sum \text{Outputs} \qquad\qquad \textbf{3.5}$$

However, in shorter periods storage and accumulation can occur. In this case, the water budget can be written as

$$P = ET + R + I + S \qquad\qquad \textbf{3.6}$$

where

P = precipitation [cm or in.]

ET = evapotranspiration or evaporation plus transpiration [cm or in.]

R = runoff [cm or in.]

I = infiltration [cm or in.]

S = storage [cm or in.]

A water balance can be used to estimate the amount of infiltration (ground water recharge) for an area, as in the following example.

EXAMPLE 3.2	USING A WATER BALANCE TO ESTIMATE EVAPOTRANSPIRATION

A $1\,\text{km}^2$ watershed has been monitored recently to estimate the summer evapotranspiration. August rainfall was 4 cm, and the runoff from the area was $5000\,\text{m}^3$. Infiltration was estimated to be 0.7 cm. Storage can be assumed negligible. What was the total evapotranspiration?

SOLUTION: We know the input to the system and two of the three outputs. We must first convert the runoff volume into depth over the $1\,km^2$ area.

$$R = \left(\frac{5000\,m^3}{1\,km^2}\right)\left(\frac{km}{1000\,m}\right)^2\left(\frac{100\,cm}{1\,m}\right)$$

$$= 0.5\,cm$$

Solving Equation 3.6 for ET we obtain

$$ET = R - (I + R + S) = 4\,cm - (0.7\,cm + 0.5\,cm + 0\,cm)$$

$$= 2.8\,cm$$

Review Questions

1. Define the following:
   a. Ecology
   b. Autotrophic
   c. Biogeochemical cycle
   d. Hydrologic cycle
   e. Ecosystem
   f. Trophic level

2. Using organisms from your area, describe several different trophic levels.

3. Obtain population records of your municipality from the university library. Using that data, estimate its population 25 years in the future.

4. Assuming that the world's population will continue to grow unabated at its present rate, estimate the year in which each person on earth will have one square meter of space.

5. From the temperature profile of a lake, shown next, graph both the temperature and density vs. water depth for the lake. Where is the **thermocline**?

DEPTH, m.	TEMPERATURE, °C
Surface, 0	22
2	22
4	20
6	8
8	6
10	6
12	6
14	6
16	6

6. In Example 3.1 the global population growth rate was estimated using the years A.D. 1950 to 1990. Using a spreadsheet, graph that population model versus the data shown at the right, which we used to generate Figure 3.6. Also use the years 0 and A.D. 1200 to estimate global population growth rate. Why is there such a discrepancy? Has the global population growth rate increased? Why? On your graph note the following events: the beginning of the machine age, the discovery of petroleum, and the first vaccination (the beginning of modern medicine).

YEAR	POPULATION (millions)
0	170
200	190
400	190
600	200
800	220
1000	265
1200	360
1400	350
1600	545
1800	900
1900	1625
1950	2500
1975	3900
1990	5500

7. What are the common forms of inorganic carbon on earth? What are the three common forms of organic carbon in geologic deposits?

8. By what percentage has global atmospheric carbon dioxide increased since 1750? Use Figure 3.10, page 39.

9. Assuming that 10% of all water in rivers and lakes is available for human consumption, that 0.001% of the water in rocks (or groundwater) is accessible, and that no other water is available, how much water is available for human consumption? (See Table 3.4, page 42) How much water is available per person on earth? Why is water such a scarce commodity in some areas?

10. The water budget, equation (3.6), does not have a term for intrusion of ocean water into an aquifer. Modify the equation by adding such a term. (This term would be an input, albeit not wanted!) Then use the modified equation to estimate the amount of **saltwater intrusion** for the following conditions that might exist in southern Florida's coastal areas. Annual precipitation is 58 in., infiltration is 14 in., runoff is 11 in., pumped withdrawal is 22 in., and evapotranspiration is 15 in. Short-term storage can be neglected.

11. What two human activities are most responsible for the increased carbon dioxide in the earth's atmosphere? How are they related?

12. The oxygen cycle was not discussed in this chapter. Using the atmosphere, biomass, minerals, and oceans, draw an oxygen cycle. Show how oxygen moves from one repository to another.

13. Describe the effects of increased carbon dioxide emissions on the carbon cycle. Will this affect the oxygen cycle? How?

14. Where is most biomass produced? What can be done to increase the production of biomass on earth and thus reduce global atmospheric carbon dioxide levels?

15. What are the major differences between the phosphorus and oxygen cycles?

16. Describe several anthropogenic sources of nitrogen in your area. How can these affect air, land, and water quality?

17. Describe several carbon sources in your area and the effect they have on the local environment.

18. Name and describe the different trophic levels in the drawing of a freshwater aquatic ecosystem in Figure 3.13.

19. For the ocean ecosystem shown in Figure 3.1, page 24, list the organisms for each trophic level.

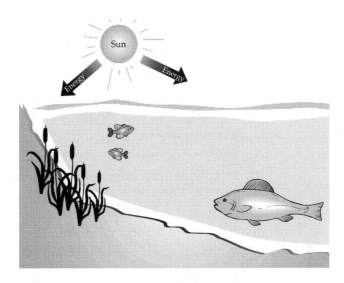

FIGURE 3.13

20. For the terrestrial ecosystem shown in Figure 3.2 list the organisms (or groups of organisms) for each trophic level.

References

1. R. L. Smith, *Elements of Ecology and Field Biology* (New York: Harper & Row, 1977).

2. R. Carson, *Silent Spring* (Boston: Houghton Mifflin, 1962).

3. G. M. Woodwell, C. F. Wurster, Jr., and P. A. Isaacson, "DDT Residues in an East Coast Estuary: A Case of Biological Concentration of a Persistent Insecticide," *Science*, vol. 156 (May 1967), pp. 821–824.

4. D. Sarokin and J. Schulkin, "The Role of Pollution in Large-Scale Population Disturbances, Part 1, Aquatic Populations," *Environmental Science and Technology*, vol. 26, no. 8 (August 1992), pp. 1476–1484.

5. J. N. Abramovitz, "Biodiversity: Inheritance from the Past, Investment in the Future," *Environmental Science and Technology*, vol. 25, no. 11 (November 1991), pp. 1817–1818.

6. C. McEvedy and R. Jones, *Atlas of World Population History* (Harmondsworth, Middlesex, England: Penguin Books, Ltd., 1978).

7. N. Keyfitz, "The Growing Human Population," *Scientific American*, vol. 261 (September 1989) pp. 118–126.

8. K. Arms, *Environmental Science* (Philadelphia: Saunders College Publishing, 1990).

9. *BP Statistical Review of World Energy* (London: The British Petroleum Company, 1991).

10. *Statistical Abstract of the United States*, 111th ed., The National Data Book (Washington, DC: U.S. Department of Commerce, Bureau of the Census. 1991).

11. A. P. Fickett, C. W. Gellings, and A. B. Lovins, "Efficient Use of Electricity," *Scientific American*, (September 1990).

12. P. H. Raven, "Our Diminishing Tropical Forests," in *Biodiversity*, E. O. Wilson, ed. (Washington, DC: National Academy Press, 1988).

13. N. Myers, "Tropical Forests and Their Species, Going, Going...?" in *Biodiversity*, E. O. Wilson, ed. (Washington, DC: National Academy Press, 1988).

14. E. F. Bruening, "Use and Misuse of Tropical Rain Forests," in *Ecosystems of the World 14B, Tropical Rain Forest Ecosystems*, pp. 611–636, M. J. A. Werger and H. Lieth, eds. (New York: Elsevier Science Publishing Co., 1989).

15. R. A. Berner and C. Lasaga, "Modeling the Geochemical Carbon Cycle," *Scientific American* (March 1989), pp. 74–81 (an excellent and easily read article on the fate and transport of organic carbon on earth).

16. A. Neftel, E. Moor, H. Oeschger, and B. Stauffer, "Evidence from Polar Ice Cores for the Increase in Atmospheric CO_2 in the Past Two Centuries," *Nature*, vol. 315 (May 2, 1985) pp. 45–47.

17. *BP Statistical Review of World Energy* (London: The British Petroleum Company, June 1991).

18. W. Stumm and J. J. Morgan, *Aquatic Chemistry, An Introduction Emphasizing Chemical Equilibria in Natural Waters* (New York: John Wiley & Sons, 1981).

4  Physical Chemistry

"A CHEMICAL COMPOUND ONCE
FORMED WOULD PERSIST FOR EVER, IF
NO ALTERATION TOOK PLACE IN
THE SURROUNDING CONDITIONS."

THOMAS HENRY HUXLEY (1825 – 1895)
English Biologist / Evolutionist

The DuPont Sabine River Works in Orange, Texas.
Photograph courtesy of DuPont.

INTRODUCTION

This chapter will familiarize you with applied physical chemistry procedures used to solve common environmental engineering problems. Topics include the gas laws and relationships, stoichiometry, equilibrium calculations for typical water systems, and two common treatment processes used in environmental engineering: adsorption and ion exchange.

GAS LAWS

There are several gas relationships. Those presented here are Henry's law, the ideal gas law, and Dalton's law of partial pressures. Although other relationships exist for gases, these three are of particular interest to the environmental engineer. Many of the other relationships can be derived from these.

Henry's Law

Henry's law states that the amount of a gas that dissolves in a liquid is proportional to the partial pressure that the gas exerts on the surface of the liquid. In equation form, this is expressed as follows:

$$C_A = K_H P_A \qquad\qquad \textbf{4.1}$$

where

C_A = concentration of A [mol/L] or [mg/L]

K_H = equilibrium constant (often called the Henry's law constant) [mol/L·atm] or [mg/L·atm]

P_A = partial pressure of A [atm]

Table 4.1 lists several gas reactions with their respective solubilities or Henry's law constants.

EXAMPLE 4.1	HENRY'S LAW — SOLUBILITY OF OXYGEN IN WATER

Although the atmosphere we breathe is approximately 20.9% oxygen, oxygen is only slightly soluble in water. In addition, the solubility decreases as the temperature increases. Thus, oxygen availability to aquatic life decreases during the summer months when the biological processes that consume oxygen are most active. Summer water temperatures of 25 to 30°C are typical for many surfacewaters in the United States. The Henry's law constant for oxygen in water is 61.2 mg/L·atm at 5°C and 40.2 mg/L·atm at 25°C. What is the solubility of oxygen at 5°C and at 25°C?

SOLUTION: At 5°C the solubility is

$$C_{O_2}(5°C) = K_{H,O_2}P_{O_2} = 61.2 \frac{mg}{L \cdot atm} \times 0.209 \text{ atm}$$

$$= 12.8 \text{ mg/L}$$

And at 25°C the solubility is

$$C_{O_2}(25°C) = K_{H,O_2}P_{O_2} = 40.2 \frac{mg}{L \cdot atm} \times 0.209 \text{ atm}$$

$$= 8.40 \frac{mg}{L}$$

Oxygen is 50% more soluble at 5°C than at 25°C in water. Organisms that breathe air have 21% oxygen to breathe, whereas fish and other aquatic organisms have only a few parts per million of oxygen to breathe!

TABLE 4.1 Selected Gas Solubility Reactions with Equilibrium Values (Henry's Law Constants) at 25°C

REACTION	NAME	K_H, mol/L·atm	$pK_H = -\log K_H$
$CO_2(g) \rightleftarrows CO_2(aq)$	Carbon dioxide	3.41×10^{-2}	1.47
$NH_3(g) \rightleftarrows NH_3(aq)$	Ammonia	57.6	−1.76
$H_2S(g) \rightleftarrows H_2S(aq)$	Hydrogen sulfide	1.02×10^{-1}	0.99
$CH_4(g) \rightleftarrows CH_4(aq)$	Methane	1.50×10^{-3}	2.82
$O_2(g) \rightleftarrows O_2(aq)$	Oxygen	1.26×10^{-3}	2.90

Air stripping is a common method of removing dissolved gases from water and wastewater. Gases commonly removed include ammonia, carbon dioxide, and hydrogen sulfide. The only gases with appreciable partial pressures in air are nitrogen and oxygen. Other gases represent only about 1% of the total pressure. Thus, since the equilibrium concentration of a dissolved gas is proportional to the partial pressure of the gas on the liquid surface, any gas except nitrogen or oxygen can be air-stripped to relatively low concentrations by bringing it into intimate contact with the air. This is normally accomplished by aerating the water, spraying it into the air, or trickling the water through a tower with a large surface area. These methods bring a large surface area of water into contact with the atmosphere, thus increasing the gas transfer rate. Figure 4.1 is a schematic of a typical air-stripping tower for the removal of carbon dioxide or other dissolved gases.

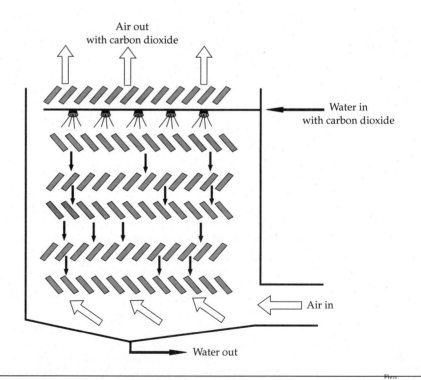

FIGURE 4.1   Typical air-stripping tower for the removal of carbon dioxide from groundwater.

| EXAMPLE 4.2 | AIR STRIPPING — REMOVAL OF CARBON DIOXIDE |

An air-stripping tower, similar to that shown in Figure 4.1, is to be used to remove dissolved carbon dioxide from a groundwater supply. If the tower lowers the level to twice the equilibrium concentration, what amount of dissolved gas will remain in the water after treatment? The partial pressure of carbon dioxide in the atmosphere is $10^{-3.5}$ atm.

SOLUTION: The first step is to determine the Henry's law constant for carbon dioxide. From Table 4.1, this is $10^{-3.5}$. The equilibrium solubility is then

$$C_{CO_2} = K_{H,CO_2}P_{CO_2} = 10^{-1.5}\,\frac{mol}{L \cdot atm} \times 10^{-3.5}\,atm$$

$$= 10^{-5}\,M = 10^{-5}\,\frac{mol}{L} \times \frac{44\,g}{mol} \times \frac{10^3\,mg}{g}$$

$$= \frac{0.44\,mg}{L}$$

The concentration after stripping is twice the equilibrium concentration, or 0.88 mg/L. Thus, the process reduces the CO_2 level to below 1 mg/L.

Ideal Gas Law

The ideal gas law states that the product of the absolute pressure of a gas and its volume is proportional to the product of its mass and absolute temperature. In equation form this is usually written

$$PV = nRT \tag{4.2}$$

where

P = absolute pressure [atm]

V = volume [L]

n = mass [mol]

T = absolute temperature [K]

R = proportionality constant or ideal gas constant [0.0821 L·atm/K·mol]

EXAMPLE 4.3	APPLICATION OF THE IDEAL GAS LAW

Anaerobic microorganisms metabolize organic matter to carbon dioxide and methane gases. Estimate the volume of gas produced (at atmospheric pressure and 25°C) from the anaerobic decomposition of 1 mol of glucose. The reaction is

$$C_6H_{12}O_6 \rightarrow 3CH_4 + 3CO_2$$

SOLUTION: Each mole of glucose produces 3 mol of methane and 3 mol of carbon dioxide gases, for a total of 6 mol. The entire volume is then

$$V = \frac{nRT}{P} = \frac{(6 \text{ mol})\left(0.0821 \frac{L \cdot atm}{K \cdot mol}\right)(298 \text{ K})}{(1 \text{ atm})}$$

$$= 147 \text{ L}$$

Note: The volume of 1 mol of any gas is the same. Thus, 1 mol of carbon dioxide gas is the same volume as 1 mol of methane gas.

Dalton's Law of Partial Pressures

Dalton's law of partial pressures states that the total pressure of a mixture of several gases is the sum of the partial pressures of the individual gases. In equation form this is simply

$$P_t = \sum_{i=1}^{n} P_i \qquad \qquad \textbf{4.3}$$

where

P_t = total pressure of the gases [atm]

P_i = pressure of the ith gas [atm]

In our atmosphere, nitrogen has a partial pressure of 0.781 atm; oxygen, 0.209 atm; argon, 0.0093 atm; and carbon dioxide, 0.00033 atm. The sum of these is 0.9996 atm. The remaining 0.0004 atm of pressure is contributed by several minor gases.

EXAMPLE 4.4	USE OF THE IDEAL GAS LAW AND DALTON'S LAW OF PARTIAL PRESSURES

Anaerobic digesters are commonly used in wastewater treatment. The biological process produces both carbon dioxide and methane gases. A laboratory worker plans to make a "synthetic" digester gas. There are currently 2 L of methane gas at 1.5 atm and 1 L of carbon dioxide gas at 1 atm in the lab. If these two samples are mixed in a 4-L tank, what will be the partial pressures of the individual gases? The total pressure?

SOLUTION: First, we must find the partial pressures of the individual gases using the ideal gas law:

$$P_1 V_1 = nRT = P_2 V_2$$

$$P_2 = P_1 \left(\frac{V_1}{V_2} \right)$$

For methane gas,

$$P_2 = (1.5 \text{ atm}) \left(\frac{2 \text{ L}}{4 \text{ L}} \right) = 0.75 \text{ atm}$$

For carbon dioxide gas,

$$P_2 = (1 \text{ atm}) \left(\frac{1 \text{ L}}{4 \text{ L}} \right) = 0.25 \text{ atm}$$

The total pressure is then the sum of the partial pressures, or

$$P_t = P_{CH_4} + P_{CO_2} = 1 \text{ atm}$$

CHEMICAL THERMODYNAMICS

An understanding of the role of chemical thermodynamics is useful in environmental engineering. Thermodynamic principles can be used to determine the equilibrium condition of a reaction or to ascertain whether a nonequilibrium reaction is proceeding forward (as written) or backward (products to reactants). Analysis of this kind can reveal the amount of heat emitted or absorbed from the reaction. As an example, for the "generic" reaction

$$aA + bB \rightleftarrows pP + rR \qquad \textbf{4.4}$$

or

$$K_{eq} = \frac{[P]^p [R]^r}{[A]^a [B]^b}$$

the equilibrium constant K_{eq} can be computed straightforwardly if the concentrations of the constituents are known. The application of basic rules will help determine whether the reaction is at equilibrium or proceeding left or right.

Heat of Reaction

The heat of a reaction is usually of less interest to environmental engineers than to mechanical or chemical engineers. However, if you have ever tried to dilute sulfuric acid with distilled water or dissolve sodium hydroxide pellets in distilled water, you probably realize there are reactions of interest to environmental engineers in which the heat of reaction is important. In several readily accessible sources you can find chemical thermodynamic parameters for different ions or molecules in solution. (Ions and molecules in this state are commonly called **species**.) Table 4.2 presents the thermodynamic properties of selected species—that is, those required for homework and example problems in this text.

The enthalpy or heat of reaction for the reaction depicted by Equation 4.4 is given by the following:

$$\Delta H^\circ_{react \times n} = (p \Delta H^\circ_P + r \Delta H^\circ_R) + (a \Delta H^\circ_A + b \Delta H^\circ_B) \qquad \textbf{4.5}$$

where

$\quad$ A, B = reactant species

$\quad$ P, R = product species

$\quad\quad$ a, b = stoichiometric coefficients of the reactants

$\quad\quad$ p, r = stoichiometric coefficients of the products

$\quad$ $\Delta H^\circ_{A,B}$ = enthalpy of reactants A and B

$\quad$ $\Delta H^\circ_{P,Q}$ = enthalpy of products P and Q

TABLE 4.2   Thermodynamic Properties of Selected Species

SPECIES	$\Delta G°$, kcal/mol	$\Delta H°$, kcal/mol
$Ca^{2+}(aq)$	-132.18	-129.77
$CaCO_3(s)$, calcite	-269.78	-288.45
$Ca(OH)_2(s)$, lime	-214.7	-235.7
$CO_2(g)$	-94.26	-94.05
$CO_2(aq)$	-92.31	-98.69
$CO_3^{2-}(aq)$	-126.2	-161.6
$Cl^-(aq)$	-31.3	-40.0
$Cl_2(aq)$	-19.1	-28.9
$OCl^-(aq)$	-8.8	-25.6
$H^+(aq)$	0	0
$HNO_3(aq)$	-26.6	-25.0
$NO_3^-(aq)$	-26.6	-49.5
$HOCl(aq)$	-19.1	-28.9
$OH^-(aq)$	-37.6	-55.0
$O_2(aq)$	3.9	-3.9
$H_2O(l)$	-56.7	-68.3
$H_2(g)$	0	0
$O_2(g)$	0	0

or, in more general terms

$$\Delta H_{\text{react}}° = \sum_{i=1}^{n} a_i \Delta H_i° - \sum_{j=1}^{m} a_j \Delta H_j°$$

4.6

where

a_i = stoichiometric coefficient of product species, i

a_j = stoichiometric coefficient of reactant species, j

ΔH_i° = enthalpy of product species, i [kcal/mol]

ΔH_j° = enthalpy of reactant species, j [kcal/mol]

Equation (4.6) can be used to solve for the amount of heat energy expelled from or absorbed by a chemical reaction. By convention, a negative enthalpy indicates an exothermic reaction, one in which heat is expelled to the surroundings. A positive enthalpy is an endothermic reaction, one in which heat is absorbed from the surroundings. One heat of reaction example will be presented here. Otherwise, in the remaining portion of this section we will focus on a concept of more importance to environmental engineers: chemical equilibria.

EXAMPLE 4.5	ENTHALPY — HEAT OF REACTION

Nitric acid is a common laboratory reagent. Commercial nitric acid is usually purchased in concentrated form as 96 to 98% acid. If 50 g of the acid is added to 1 L of water, how much heat is released?

SOLUTION: The reaction is

$$HNO_3(aq) \rightleftarrows H^+ + NO_3^-$$

The heat of reaction is then

$$\Delta H_{react}^\circ = (1\,\Delta H_{H^+}^\circ + 1\,\Delta H_{NO_3^-}^0)_i + (1\,\Delta H_{HNO_3}^\circ)_j$$

$$= \left[0\,\frac{kcal}{mol} + \left(-49.5\,\frac{kcal}{mol} \right) \right] - \left[-25.0\,\frac{kcal}{mol} \right]$$

$$= -24.5\,\frac{kcal}{mol}$$

The negative heat of reaction indicates that the reaction is exothermic. For each mole of acid ionized to H^+ and NO_3^-, 24.5 kcal of heat is released. Using specific heats (from freshman chemistry) yields a temperature increase of about 19°C for the solution.

Gibbs Free Energy

The concept of Gibbs free energy, so named after American mathematician and physicist J. Willard Gibbs, can be used to determine the equilibrium position of a chemical reaction, or whether reactants and reaction products are proceeding left or right. Reactants and products will advance toward a minimum Gibbs free energy much as a roller coaster will come to rest at the lowest point between two peaks in its track if gravity alone is operating on it. Figure 4.2 depicts this analogy.

The change in Gibbs free energy for a process occurring at constant temperature and pressure is related to the change in entropy and enthalpy by the following expression:

$$\Delta G = \Delta H - T \Delta S \qquad \textbf{4.7}$$

where

ΔG = Gibbs free energy [kcal/mol]

ΔH = enthalpy [kcal/mol]

ΔS = entropy [kcal/K·mol]

T = temperature [K]

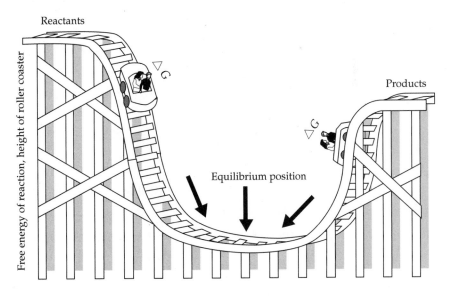

FIGURE 4.2 The energy ``roller coaster'' illustrates the equilibrium position of a chemical reaction.

A reaction with a negative ΔG will proceed from reactants to products. A reaction with a positive ΔG will proceed from products to reactants. Where ΔG is zero, the reaction is at equilibrium.

Standard values of ΔG for elements and compounds in various states have been determined, and these are instrumental in calculating the $\Delta G°$ for a given mixture or solution. For the reaction shown in Equation 4.4, the following relationship can be used to find the standard Gibbs free energy, $\Delta G°$, for the reaction:

$$\Delta G°_{react} = (p\Delta G°_P + r\Delta G°_R) - (a\Delta G°_A + b\Delta G°_B) \qquad \textbf{4.8}$$

where

A, B = reactant species

P, R = product species

a, b = stoichiometric coefficients of the reactants

p, r = stoichiometric coefficients of the reactants

$\Delta G°_{A,B}$ = Gibbs free energy of reactants A and B

$\Delta G°_{P,Q}$ = Gibbs free energy of reactants P and Q

or, in more general terms

$$\Delta G°_{react} = \left(\sum_{i=1}^{n} a_i \Delta G°_i\right) - \left(\sum_{j=1}^{m} a_j \Delta G°_j\right) \qquad \textbf{4.9}$$

where

a_i = stoichiometric coefficient of product species, i

a_j = stoichiometric coefficient of reactant species, j

$\Delta G°_i$ = standard Gibbs free energy of product species, i [kcal/mol]

$\Delta G°_j$ = standard Gibbs free energy of reactant species, j [kcal/mol]

However, $\Delta G°$ is only half the information we need. It will tell us if the reactants and products, at their standard activities (see Table 4.3), will proceed forward or backward, or are at equilibrium. Most systems of environmental importance, however, are not at $1\,M$ concentration. So what we actually need is ΔG for the reaction, not $\Delta G°$. For the generic reaction represented by equation 4.4, ΔG and $\Delta G°$ are related by the following expression:

TABLE 4.3 Standard States for Gibbs Free Energy

PARAMETER	STANDARD STATE OR CONDITION
Temperature	25°C
Gas	1 atm
Solid	Pure solid
Liquid	Pure liquid
Solution	1 M (m/L)
Element*	0

* In its normal state.

$$\Delta G = \Delta G^\circ + RT \ln \frac{P^p R^r}{A^a B^b}$$ **4.10**

The natural log term, or reaction quotient, is often denoted Q, or

$$Q = \frac{P^p R^r}{A^a B^b}$$ **4.11**

In addition, when ΔG is zero, the reaction is at equilibrium. Thus

$$\Delta G^\circ = -RT \ln K_{eq}$$ **4.12**

Therefore, thermodynamics will also lead us to the determination of the reaction equilibrium constant, K_{eq}.

We now have the necessary tools to determine whether a reaction is at equilibrium, proceeding as written, or going in reverse.

EXAMPLE 4.6	GIBBS FREE ENERGY — DIRECTION OF A REACTION

Lime, or calcium hydroxide, is commonly used to improve sedimentation processes or precipitate toxic metals. A 500-L reaction vessel contains an excess amount of solid calcium hydroxide. The solution contains 10^{-8} M of Ca^{2+} and 10^{-8} M of OH^-. Is the lime still dissolving, is the solution at equilibrium, or is lime precipitating?

SOLUTION: The reaction is

$$Ca(OH)_2(s) \rightleftarrows Ca^{2+} + 2OH^-$$

The standard Gibbs free energy is then

$$\Delta G^\circ_{rxn} = (2 \times \Delta G^\circ_{OH^-} + \Delta G^\circ_{Ca^{2+}}) - \Delta G^\circ_{Ca(OH)_2}$$

$$= [(-132.3 \text{ kcal/mol}) + 2(-37.6 \text{ kcal/mol})$$
$$- (-214.7 \text{ kcal/mol})]$$

$$= +7.22 \text{ kcal/mol}$$

Note, however, that our solution does not contain species with concentrations of $1\,M$, so we must now calculate ΔG_j.

$$\Delta G_j = \Delta G^\circ + RT \ln[Ca^2][OH^-]^2$$

$$= \left(7.22 \frac{\text{kcal}}{\text{mol}}\right) + \left(1.987 \frac{\text{cal}}{\text{K} \cdot \text{mol}}\right)\left(\frac{\text{kcal}}{1000 \text{ cal}}\right)$$

$$\times (298 \text{ K})[\ln(10^{-8})(10^{-8})^2]$$

$$= -13.2 \text{ kcal/mol}$$

Since ΔG is negative, the reaction is proceeding as written. In other words the lime is continuing to dissolve. (It is not a saturated or supersaturated solution.)

EXAMPLE 4.7   **GIBBS FREE ENERGY — OXYGEN SOLUBILITY**

A polluted stream with a temperature of 25°C has a dissolved oxygen concentration of 4 mg/L. Use Gibbs free energy to determine if oxygen from the atmosphere is dissolving in the water or is at equilibrium with it, or whether oxygen from the stream is going into the atmosphere.

SOLUTION: The first step is to write the reaction. We have dissolved molecular oxygen [O_2(aq)] converting to atmospheric molecular oxygen [O_2(g)], or

$$O_2(aq) \rightleftarrows O_2(g)$$

The standard Gibbs free energy for the reaction (from Table 4.2) is

$$\Delta G^\circ_{react} = \Delta G^\circ_{O_2(g)} - \Delta G^\circ_{O_2(aq)} = \left(0\ \frac{kcal}{mol}\right) - \left(+3.9\ \frac{kcal}{mol}\right)$$

$$= -3.9\ \frac{kcal}{mol}$$

Thus, at standard conditions (1 M dissolved oxygen, 1 atm partial pressure gaseous oxygen) aqueous oxygen would evolve into the atmosphere. However, these are not exactly typical environmental conditions. The level of dissolved oxygen in our example is only 4 mg/L, and the partial pressure of oxygen in the atmosphere is 0.209 atm. We need to determine ΔG. Recall that the standard method of expressing the activity of a gas is as a partial pressure, P_{O_2}. Thus,

$$\Delta G = \Delta G^\circ + RT \ln \frac{P_{O_2}}{[O_2(aq)]}$$

The concentration of oxygen in the water is

$$[O_2(aq)] = 4\ \frac{mg\ O_2}{L} \times \frac{g\ O_2}{1000\ mg\ O_2} \times \frac{mol\ O_2}{32\ g\ O_2} = 1.25 \times 10^{-4}\ M$$

The free energy is then

$$\Delta G = 3.9\ \frac{kcal}{mol} \times \frac{1000\ cal}{kcal}$$

$$+ \left(1.987\ \frac{cal}{K \cdot mol}\right)(298\ K)\ln\left(\frac{0.209\ atm}{1.25 \times 10^{-4}\ M}\right)$$

$$= 491\ \frac{cal}{mol} > 0$$

Since ΔG is positive, the reaction is proceeding in reverse (from right to left). Atmospheric oxygen is dissolving in the water.

See Problem 5 at the end of the chapter for calculation of the saturation concentration of dissolved oxygen.

STOICHIOMETRY

Stoichiometry deals with numerical relationships between reactants and products in chemical reactions. Stoichiometric analysis can be used to determine the product yield for a given amount of reactant converted. When water is formed from the combustion of hydrogen and oxygen, for example, the gases combine in fixed atomic and molecular ratios. Two hydrogen atoms and one oxygen atom form one water molecule. However, since the gases exist as diatomic molecules themselves, the actual formula is

$$2H_2 + O_2 \rightleftharpoons 2H_2O \qquad \qquad \textbf{4.13}$$

Two hydrogen molecules plus one oxygen molecule form two water molecules. On a molar basis, 2 mol of hydrogen (2×1.008 g) combine with 1 mol of oxygen (1×16.00 g) and form 2 mol of water (2×18.016 g), or

$$
\begin{array}{ccccc}
2H_2 & + & O_2 & \rightleftharpoons & 2H_2O \\
2 \text{ mol } H_2 & + & 1 \text{ mol } O_2 & \rightleftharpoons & 2 \text{ mol } H_2O \\
\end{array} \qquad \textbf{4.14}
$$
$$(2 \times 2 \times 1.008 \text{ g}) + (1 \times 2 \times 16.00 \text{ g}) = (2 \times 18.016 \text{ g})$$

EXAMPLE 4.8	STOICHIOMETRY — NEUTRALIZATION OF HYDROCHLORIC ACID WITH LIME

A tractor trailer truck with a full load of hydrochloric acid (40,000 lbs) crashes into a bridge, spilling the acid into the stream. How much lime [$Ca(OH)_2$] is required to neutralize the acid? The neutralization reaction is

$$2HCl + Ca(OH)_2 \rightleftharpoons CaCl_2 + 2H_2O$$

SOLUTION: Determine the number of moles , n, of hydrochloric acid.

$$n_{HCl} = 40{,}000 \text{ lbs} \times \frac{454 \text{ g}}{1 \text{ lb}} \times \frac{1 \text{ mol}}{36.5 \text{ g}} = 5.0 \times 10^5 \text{ mol HCl}$$

From the stoichiometric formula, 1 mol of lime neutralizes 2 mol of hydrochloric acid. The required amount of lime is then

$$n_{Ca(OH)_2} = 5.0 \times 10^5 \text{ mol HCl} \times \frac{1 \text{ mol Ca(OH)}_2}{2 \text{ mol HCl}}$$

$$= 2.5 \times 10^5 \text{ mol Ca(OH)}_2$$

The mass of lime required is

$$m_{Ca(OH)_2} = 2.5 \times 10^5 \text{ mol Ca(OH)}_2 \times \frac{74 \text{ g Ca(OH)}_2}{1 \text{ mol Ca(OH)}_2}$$

$$= 1.8 \times 10^5 \text{ Kg}$$

$$= 41{,}000 \text{ lb Ca(OH)}_2$$

EXAMPLE 4.9	STOICHIOMETRY — OXIDATION OF ACETIC ACID TO CARBON DIOXIDE AND WATER

Acetic acid [CH_3COOH] or vinegar is oxidized to carbon dioxide and water by microbial action. A vinegar manufacturing plant has 50 mg/L of acetic acid in its wastewater effluent. The plant flow is 500 m^3 per day. How much oxygen is required each day to oxidize the vinegar?

SOLUTION: First, the equation must be balanced. The end products are carbon dioxide and water, and the reactants are acetic acid and oxygen. The unbalanced equation is

$$\underline{}CH_3COOH + \underline{}O_2 \rightleftarrows \underline{}CO_2 + \underline{}H_2O$$

The balanced equation is then

$$CH_3COOH + 2O_2 \rightleftarrows 2CO_2 + 2H_2O$$

Thus, 2 mol of oxygen are required to oxidize the acid to carbon dioxide and water. The amount of acid required for oxidization each day is

$$n_{HAc} = 50 \frac{\text{mg HAc}}{L} \times \frac{g}{1000 \text{ mg}} \times 500 \frac{m^3}{\text{day}} \times \frac{1000 \text{ L}}{1 \text{ m}^3}$$

$$\times \frac{1 \text{ mol HAc}}{60 \text{ g HAc}}$$

$$= 417 \frac{\text{mol}}{\text{day}}$$

Since 2 mol of oxygen are required for each mole of acetic acid, the required oxygen is

$$n_{O_2} = 417 \frac{\text{mol HAc}}{\text{day}} \times \frac{2 \text{ mol O}_2}{\text{mol HAc}} = 834 \frac{\text{mol O}_2}{\text{day}}$$

The mass of oxygen is then

$$m_{O_2} = 834 \frac{\text{mol O}_2}{\text{day}} \times \frac{32 \text{ g O}_2}{\text{mol O}_2} = 26,700 \frac{\text{g O}_2}{\text{day}}$$

$$= 26.7 \frac{\text{Kg O}_2}{\text{day}}$$

Theoretical Oxygen Demand

An environmental engineering application of stoichiometry is the estimation of the amount of oxygen a known organic chemical will consume as it is converted to carbon dioxide and water. This is frequently termed **theoretical oxygen demand**, or **ThOD**. It is simply the amount of oxygen required to convert the material to carbon dioxide and water. In situations where the organic matter contains amine compounds [$-NH_2$], the end product of the nitrogen is ammonia [NH_3]. The process of determining the ThOD is presented in the following example.

EXAMPLE 4.10	THEORETICAL OXYGEN DEMAND

Ethanol, or ethyl, alcohol is used in beverages, as a gasoline additive, and in other industrial applications. Because small amounts of ethanol and sugar are used in the biological process to produce methanol, both of these compounds inevitably end up in the wastewater of methanol plants. Calculate the theoretical oxygen demand for a wastewater containing

    a. 25 mg/L ethanol [CH_3CH_2OH].
    b. 45 mg/L sucrose [$C_6H_{12}O_6$].
    c. a mixture of 25 mg/L ethanol and 45 mg/L sucrose.

SOLUTION: **a.** The first step is to write the balanced equation of the oxidation of ethanol (often written EtOH) to end products of carbon dioxide and water. The unbalanced equation is

$$CH_3CH_2OH + \_\_O_2 \rightarrow \_\_CO_2 + \_\_H_2O$$

The balanced equation, with molecular weights calculated, is then

$$CH_3CH_2OH + 3\tfrac{1}{2}O_2 \rightarrow 2CO_2 + 3H_2O$$
$$44 \qquad\qquad 96$$

Thus, 96 mg of O_2 is required to oxidize 44 mg of EtOH. The ThOD is then

$$ThOD = 25\ \frac{mg\ EtOH}{L} \times \frac{96\ mg\ O_2}{44\ mg\ EtOH}$$

$$= 55\ mg\ O_2/L$$

b. Here, as for the ethanol, the first step is to write the balanced equation for the oxidation of sucrose (table sugar) to end products of carbon dioxide and water. The unbalanced equation is

$$C_6H_{12}O_6 + \_\_O_2 \rightarrow \_\_CO_2 + \_\_H_2O$$

And the balanced equation, with appropriate molecular weights, is then

$$C_6H_{12}O_6 + 6O_2 \rightarrow 6CO_2 + 6H_2O$$
$$180 \qquad 192$$

192 mg of oxygen is required to oxidize 180 mg of sucrose. The ThOD is then

$$ThOD = 45\ \frac{mg\ C_6H_{12}O_6}{L} \times \frac{192\ mg\ O_2}{180\ mg\ C_6H_{12}O_6}$$

$$= 48\ mg\ O_2/L$$

c. The ThOD of the mixture is simply the sum of the ThODs of the individual components, or

$$ThOD_{tot} = 55\ mg\ O_2/L + 48\ mg\ O_2/L$$

$$= 109\ mg\ O_2/L$$

EQUILIBRIUM CHEMISTRY

Equilibrium chemistry can be used to analyze a variety of different aqueous reactions of interest to the environmental engineer — or the environmental engineering student! Reactions of environmental significance include determining the amount of a base to add to an acid spill, the amount of acid to neutralize a basic process wastewater, and the solubility of a metal in a chemical waste stream; and estimating the removal of phosphorus in a wastewater treated with lime or the solubility of mercury complexed in seawater.

An important point to remember when using equilibrium chemistry calculations is that you are assuming that the system is at equilibrium or is going toward it. You cannot determine the time required to reach equilibrium using equilibrium chemistry, however. This must be known, determined in the laboratory, or calculated using chemical kinetics. (Chemical kinetics is discussed in Chapter 8.) Some reactions reach equilibrium rapidly. Acid–base reactions are among the fastest — most moving as quickly as the mass transport will allow. Precipitation–dissolution reactions are typically slow. Some carbonate reactions involving solids take years to approach equilibrium.

Although there are several different methods of solving aqueous systems, the approach used here will be to write (and then solve) a system of simultaneous equations. This is the usual method employed in computer solutions. The first step is to determine the species involved in the reactions — that is, what are the reactants and the potential products? When the species are known, three types of equations must be written, then solved simultaneously. They are equilibrium equations for each reaction, mass balances for each constituent, and an electroneutrality equation. The following discussion takes each step in turn.

Determination of Species

Determination of the species present in the system is often the most difficult task. You may know them from your past experience, from using reference books, or from making laboratory determinations. The main point is, under normal circumstances, you cannot predict the outcome of a reaction by simply knowing the reactants and reaction conditions. You must have some information (prior knowledge) to indicate the final products as well. Only experience, laboratory work, or extensive study of a particular system will help you in this process.

EXAMPLE 4.11	DETERMINATION OF SPECIES

Determine the species present if the following compounds are dissolved in water in both open and closed systems:

    a. Sodium carbonate [Na_2CO_3]
    b. Sodium bicarbonate [$NaHCO_3$]
    c. Sodium phosphate [Na_3PO_4]

SOLUTION: **a.** Sodium carbonate. When carbonate is present in water, it will convert between the different carbonate-containing ions in an effort to reach equilibrium. Thus, simply adding sodium carbonate does not mean that other carbonate species will not be present. If sodium carbonate is added to water, at least the following reactions should be considered:

$$Na_2CO_3(s) \rightarrow 2Na^{2+} + CO_3^{2-}$$

Since water is present, both [H^+] and [OH^-] ions will also be present. The carbonate will react with the [H^+] ions as follows:

$$H^+ + CO_3^{2-} \rightleftarrows HCO_3^-$$
$$H^+ + HCO_3^- \rightleftarrows H_2CO_3(aq)$$
$$H_2CO_3(aq) \rightleftarrows H_2O + CO_2(aq)$$

So for a closed system the possible species are at least: $Na_2CO_3(s)$, Na^+, CO_3^{2-}, HCO_3^-, $H_2CO_3(aq)$, $CO_2(aq)$, H^+, and OH^-. In addition, if the system is open to the atmosphere, the $CO_2(aq)$ can come out of solution into the atmosphere in accordance with Henry's law::

$$CO_2(aq) \rightleftarrows CO_2(g)$$

b. Sodium bicarbonate. The reactions would be

$$NaHCO_3(s) \rightarrow Na^+ + HCO_3^-$$
$$HCO_3^- \rightleftarrows H^+ + CO_3^{2-}$$
$$H^+ + HCO_3^{2-} \rightleftarrows H_2CO_3(aq)$$
$$H_2CO_3(aq) \rightleftarrows H_2O + CO_2(aq)$$

The species present when sodium bicarbonate is added to water are the same as those for sodium carbonate. However, the concentrations will be different. If the system is open to the atmosphere, the aqueous CO_2 could also pass to the atmosphere, adding the $CO_2(g)$ species.

c. Sodium phosphate. When phosphate dissolves in water, it will inter-convert to different species by combining with the [H^+] ion. So the following reactions will occur:

$$Na_3PO_4 \rightarrow 3Na^+ + PO_4^{3-}$$
$$H^+ + PO_4^{3-} \rightleftharpoons HPO_4^{2-}$$
$$H^+ + HPO_4^{2-} \rightleftharpoons H_2PO_4^-$$
$$H^+ + H_2PO_4^- \rightleftharpoons H_3PO_4(aq)$$

The species would then be H^+, OH^-, Na^+, PO_4^{3-}, HPO_4^{2-}, $H_2PO_4^-$, and H_3PO_4. Unless we are to consider species dissolving from the atmosphere, there would be no difference between an open and closed system in this case since phosphate has no gaseous phase.

Equilibrium Reactions

You should be familiar with equilibrium reactions from freshman chemistry. An equilibrium reaction can be expressed in terms of the products and reactants. The equilibrium constant K_{eq} is the product of the products raised to their stoichiometric coefficients divided by the product of the reactants raised to their respective stoichiometric coefficients. For the "generic" reaction:

$$aA + bB \rightleftharpoons pP + rR \qquad\qquad\qquad \textbf{4.15}$$

the equilibrium expression is

$$K_{eq} = \frac{\{C\}^c \{D\}^d}{\{A\}^a \{B\}^b} \qquad\qquad\qquad \textbf{4.16}$$

where

a, b, p, r = stoichiometric coefficients of the respective reactants

$\{A\}, \{B\}, \{P\}, \{R\}$ = activity of the reactants and products

The chemical activity of a substance is related to its concentration by the relationship

$$\{A\} = \gamma_A [A] \qquad\qquad\qquad \textbf{4.17}$$

where

$\{A\}$ = activity of species A [mol/L]

$[A]$ = concentration of species A [mol/L]

γ_A = activity coefficient of A, unitless

This activity coefficient γ is dependent on ionic strength — the concentration of ions in solution. For dilute aqueous solutions, γ is near unity. Thus, the activity and concentraton are approximately the same. For most freshwater systems, γ may be neglected. For many industrial processes and for marine environments, γ must be calculated and activities rather than concentrations used in calculations. Otherwise, unacceptable errors may result. Those interested in calculating γ or activity should consult other references [1, 2]. For the remainder of this text, we will assume that the ionic strength is sufficiently low that concentration and activity are, for all practical purposes, equal.

There are various conventions for expressing the activity in equilibrium reactions. Like most chemistry and engineering publications, this text uses the following conventions:

▶ the activity of the solvent, water, is set equal to unity;
▶ the activity of a solid in equilibrium with a solution is unity;
▶ the activity of a gas is the partial pressure the gas exerts on the liquid surface; and
▶ the activity of a molecule or ion is related to the concentration by $\{A\} = \gamma_A [A]$, where γ_A is the activity coefficient of species A.

Equilibrium reactions of environmental significance include acid–base reactions, precipitation–dissolution reactions, gas solubility reactions, and complexation reactions. Each will be discussed in turn.

Acid–Base Reactions. Acid–base reactions involve either the gain or loss of a proton, H^+, or the gain or loss of a hydroxyl, OH^-. The self-ionization of water is considered an acid–base reaction. Water does not exist only as H_2O, for a small portion of the water molecules ionize to form H^+ and OH^-. The reaction is

$$H_2O \rightleftarrows H^+ + OH^-$$ **4.18**

The equilibrium equation is then

$$K_w = \frac{\{H^+\}\{OH^-\}}{\{H_2O\}} = \{H^+\}\{OH^-\}$$ **4.19**

At 25°C the value of K_w is 10^{-14}. Thus, if the activity of H^+ increases, the activity of OH^- must decrease in order for the product $\{H^+\}\{OH^-\}$ to remain constant.

Many concentrations or activities are expressed as pC. A lower case p preceding a species indicates the negative logarithm of the activity or concentration of that particular species. The most common is pH, or the negative logarithm of the H^+ activity. However, the terminology is used for hydroxide activity, pOH, as well as for others.

EXAMPLE 4.12	THE pH–pOH RELATIONSHIP

Find the pH and pOH of water at 25°C if the concentration of $[H^+]$ ions is $10^{-5} M$.

SOLUTION: The pH can be determined by taking the negative log of the H^+ activity, or

$$pH = -\log\{H^+\} = -\log(10^{-5}) = 5$$

The pOH may be determined one of two ways.

a. Using the ion product of water, we have

$$\{OH^-\} = \frac{K_w}{\{H\}} = \frac{10^{-14}}{10^{-5}} = 10^{-9}$$

$$pOH = -\log\{OH^-\} = -\log(10^{-9}) = 9$$

b. The alternative solution takes the negative logarithm of both sides of Equation 4.19,

$$14 = -\log\{H^+\} - \log\{OH^-\}$$

or

$$14 = pH + pOH$$

Since the pH has already been calculated as 5, the pOH is

$$pOH = 14 - pH = 14 - 5 = 9$$

The reaction for a typical monoprotic (one proton or H^+) acid is

$$HA \rightleftharpoons H^+ + A^-$$
4.20

And the equilibrium equation then becomes

$$K_A = \frac{\{H^+\}\{A^-\}}{\{HA\}}$$
4.21

Similarly, for a diprotic acid, the first ionization reaction is

$$H_2A = H^+ + HA^-$$
4.22

And the resulting equilibrium reaction is

$$K_{A1} = \frac{\{H^+\}\{HA^-\}}{\{H_2A\}}$$ **4.23**

The second ionization reaction is

$$HA^- \rightleftarrows H^+ + A^{2-}$$ **4.24**

and

$$K_{A2} = \frac{\{H^+\}\{A^{2-}\}}{\{HA^-\}}$$ **4.25**

Thus, a diprotic acid has two reactions (one for the loss of each proton) and two equilibrium reactions. The reactions for a triprotic acid follow similar trends. Selected acid–base reactions and their equilibrium constants are shown in Table 4.4.

TABLE 4.4 Selected Acid–Base Reactions with Approximate Equilibrium Values at 25°C

REACTION	NAME	K_a	$pK_a = -\log K_a$
$HCl \rightleftarrows H^+ + Cl^-$	Hydrochloric	1000	-3
$H_2SO_4 \rightleftarrows H^+ + HSO_4^-$	Sulfuric, H1	1000	-3
$HNO_3 \rightleftarrows H^+ + NO_3^-$	Nitric	$\simeq 1$	$\simeq 0$
$HSO_4^- \rightleftarrows H^+ + SO_4^{2-}$	Sulfuric, H2	1×10^{-2}	2
$H_3PO_4 \rightleftarrows H^+ + H_2PO_4^-$	Phosphoric, H1	7.94×10^{-3}	2.1
$HAc \rightleftarrows H^+ + Ac^-$	Acetic	2.00×10^{-5}	4.7
$H_2CO_3 \rightleftarrows H^+ + HCO_3^-$	Carbonic, H1	5.01×10^{-7}	6.3
$H_2S \rightleftarrows H^+ + HS^-$	Hydrosulfuric, H1	7.94×10^{-8}	7.1
$H_2PO_4^- \rightleftarrows H^+ + HPO_4^{2-}$	Phosphoric, H2	6.31×10^{-8}	7.2
$HOCl \rightleftarrows H^+ + OCl^-$	Hypochlorous	3.16×10^{-8}	7.5
$NH_4^+ \rightleftarrows H^+ + NH_3$	Ammonium	5.01×10^{-10}	9.3
$HCO_3^- \rightleftarrows H^+ + CO_3^{2-}$	Carbonic, H2	5.01×10^{-11}	10.3
$HPO_4^{2-} \rightleftarrows H^+ PO_4^{3-}$	Phosphoric, H3	5.01×10^{-13}	12.3

"Strong acid" is a relative term indicating any acid that donates or loses most of its protons in the pH range of interest. If we are studying natural waters, the pH range would be approximately 6 to 9. Any acid that loses most of its protons in this range would be considered a **strong acid**. If we are considering an industrial process that operates in the pH range of 3 to 5, then a strong acid would be any that loses most of its protons in this range. Although there are others, three strong acids common in both industrial processes and environmental engineering laboratories are nitric, hydrochloric, and sulfuric. Conversely, a **weak acid** is one that does not ionize completely (or donate all its protons) under the conditions of interest. Examples include carbonic acid, acetic acid and hypochlorus acid.

EXAMPLE 4.13	ACID IONIZATION REACTIONS

What fractions of the following acids are ionized at pH 7?

    a. Nitric
    b. Hydrochloric
    c. Sulfuric
    d. Hypochlorous

SOLUTION: **a.** Nitric acid. From Table 4.4, the pK_a of nitric acid is approximately 0. If we substitute the K_a and $[H^+]$ concentration into the equilibrium equation, we can obtain the ratio of the acid to the anion.

$$K_a = \frac{[H^+][NO_3^-]}{[HNO_3]}$$

$$\frac{[HNO_3]}{[NO_3^-]} = \frac{[H^+]}{K_a} = \frac{10^{-7}}{1} = 10^{-7}$$

The fraction ionized, F, is then

$$F = \frac{[NO_3^-]}{[HNO_3] + [NO_3^-]} = \frac{[NO_3^-]}{10^{-7}[NO_3^-] + [NO_3^-]} \simeq 1$$

The acid is almost completely ionized (almost all protons are released into the water).

    b. Hydrochloric acid. Similarly, from Table 4.4, the pK_a of hydrochloric acid is approximately -3. If we substitute the K_a and $[H^+]$

concentration into the equilibrium equation, we can obtain the ratio of the acid to the anion.

$$K_a = \frac{[H^+][Cl^-]}{[HCl]}$$

$$\frac{[HCl]}{[Cl^-]} = \frac{[H^+]}{K_a} = \frac{10^{-7}}{10^3} = 10^{-10}$$

The fraction ionized, F, is then

$$F = \frac{[Cl^-]}{[HCl] + [Cl^-]} = \frac{[Cl^-]}{10^{-10}[Cl^-] + [Cl^-]} \simeq 1$$

Again, almost all the acid has been ionized to the anion.

c. Sulfuric acid. This presents us with a slightly different problem. A diprotic acid, sulfuric has two protons to donate. Will it donate only one or both at normal pH? We can use the previous method to answer the question, but we must consider both reactions.

$$K_{a1} = \frac{[H^+][HSO_4^-]}{[H_2SO_4^-]}$$

$$\frac{[H_2SO_4]}{[HSO_4^-]} = \frac{[H^+]}{K_{a1}} = \frac{10^{-7}}{10^3} = 10^{-10}$$

The fraction ionized, F_1, is then

$$F_1 = \frac{[HSO_4^-]}{[H_2SO_4] + [HSO_4^-]} = \frac{[HSO_4^-]}{10^{-10}[HSO_4^-] + [HSO_4^-]} \simeq 1$$

Thus, sulfuric acid loses the first proton at normal pH. What about the second proton?

$$K_{a2} = \frac{[H^+][SO_4^{2-}]}{[HSO_4^-]}$$

$$\frac{[HSO_4^-]}{[SO_4^{2-}]} = \frac{[H^+]}{K_{a2}} = \frac{10^{-7}}{10^{-2}} = 10^{-5}$$

The fraction ionized, F_2, is then

$$F_2 = \frac{[SO_4^{2-}]}{[HSO_4^-] + [SO_4^{2-}]} = \frac{[SO_4^{2-}]}{[10^{-5}][SO_4^{2-}] + [SO_4^{2-}]} \simeq 1$$

The ratio of bisulfate (an acid) to sulfate anion is 10^{-5}. Again, the acid is almost completely ionized (almost all protons released into the water). Thus, in normal pH ranges, sulfuric acid ionizes almost completely, losing both protons.

d. Hypochlorous acid. From Table 4.4, the pK_a of hypochlorous acid is approximately 7.5. If we substitute the K_a and $[H^+]$ concentration into the equilibrium equation, we can obtain the ratio of the acid to the anion.

$$K_a = \frac{[H^+][OCl^-]}{[HOCl]}$$

$$\frac{[HOCl]}{[OCl^-]} = \frac{[H^+]}{K_a} = \frac{10^{-7}}{10^{-7.5}} = 3.16$$

The fraction ionized, F, is then

$$F = \frac{[OCl^-]}{[HOCl] + [OCl^-]} = \frac{[OCl^-]}{3.16[OCl^-] + [OCl^-]} = 0.24$$

Since only 24% of the acid ionizes, hypochlorous acid is a weak acid.

Precipitation–Dissolution Reactions.   Reactions involving solid phases are important in both natural and engineered systems. Carbonates and their effect on the composition of groundwater is one important example. Another is the use of precipitation reactions to remove materials from water and wastewater. Many toxic metals can be removed to leave low concentrations using precipitation. Several precipitation reactions of environmental significance are shown in Table 4.5.

Metals often precipitate as hydroxides. For a monobasic (1 OH^-) metal, M, the reaction is

$$MOH = M^+ + OH^- \qquad\qquad \textbf{4.26}$$

TABLE 4.5 Selected Precipitation Reactions with Approximate Equilibrium Values at 25°C

REACTION	NAME	K_{sp}	$pK_{sp} = -\log K_{sp}$
$CaCO_3 \rightleftarrows Ca^{2+} + CO_3^{2-}$	Calcium carbonate	5×10^{-9}	8.3
$Ca(OH)_2 \rightleftarrows Ca^{2+} + 2OH^-$	Calcium hydroxide	8×10^{-6}	5.1
$Ca_3(PO_4)_2 \rightleftarrows 3Ca^{2+} + 2PO_4^{3-}$	Calcium phosphate	1×10^{-27}	27
$Cr(OH)_3 \rightleftarrows Cr^{3+} + 3OH^-$	Chromium hydroxide	6×10^{-31}	30
$Cu(OH)_2 \rightleftarrows Cu^{2+} + 2OH^-$	Copper hydroxide	2×10^{-19}	18.7
$Fe(OH)_3 \rightleftarrows Fe^{3+} + 3OH^-$	Ferric hydroxide	6×10^{-36}	37
$Mg(CO)_3 \rightleftarrows Mg^{2+} + CO_3^{2-}$	Magnesium carbonate	4×10^{-5}	4.4
$Mg(OH)_2 \rightleftarrows Mg^{2+} + 2OH^-$	Magnesium hydroxide	9×10^{-12}	11
$Ni(OH)_2 \rightleftarrows Ni^{2+} + 2OH^-$	Nickel hydroxide	2×10^{-16}	16
$Zn(OH)_2 \rightleftarrows Zn^{2+} + 2OH^-$	Zinc hydroxide	3×10^{-17}	17

SOURCE: Sawyer and McCarty, Chemistry for Environmental Engineering, 3rd ed. (New York: McGraw-Hill, 1978).

and the equilibrium equation is

$$K_{sp} = \{M^+\}\{OH^-\} \qquad \textbf{4.27}$$

(Recall that the activity of a solid is unity.) For a dibasic metal the equilibrium reaction would be

$$M(OH)_2 = M^{2+} + 2OH^- \qquad \textbf{4.28}$$

and the equilibrium equation is

$$K_{sp} = \{M^{2+}\}\{OH^-\}^2 \qquad \textbf{4.29}$$

Calcium hydroxide (slaked lime) and zinc hydroxide are examples of dibasic metal hydroxides that behave as noted.

Gas Solubility Reactions. By convention, the concentration or activity of a gas is expressed as the partial pressure of the gas. For this general equilibrium reaction between a gas and its aqueous species

$$A(g) \rightleftarrows A(aq) \qquad \textbf{4.30}$$

the equilibrium expression (or Henry's law) is

$$K_H = \frac{\{A(aq)\}}{p_A}$$ **4.31**

where

K_H = equilibrium constant (often called Henry's law constant)

P_A = partial pressure of A [atm]

Table 4.1 (page 51) lists several gas reactions with their respective solubility constants.

Complexation. Some ions or molecules dissolved in water will combine with a variety of other ions or molecules to form several different species. This process is known as **complexation**. It can increase the solubility of the material significantly and can also alter its other chemical properties. And when complexation is not taken into account, the use of only the K_{sp} or solubility product may result in a gross underestimation of the actual solubility. Complexation is also used in several different laboratory determinations, including hardness and chloride analyses.

An example of the effects of complexation is mercury discharges into seawater. In the past, mercury has been associated with several serious pollution episodes, including those in the Great Lakes of the United States and Canada and in seawater off the coast of Japan. In seawater, mercury complexes with chloride in the water, forming several mercury-chloride species

$$Hg^{2+} + Cl^- \rightleftarrows HgCl^+; \quad K_1 = \frac{\{HgCl^+\}}{\{Hg^{2+}\}\{Cl^-\}}$$ **4.32**

$$HgCl^+ + Cl^- \rightleftarrows HgCl_2^0; \quad K_2 = \frac{\{HgCl_2^0\}}{\{HgCl^+\}\{Cl^-\}}$$ **4.33**

$$HgCl_2^0 + Cl^- \rightleftarrows HgCl_3^-; \quad K_3 = \frac{\{HgCl_3^-\}}{\{HgCl_2^0\}\{Cl^-\}}$$ **4.34**

$$HgCl_3^- + Cl^- \rightleftarrows HgCl_4^{2-}; \quad K_4 = \frac{\{HgCl_4^{2-}\}}{\{HgCl_3^-\}\{Cl^-\}}$$ **4.35**

The $\{HgCl_2^0\}$ indicates an uncharged aqueous species, not a solid or precipitate. The complexation of mercury with chloride greatly increases its solubility. This in turn means that more mercury is available in the water to be absorbed by aquatic life.

Many other compounds will sequentially complex with other ions in solution. Examples include the complexation of several metals [aluminum, copper, iron(III), lead, manganese, mercury, and zinc] with hydroxide species, and the complexation of mercury with ammonia. Other species will complex with organic **ligands** such as ethylaminediaminetetraacetic acid (or EDTA).

Cyanide is a common electroplating component. It is used in zinc, copper, and brass plating. It can be removed from drainage and rinsewater coming from plated parts by chemical destruction, ion exchange (discussed later in this chapter), and other means. However, chemical destruction requires that the cyanide not be strongly complexed. Unfortunately, iron forms strong complexes with cyanide, preventing its destruction by chemical means. Although iron is not present in most plating tanks (they are lined or made of nonreactive stainless steel), it can dissolve in the water if cast iron, rather than stainless steel, pipes are used. Example 4.7 at the end of this section on equilibrium chemistry demonstrates complexation calculations using iron–cyanide reactions.

Electroneutrality

Any aqueous solution is electrically neutral. The net sum of the positive charges minus the net sum of the negative charges is zero, or the net sum of the positive ionic charges equals the net sum of the anionic charges. In equation form, this is

$$\left(\sum_{i=1}^{n} n_A Z_i C_i \right)_{cation} - \left(\sum_{j=1}^{p} n_A Z_j C_j \right)_{anion} = 0 \qquad \textbf{4.36}$$

where

n_A = Avogadro's number, 6.02×10^{23} [atoms/mol]

Z_i = charge per cation of the ith species [charge/ion]

Z_j = charge per anion of the jth species [charge/ion]

C_i = concentration of the ith cationic species [mol/L]

C_j = concentration of the jth anionic species [mol/L]

Since Avogadro's number cancels from each side, the equation becomes

$$\left(\sum_{i=1}^{n} Z_i C_i\right)_{\text{cation}} - \left(\sum_{j=1}^{p} Z_j C_j\right)_{\text{anion}} = 0 \qquad \textbf{4.37}$$

It is important to note that since the electroneutrality equation is for the entire solution, there can be only one electroneutrality equation for any given solution!

| **EXAMPLE 4.14** | **ELECTRONEUTRALITY** |

Write the electroneutrality equation for aqueous solutions of the following:

    a. NaOH
    b. CH$_3$COOH
    c. Na$_2$CO$_3$
    d. Na$_2$CO$_3$ + NaHCO$_3$

SOLUTION: **a.** NaOH. When NaOH dissolves into water, the species are H$_2$O, H$^+$, OH$^-$, Na$^+$, and NaOH(aq) (negligible activity). Note that water and its ionization products are included (as always in aqueous chemistry). The cations are H$^+$ and Na$^+$. The only anion is OH$^-$. The nonionic (uncharged) species H$_2$O and NaOH(aq) do not enter into the electroneutrality equation. The equation is

$$Z_{\text{Na}^+} C_{\text{Na}^+} + Z_{\text{H}^+} C_{\text{H}^+} - Z_{\text{OH}^-} C_{\text{OH}^-} = 0$$

$$1\frac{\text{charge}}{\text{mol}} C_{\text{Na}^+} + 1\frac{\text{charge}}{\text{mol}} C_{\text{H}^+} - 1\frac{\text{charge}}{\text{mol}} C_{\text{OH}^-} = 0$$

b. CH$_3$COOH. Acetic acid ionizes in water to form H$^+$ and CH$_3$COO$^-$ (acetate) ions. (The acetate ion is often written Ac$^-$ and acetic acid HAc.) The electroneutrality equation is then

$$Z_{\text{H}^+} C_{\text{H}^+} - (Z_{\text{Ac}^-} C_{\text{Ac}^-} + Z_{\text{OH}^-} C_{\text{OH}^-}) = 0$$

$$1\frac{\text{charge}}{\text{mol}} C_{\text{H}^+} - \left(1\frac{\text{charge}}{\text{mol}} C_{\text{Ac}^-} + 1\frac{\text{charge}}{\text{mol}} C_{\text{OH}^-}\right) = 0$$

c. Na$_2$CO$_3$. When sodium carbonate dissolves in water, sodium ions [Na$^+$] and carbonate ions [CO$_3^{2-}$] are formed. However, the carbonate ions will react with the [H^+] ions already present in the water. This forms bicarbonate ions [HCO_3^-] and carbonic acid, H_2CO_3. The resulting electroneutrality equation is

$$(Z_{H^+}C_{H^+} + Z_{Na^+}C_{Na^+})$$
$$-(Z_{CO_3^{2-}}C_{CO_3^{2-}} + Z_{HCO_3^-}C_{HCO_3^-} + Z_{OH^-}C_{OH^-}) = 0$$

$$1\frac{charge}{ion}C_{H^+} + 1\frac{charge}{ion}C_{Na^+}$$

$$-\left(2\frac{charge}{ion}C_{CO_3^{2-}} + 1\frac{charge}{ion}C_{HCO_3^-} + 1\frac{charge}{ion}C_{OH^-}\right) = 0$$

d. $Na_2CO_3 + NaHCO_3$. When sodium bicarbonate dissolves in water, the same species are formed as when sodium carbonate dissolves. The resulting concentrations are different, but the species are not. So the electroneutrality equation is

$$(Z_{H^+}C_{H^+} + Z_{Na^+}C_{Na^+})$$
$$-(Z_{CO_3^{2-}}C_{CO_3^{2-}} + Z_{HCO_3^-}C_{HCO_3^-} + Z_{OH^-}C_{OH^-}) = 0$$

$$1\frac{charge}{ion}C_{H^+} + 1\frac{charge}{ion}C_{Na^+}$$

$$-\left(2\frac{charge}{ion}C_{CO_3^{2-}} + 1\frac{charge}{ion}C_{HCO_3^-} + 1\frac{charge}{ion}C_{OH^-}\right) = 0$$

Mass Balances

A mass balance is a written statement or equation that describes the different species in which a **component** may exist. As an example, carbonate may exist as in water as CO_3^{2-}, HCO_3^-, H_2CO_3, and CO_2(aq). Depending on equilibrium conditions, it may interconvert among these different species. So a mass balance on carbonate must take this into consideration. A generalized expression for the mass balance of a component is

$$C_{t,i} = \sum_{j=1}^{n} C_j \qquad\qquad \textbf{4.38}$$

where

$$C_{t,i} = \text{total concentration of component } i \text{ [mol/L, } M]$$

$$C_j = \text{concentration of species } j \text{ of component } i \text{ [mol/L, } M]$$

EXAMPLE 4.15	MASS BALANCES

Write a mass balance for the following:

a. 0.01 M sodium chloride [NaCl]
b. 0.05 M phosphoric acid [H$_3$PO$_4$]
c. 0.025 M phosphoric acid [H$_3$PO$_4$] and 0.025 M sodium hydrogen phosphate [NaH$_2$PO$_4$]
d. 0.005 M sodium acetate [CH$_3$COONa] and 0.005 M acetic acid [CH$_3$COOH]

SOLUTION:

a. 0.01 M sodium chloride. What can form when sodium chloride is dissolved in water? You cannot answer this question without some previous experience or the help of a reference. The answer is that neither sodium nor chloride will form any significant other species — just sodium ions and chloride ions. Thus, the mass balances are

$$C_{t,\text{Na}} = [\text{Na}^+] = 0.01 \, M$$

$$C_{t,\text{Cl}} = [\text{Cl}^-] = 0.01 \, M$$

b. 0.05 M phosphoric acid [H$_3$PO$_4$]. When phosphoric acid dissolves in water, some molecules will retain all three protons and remain as [H$_3$PO$_4$]. Others will lose one, two, or three protons, becoming the phosphate anions. Thus, all four species must be considered, or

$$C_{t,\text{PO}_4} = [\text{H}_3\text{PO}_4^-] + [\text{H}_2\text{PO}_4^-] + [\text{HPO}_4^{2-}] + [\text{PO}_4^{3-}] = 0.05 \, M$$

Since [H$^+$] ions are in the water prior to the addition of the phosphoric acid, no mass balance can be written on [H$^+$].

c. 0.025 M phosphoric acid [H$_3$PO$_4$] and 0.025 M sodium hydrogen phosphate [Na$_2$HPO$_4$]. We take a similar approach to writing this mass balance, except that we must also consider sodium. Note that there are two sodium atoms in each molecule of sodium hydrogen phosphate. Thus, the total molar concentration of sodium is 0.05 M.

The mass balance is then

$$C_{t,\mathrm{Na}} = [\mathrm{Na}^+] = 0.05\,M$$

As in part (b), 0.05 M of phosphate is added to the water. Thus, the mass balance for phosphate is

$$C_{t,\mathrm{PO_4}} = [\mathrm{H_3PO_4}] + [\mathrm{H_2PO_4^-}] + [\mathrm{HPO_4^{2-}}] + [\mathrm{PO_4^{3-}}] = 0.05\,M$$

Note: *It is unimportant for the mass balane equation whether the various phosphate species originated from the phosphoric acid or from the sodium hydrogen phosphate.*

d. 0.005 M sodium acetate [$\mathrm{CH_3COONa}$] and 0.005 M acetic acid [$\mathrm{CH_3COOH}$]. In this situation, the total sodium concentration would be 0.005 M, or

$$C_{t,\mathrm{Na}} = [\mathrm{Na}^+] = 0.005\,M$$

The total acetate concentration would be 0.01 M. However, the acetate can exist as either acetic acid or the acetate ion, depending on the pH of the solution. Thus, the mass balance for the acetate is

$$C_{t,\mathrm{Ac}} = [\mathrm{HAc}] + [\mathrm{Ac}^-] = 0.01\,M$$

Equilibrium Calculations

We now have the fundamental tools to write and solve the necessary equations to determine the concentrations of the various species in an aqueous system. The first step in the process is to identify all possible species in the system. Once we know reactants and products, we can write the various equilibrium reactions and determine their constants—again by reference to appropriate tables or books [3, 4, 5]. We write out each possible reaction. Then we write the types of equations necessary: equilibrium, mass balance, and electroneutrality. The following examples illustrate the process.

EXAMPLE 4.16	EQUILIBRIUM CALCULATIONS

Write the necessary equations to solve for all species in the following aqueous solutions:

    a. $5 \times 10^{-4}\,M$ phosphoric acid [$\mathrm{H_3PO_4}$].
    b. $5 \times 10^{-4}\,M$ sodium hydrogen phosphate [$\mathrm{NaH_2PO_4}$].

SOLUTION: **a.** $0.05\,M$ phosphoric acid $[H_3PO_4]$

Step 1. Determination of species:

The possible species formed from phosphoric acid are $[H_3PO_4]$, $[H_2PO_4^-]$, $[HPO_4^{2-}]$, $[PO_4^{3-}]$, and $[H^+]$. In addition, water contributes the hydroxide ion $[OH^-]$. So there are six unknown species, and six equations are required.

Step 2. Equilibrium equations:

There are three equilibrium reactions for the phosphoric acid. They are

$$H_3PO_4 \rightleftharpoons H^+ + H_2PO_4^-$$

$$H_2PO_4^- \rightleftharpoons H^+ + HPO_4^{2-}$$

$$HPO_4^{2-} \rightleftharpoons H^+ + PO_4^{3-}$$

The resulting three equilibrium equations are

$$K_{a1} = \frac{[H^+][H_2PO_4^-]}{[H_3PO_4]}$$

$$K_{a2} = \frac{[H^+][HPO_4^{2-}]}{[H_2PO_4^-]}$$

$$K_{a3} = \frac{[H^+][PO_4^{3-}]}{[HPO_4^{2-}]}$$

We now have three of the required equations. Since this is an aqueous system, we can use the self-ionization of water:

$$H_2O \rightleftharpoons H^+ + OH^-$$

and the equilibrium equation is then

$$K_w = [H^+][OH^-]$$

Step 3. Mass balances:

There is only one mass balance possible in this system, that for phosphate:

$$C_{t,PO_4} = [H_3PO_4] + [H_2PO_4^-] + [HPO_4^{2-}] + [PO_4^{3-}]$$

Step 4. Electroneutrality:

The electroneutrality equation is

$$[H^+] = [H_2PO_4^-] + 2[HPO_4^{2-}] + 3[PO_4^{3-}] + [OH^-]$$

Thus, we have six equations and six unknowns. These may now be solved for the different species, as well as the pH. Computer solution of this system yields:

SPECIES	CONCENTRATION, mol/L
H_3PO_4	3.16×10^{-5}
$H_2PO_4^-$	4.68×10^{-4}
HPO_4^{2-}	6.3×10^{-8}
PO_4^{3-}	6.3×10^{-17}
H^+	5.01×10^{-4}
OH^-	1.99×10^{-11}

b. $5 \times 10^{-4} M$ sodium hydrogen phosphate [NaH_2PO_4]

Step 1. Determination of species:

The solution to this problem differs very little from the previous one. Sodium contributes one additional species and requires one additional equation, the trivial mass balance for sodium. The electroneutrality equation also changes. The answers, however, are quite different.

The possible species formed from sodium hydrogenphosphate are [H_3PO_4], [$H_2PO_4^-$], [HPO_4^{2-}], [PO_4^{3-}], [H^+], and in addition, [Na^+]. As before, water also contributes the hydroxide ion [OH^-]. So, there are seven unknown species and seven equations are required.

Step 2. Equilibrium equations:

There are three equilibrium reactions for the phosphoric acid. The presence of sodium in the solution does not change these equilibrium equations.

$$H_3PO_4 \rightleftarrows H^+ + H_2PO_4^-$$
$$H_2PO_4^- \rightleftarrows H^+ + HPO_4^{2-}$$
$$HPO_4^{2-} \rightleftarrows H^+ + PO_4^{3-}$$

The resulting three equilibrium equations are

$$K_{a1} = \frac{[H^+][H_2PO_4^-]}{[H_3PO_4]}$$

$$K_{a2} = \frac{[H^+][HPO_4^{2-}]}{[H_2PO_4^-]}$$

$$K_{a3} = \frac{[H^+][PO_4^{3-}]}{[HPO_4^{2-}]}$$

We now have three of the required equations. Since this is an aqueous system, we can use the self-ionization of water:

$$H_2O \rightleftharpoons H^+ + OH^-$$

and the equilibrium equation is then

$$K_w = [H^+][OH^-]$$

Step 3. Mass balances:

The additional equation occurs here. There is one mass balance for phosphate and a second for sodium:

$$C_{t,PO_4} = [H_3PO_4] + [H_2PO_4^-] + [HPO_4^{2-}] + [PO_4^{3-}]$$
$$= [Na^+]$$

Step 4. Electroneutrality:

The electroneutrality equation is different due to the presence of the sodium ion:

$$[H^+] + [Na^+] = [H_2PO_4^-] + 2[HPO_4^{2-}] + 3[PO_4^{3-}] + [OH^-]$$

We now have seven equations and seven unknowns. These may be solved for the different species, as well as the pH. Solution of these equations yields:

SPECIES	CONCENTRATION, mol/L
H_3PO_4	4.26×10^{-7}
$H_2PO_4^-$	4.81×10^{-4}
HPO_4^{2-}	6.64×10^{-6}
PO_4^{3-}	9.36×10^{-13}
Na^+	5.00×10^{-4}
H^+	4.52×10^{-6}
OH^-	2.14×10^{-9}

EXAMPLE 14.17	COMPLEXATION — CYANIDE AND IRON(III)

A zinc electroplating rinse water has the following composition:

Total sodium:	$10^{-4}\,M$
Total iron(III):	$10^{-3}\,M$
Total zinc:	$10^{-3}\,M$
Total carbonate:	$10^{-3}\,M$
Total cyanide:	$10^{-3}\,M$

At pH 10, what fraction of the cyanide ions is complexed with iron(III)? You may assume that the system is closed. The following reactions are applicable:

$$Fe^{3+} + 6CH^- \rightleftharpoons Fe(CN)_6^{3-} \qquad pK = 43.57$$

$$Zn^{2+} + 4CH^- \rightleftharpoons Zn(CN)_4^{2-} \qquad pK = 17.18$$

SOLUTION: We must first determine the possible reactions and species. Descriptions of the reactions can be found in several aquatic chemistry texts and other chemical references. The solubility reactions are

$$Na_2CO_3(s) \rightleftharpoons 2Na^+ + CO_3^{2-} \qquad pK_{sp} = 12.7$$

$$ZnCO_3(s) \rightleftharpoons Zn^{2+} + CO_3^{2-} \qquad pK_{sp} = 10.76$$

$$Zn(OH)_2(s) \rightleftharpoons Zn^{2+} + 2OH^- \qquad pK_{sp} = -11.81$$

$$Fe(OH)_3(s) \rightleftharpoons Fe^{3+} + 3OH^- \qquad pK_{sp} = -2.93$$

Self-ionization of water:

$$H_2O \rightleftharpoons H^+ + OH^- \qquad pK_w = 14$$

Carbonate reactions:

$$H_2CO_3(aq) \rightleftharpoons H^+ + HCO_3^- \qquad pK_{a1} = 6.3$$

$$HCO_3^- \rightleftharpoons H^+ + CO_3^{2-} \qquad pK_{a2} = 10.3$$

Complexation reactions:

$$Fe^{3+} + 6CN^- \rightleftharpoons Fe(CN)_6^{3-} \qquad pK = 43.57$$

$$Zn^{2+} + 4CN^- \rightleftharpoons Zn(CN)_4^{2-} \qquad pK = 17.18$$

Cyanide:

$$H^+ + CN^- \rightleftharpoons HCN(aq) \qquad pK_a = 9.3$$

The species are then:

$[H^+]$, $[Na^+]$, $[Zn^{2+}]$, $[Fe^{3+}]$, $[HO^-]$, $[CO_3^{2-}]$, $[HCO_3^-]$, $[H_2CO_3]$, $[CN^-]$, $[HCN(aq)]$, $[Fe(CN)_6^{3-}]$, $[Zn(CN)_4^{2-}]$, $[Na_2CO_3(s)]$, $[Fe(OH)_3(s)]$, $[Zn(OH)_2(s)]$, and $[ZnCO_3(s)]$.

There are 16 in all. So we need 16 equations to solve the problem. The equilibrium reactions provide 10 equations. And there are 5 mass balance equations:

$$C_{t,Na} \rightleftharpoons [Na^+] + [Na_2CO_3(s)] \rightleftharpoons 10^{-4} M$$

$$C_{t,Fe} \rightleftharpoons [Fe^{3+}] + [Fe(OH)_3(s)] + [Fe(CN)_6^{3-}] \rightleftharpoons 10^{-3} M$$

$$C_{t,Zn} \rightleftharpoons [Zn^{2+}] + [ZnCO_3(s)] + [Zn(OH)_2(s)] + [Zn(CN)_4^{2-}] \rightleftharpoons 10^{-3} M$$

$$C_{t,CO_3} \rightleftharpoons [CO_3^{2-}] + [HCO_3^-] + [H_2CO_3(aq)] + [Na_2CO_3(s)]$$
$$+ [ZnCO_3(s)] \rightleftharpoons 10^{-4} M$$

$$C_{t,CN} \rightleftharpoons [CN^-] + [HCN(aq)] + [Fe(CN)_6^{3-}] + [Zn(CN)_4^{2-}] \rightleftharpoons 10^{-4} M$$

And finally electroneutrality:

$$[H^+] + [Na^+] + 2[Zn^{2+}] + 3[Fe^{3+}] \rightleftharpoons [OH^-] + [HCO_3^-]$$
$$+ 2[CO_3^{2-}] + 2[Zn(CN)_4^{2-}] + 3[Fe(CN)_6^{3-}]$$

Solution of these equations using an equilibrium chemistry computer program yields:

SPECIES	EQUILIBRIUM CONCENTRATION, M	COMMENTS
$[H^+]$	10^{-10}	
$[OH^-]$	10^{-4}	
$[HCN(aq)]$	$10^{-4.16}$	
$[CN^-]$	$10^{-3.43}$	
$[Na^+]$	$10^{-4.00}$	
$[Na_2CO_3(s)]$	0	Does not precipitate
$[Fe^{3+}]$	$10^{-26.9}$	
$[Fe(OH)_3(s)]$	$10^{-3.04}$	Precipitates
$[Fe(CN)_6^{3-}]$	$10^{-4.05}$	
$[Zn^{2+}]$	$10^{-8.50}$	
$[Zn(OH)_2(s)]$	0	Does not precipitate
$[ZnCO_3(s)]$	$10^{-3.70}$	Precipitates
$[Zn(CN)_4^{2-}]$	$10^{-5.10}$	
$[CO_3^{2-}]$	$10^{-3.66}$	
$[HCO_3^-]$	$10^{-3.41}$	
$[H_2CO_3(aq)]$	$10^{-7.04}$	

The Carbonate System

Carbon dioxide and carbonates are crucial in water chemistry. Carbon dioxide is a minor, although important, constituent of the earth's atmosphere. Carbon dioxide and carbonates are in all surfacewaters, and carbonates are contained in many minerals. Groundwater contains significant levels of carbonates when it has been exposed to magnesium and calcium carbonate minerals. Also, carbonates buffer many natural waters against abrupt pH changes. For these reasons and others, carbonates play an important role in all natural water systems.

Dissolved carbon dioxide molecules combine with water molecules to form carbonic acid. The reaction is as follows:

$$CO_2(aq) + H_2O \rightleftharpoons H_2CO_3 \qquad \textbf{4.39}$$

It is so difficult to distinguish between the aqueous-phase carbon dioxide and carbonic acid that many calculations and measurements do not differentiate the two forms. This combined species is written $[H_2CO_3^*]$ and usually expressed as "H-2-C-O-3-Star." Thus, the reaction is

$$[CO_2(aq)] + [H_2CO_3] = [H_2CO_3^*] \qquad \textbf{4.40}$$

Relationship Between pH and pK_a

A useful relationship for describing the ratio of the anion (or conjugate base) to the acid $[A^-]/[HA]$ can be established by taking the logarithm of each side of a generalized acid reaction. For this reaction

$$HA \rightleftharpoons H^+ + A^- \qquad \textbf{4.41}$$

or

$$K_a = \frac{[H^+][A^-]}{[HA]} \qquad \textbf{4.42}$$

the equilibrium equation is written by taking the logarithm of

$$\log K_a = \log [H^+] + \log \frac{[A^-]}{[HA]} \qquad \textbf{4.43}$$

or

$$pH = pK_a + \log \frac{[A^-]}{[HA]} \qquad \textbf{4.44}$$

Equation 4.44, often called the Henderson–Hasselbach equation, provides a useful method of estimating the ratio of anion to acid — when the pH and the acid's pK_a value are known.

The pH at which an acid is 50% ionized — that is, where the $[A^-]$ concentration equals the $[HA]$ concentration — is related to the pK_a of the acid. If we use $[A^-]/[HA] = 1$ in Equation 4.44, we obtain

$$pH = pK_a + \log \frac{[A^-]}{[HA]} = pK_a + \log(1) \qquad \textbf{4.45}$$

or

$$pH = pK_a \qquad \textbf{4.46}$$

Therefore, when the $pH = pK_a$, the anion (or conjugate base) and the acid concentrations are equal, or $[A^-] = [HA]$. The following example demonstrates the use of this relationship.

EXAMPLE 4.18	**ACID IONIZATION**

The K_a for hypochlorous acid is $10^{-7.5}$ and the pK_a is 7.5.
a. Find the pH at which the acid and conjugate base concentrations are equal—that is $[HA] = [A^-]$.
b. Determine the pH at which the conjugate base is only 10% of the acid concentration, or $[A^-] = 0.1\,[HA]$.

SOLUTION: **a.** From Equation 4.46 the pH must equal the pK_a, or $pH = 7.5$.
b. If we substitute $[A^-] = 0.1[HA]$ into Equation 4.44, we obtain

$$pH = pK_a + \log \frac{0.1[HA]}{[HA]} = 7.5 + \log(0.1)$$

or

$$pH = 6.5$$

UNITS OF EXPRESSION

There are several ways to denote the concentrations of solutions. The most common units are probably moles/liter (M) and milligrams/liter (mg/L), but other units of expression have advantages in certain applications, and some materials are traditionally expressed in units that are less common. Equivalents per liter (normality) or milliequivalents per liter (meq/L) are ordinarily used in many laboratories, including environmental laboratories. Equivalents, or mg/L, of $CaCO_3$ are often used to express hardness and alkalinity concentrations. Other atypical approaches include using only the nitrogen content to express the concentration of several nitrogen-containing species, and the similar use of phosphorus to denote different phosphorus-containing compounds. Each of these practices will be briefly explained.

Normality

Equivalent concentrations are often used in the laboratory because equivalent amounts of two substances will react completely. One **equivalent** is the mass of a compound that will produce 1 mol of available reacting substance. Thus, for an acid, this would be the mass that produces 1 mol of H^+. For a base, it is the mass that will yield 1 mol of OH^-. The equivalent concentration of a compound is dependent on the reaction in which it is used. For example, if sulfuric acid $[H_2SO_4]$ is used to neutralize sodium hydroxide [NaOH], the reaction is

$$H_2SO_4 + 2NaOH \rightarrow Na_2SO_4 + 2H_2O \qquad\qquad \textbf{4.47}$$

In this reaction each mole of sulfuric acid produces 2 mol of H^+ ions. A generalized relationship for the situation is then

$$E = Zn \qquad\qquad \textbf{4.48}$$

where

E = equivalents [eq]

Z = equivalents per mole [eq/mol]

n = moles [mol]

In this equation each mole of sulfuric acid supplies 2 mol of H^+ to react with the OH^- in the sodium hydroxide. Thus, sulfuric acid has two equivalents per mole, or $Z = 2$ eq/mol. However, the sodium hydroxide supplies only one OH^- for each mole. Therefore, the Z for sodium hydroxide is 1 eq/mol.

For concentration terms, we can simply divide Equation 4.48 by the solution volume, obtaining

$$C_E = ZC_n$$

or

$$\left[\frac{\text{equivalents}}{\text{liter}}\right] = \left[\frac{\text{equivalents}}{\text{mole}}\right] \times \left[\frac{\text{moles}}{\text{liter}}\right] \qquad\qquad \textbf{4.49}$$

where

C_E = concentration in equivalents per liter, or normality [N]

Z = relationship between equivalents and moles [mole/eq]

C_n = concentration in moles per liter [M]

The following examples illustrate the use of equivalent concentrations.

EXAMPLE 4.19	EQUIVALENT CONCENTRATIONS

Determine the Z value and then calculate the normality of the following solutions:

a. 36.5 g/L hydrochloric acid [HCl]
b. 80 g/L sodium hydroxide [NaOH]
c. 9.8 g/L sulfuric acid [H_2SO_4]
d. 9.0 g/L acetic acid [CH_3COOH]

SOLUTION: **a.** The molecular weight of hydrochloric acid is 36.5 g. The concentration of HCl is therefore

$$C_n = \frac{mol}{36.5\,g} \times \frac{36.5\,g}{L} = 1\,M$$

Since hydrochloric acid supplies one proton, the Z term is 1 eqmol. The normality is then

$$C_E = ZC_m = 1\,\frac{eq}{mol} \times 1\,\frac{mol}{L}$$

$$= 1\,\frac{eq}{L} = 1\,N$$

b. The molecular weight of sodium hydroxide is 40 g. Thus,

$$C_n = \frac{mol}{40\,g} \times \frac{80\,g}{L} = 2\,M$$

NaOH donates one OH$^-$. The Z term is 1 eq/mol. The normality of the solution is thus

$$C_E = ZC_n = 1 \frac{eq}{mol} \times 2 \frac{mol}{L}$$

$$= 2 \frac{eq}{L} = 2\,N$$

c. With sulfuric acids' molecular weight of 98 g/mol, the concentration of H_2SO_4 is

$$C_n = \frac{mol}{98\,g} \times \frac{9.8\,g}{L} = 0.1\,M$$

Under conditions of environmental interest, sulfuric acid will donate both protons. The Z term is then 2 eq/mol, and the normality of the solution is

$$C_E = ZC_n = 2 \frac{eq}{mol} \times 0.1 \frac{mol}{L}$$

$$= 0.2 \frac{eq}{L} = 0.2\,N$$

d. Since molecular weight of acetic acid is 60 g/mol, its concentration is

$$C_n = \frac{mol}{60\,g} \times \frac{9\,g}{L} = 0.15\,M$$

To determine the Z value, you must realize that acetic acid will donate only one proton. The reaction for acetic acid, then, is

$$CH_3COOH \rightarrow CH_3COO^- + H^+$$

The Z term is thus 1 eq/mol. The normality of the solution is

$$C_E = ZC_n = 1 \frac{eq}{mol} \times 0.15 \frac{mol}{L}$$

$$= 0.15 \frac{eq}{L} = 0.15\,N$$

| **EXAMPLE 4.20** | **NORMAL SOLUTIONS** |

Determine the mass of base or acid in each of the following solutions:

  a. 500 mL of 10 N sodium hydroxide [NaOH]
  b. 2 L of 0.1 N phosphoric acid [H_3PO_4]

SOLUTION: **a.** To determine the mass of sodium hydroxide in 500 mL of 10 N NaOH, we solve Equation 4.49 for C_n.

$$C_n = \frac{C_E}{Z} = \frac{10 \text{ eq/L}}{1 \text{ eq/mol}} = 10 \frac{\text{mol}}{\text{L}}$$

Recall that concentration times volume determines mass. Hence,

$$m = CV = 10 \frac{\text{mol}}{\text{L}} \times \frac{40 \text{ g}}{\text{mol}} \times 500 \text{ mL} \times \frac{1 \text{ L}}{1000 \text{ mL}}$$

$$= 200 \text{ g NaOH}$$

b. In that each mole of phosphoric acid releases three protons, the concentration in moles per liter is

$$C_n = \frac{C_E}{Z} = \frac{0.1 \text{ eq/L}}{3 \text{ eq/mol}} = 0.033 \text{ mol/L}$$

And the mass contained in the 2-L solution is then

$$m = CV = 0.033 \frac{\text{mol}}{\text{L}} \times \frac{98 \text{ g}}{\text{mol}} \times 2 \text{ L}$$

$$= 6.5 \text{ g } H_3PO_4$$

Concentrations Using $CaCO_3$

Both alkalinity and hardness are usually expressed in either meq/L or mg/L $CaCO_3$. Alkalinity is the acid-neutralizing capacity of a sample of water. In equation form it is

$$\text{Alk} = [HCO_3^-] + 2[CO_3^{2-}] + [OH^-] - [H^+] \qquad \textbf{4.50}$$

where concentrations are given in moles per liter. (The carbonate can accept two protons.) However, if alkalinity is expressed as meq/L or mg/L $CaCO_3$, the units reflect the reactivity of the carbonate ion. The equation then becomes

$$Alk = [HCO_3^-] + [CO_3^{2-}] + [OH^-] - [H^+] \qquad \textbf{4.51}$$

Hardness is the sum of the divalent cations in water. Both alkalinity and hardness will be discussed in more detail in Chapter 8, Water Quality. Our objective here is to introduce the procedure to express concentrations in units of mg/L $CaCO_3$.

To express a component as calcium carbonate, multiply its concentration by the ratio of the equivalent weight of calcium carbonate divided by the component's equivalent weight. The following two examples demonstrate this approach.

EXAMPLE 4.21	ALKALINITY EXPRESSED AS mg/L CaCO₃ AND AS meq/L

A water with pH 9.5 has the following composition. Calculate the alkalinity of the water expressed as mg/L $CaCO_3$ and as meq/L.

ION	CONCENTRATION mg/L
CO_3^{2-}	10
HCO_3^-	65

SOLUTION:

a. The pH is 9.5, so $[H^+] = 10^{-9.5} M$. To express this as $CaCO_3$, we multiply by the equivalent mass of $CaCO_3$:

$$[H^+] = 10^{-9.5} \frac{\text{mol } H^+}{L} \times \frac{1 \text{ eq}}{\text{mol } H^+} \times \frac{50 \text{ g } CaCO_3}{\text{eq}}$$

$$\times \frac{10^3 \text{ mg } CaCO_3}{\text{g } CaCO_3}$$

$$= 0.000016 \text{ mg/L as } CaCO_3 \simeq 0 \text{ mg/L as } CaCO_3$$

For OH^-,

$$[OH^-] = 10^{-4.5} \frac{\text{mol } OH^-}{L} \times \frac{1 \text{ eq}}{\text{mol } OH^-} \times \frac{50 \text{ g CaCO}_3}{\text{eq}}$$

$$\times \frac{10^3 \text{ mg CaCO}_3}{\text{g CaCO}_3}$$

$$= 1.6 \text{ mg/L as CaCO}_3$$

For HCO_3^-,

$$[HCO_3^-] = 65 \frac{\text{mg } HCO_3^-}{L} \times \frac{\text{mol } HCO_3^-}{61 \text{ g } HCO_3^-} \times \frac{\text{g } HCO_3^-}{10^3 \text{ mg } HCO_3^-}$$

$$\times \frac{1 \text{ eq}}{\text{mol } HCO_3^-} \times \frac{50 \text{ g CaCO}_3}{\text{eq}} \times \frac{10^3 \text{ mg CaCO}_3}{\text{g CaCO}_3}$$

$$= 53 \text{ mg/L as CaCO}_3$$

Note that for CO_3^{2-} there are 2 eq/mol. This is because carbonate consumes or neutralizes two protons. Thus,

$$[CO_3^{2-}] = 10 \frac{\text{mg } CO_3^{2-}}{L} \times \frac{\text{mol } CO_3^{2-}}{60 \text{ g } CO_3^{2-}} \times \frac{\text{g } CO_3^{2-}}{10^3 \text{ mg } CO_3^{2-}}$$

$$\times \frac{2 \text{ eq}}{\text{mol } CO_3^{2-}} \times \frac{50 \text{ g CaCO}_3}{\text{eq}} \times \frac{10^3 \text{ mg CaCO}_3}{\text{g CaCO}_3}$$

$$[CO_3^{2}] = 17 \text{ mg/L as CaCO}_3$$

Now, using Equation 4.51, we derive the total alkalinity:

$$\text{Alk} = [HCO_3] + [CO_3^{2-}] + [OH^-] - [H^+]$$

$$\text{Alk as CaCO}_3 = 53 \frac{\text{mg}}{L} + 17 \frac{\text{mg}}{L} + 2 \frac{\text{mg}}{L} - 0 \frac{\text{mg}}{L}$$

$$= 72 \frac{\text{mg}}{L}$$

b. To express the alkalinity in meq/L, we first determine the concentration for both $[H^+]$ and $[OH^-]$ in meq/L.

For H^+,

$$[H^+] = 10^{-9.5} \frac{\text{mol } H^+}{L} \times \frac{1 \text{ eq}}{\text{mol } H^+}$$

$$= 10^{-9.5} \frac{\text{eq}}{L} = 3.2 \times 10^{-10} N$$

For OH^-,

$$[OH^-] = 10^{-4.5} \frac{\text{mol } OH^-}{L} \times \frac{1 \text{ eq}}{\text{mol } OH^-}$$

$$= 10^{-4.5} \frac{\text{eq}}{L} = 3.2 \times 10^{-5} N$$

For HCO_3^-

$$[HCO_3^-] = 65 \frac{\text{mg } HCO_3^-}{L} \times \frac{\text{mol } HCO_3^-}{61 \text{ g } HCO_3^-} \times \frac{\text{g } HCO_3^-}{10^3 \text{ mg } HCO_3^-}$$

$$\times \frac{1 \text{ eq}}{\text{mol } HCO_3^-}$$

$$= 1.1 \times 10^{-3} N$$

Note that for CO_3^{2-} there are 2 eq/mol. This, again, results from carbonate's capacity to consume or neutralize two protons. So

$$[CO_3^{2-}] = 10 \frac{\text{mg } CO_3^{2-}}{L} \times \frac{\text{mol } CO_3^{2-}}{60 \text{ g } CO_3^{2-}} \times \frac{\text{g } CO_3^{2-}}{10^3 \text{ mg } CO_3^{2-}}$$

$$\times \frac{2 \text{ eq}}{\text{mol } CO_3^{2-}}$$

$$= 3.3 \times 10^{-4} N$$

Again, using Equation 4.51, we obtain the total alkalinity:

$$Alk = [HCO_3^-] + [CO_3^{2-}] + [OH^-] - [H^+]$$

$$= 1.1 \times 10^{-3} N + 3.3 \times 10^{-4} N + 3.2 \times 10^{-5} N - 3.2 \times 10^{-10} N$$

$$= 1.4 \times 10^{-3} N$$

EXAMPLE 4.22	HARDNESS EXPRESSED AS mg/L CaCO₃ AND AS meq/L

A sample of water has 25 mg/L of Ca^{2+} and 15 mg/L of Mg^{2+}. Calculate its hardness as mg/L $CaCO_3$ and as meq/L.

SOLUTION: **a.** First we must express each concentration as mg/L $CaCO_3$. For Ca^{2+} it is

$$[Ca^{2+}] = 25\,\frac{mg\,Ca^{2+}}{L} \times \frac{g\,Ca^{2+}}{10^3\,mg\,Ca^{2+}} \times \frac{mol\,Ca^{2+}}{40\,g\,Ca^{2+}} \times \frac{2\,eq}{mol\,Ca^{2+}}$$

$$\times \frac{50\,g\,CaCO_3}{eq} \times \frac{10^3\,mg\,CaCO_3}{g\,CaCO_3}$$

$$= 62.5\,\frac{mg}{L}\text{ as }CaCO_3$$

Similarly, for Mg^{2+}, it is

$$[Mg^{2+}] = 15\,\frac{mg\,Mg^{2+}}{L} \times \frac{g\,Mg^{2+}}{10^3\,mg\,Mg^{2+}} \times \frac{mol\,Mg^{2+}}{24.3\,g\,Mg^{2+}}$$

$$\times \frac{2\,eq}{mol\,Mg^{2+}} \times \frac{50\,g\,CaCO_3}{eq} \times \frac{10^3\,mg\,CaCO_3}{g\,CaCO_3}$$

$$= 61.7\,\frac{mg}{L}\text{ as }CaCO_3$$

b. The total hardness, then, is the sum of the calcium and magnesium concentrations expressed as $CaCO_3$, or

$$\text{Total hardness} = 124\,\frac{mg}{L}\text{ as }CaCO_3$$

Concentrations as N or P

Nitrogen exists in several inorganic forms in the natural environment. It can be readily interconverted among the different forms by a variety of microorganisms. Phosphorus also occurs in different forms, and it can also be interconverted. It is easier to express the different forms not as the total concentration of the compound containing the element but as only the N or P fraction. The advantage of this is that as the compound is converted, the concentration then remains the same. As an example, if we express the concentration of ammonia as ammonia nitrogen instead of ammonia, as it is converted to nitrate, one milligram of ammonia nitrogen produces one milligram of nitrate nitrogen. When ammonia is expressed as nitrogen, it is written x.x mg NH_3-N/L or x.x mg/L (as N). Nitrate would be y.y mg NO_3^--N/L or y.y mg/L (as N). To convert a concentration from mg/L of the compound to mg/L as N, multiply by the ratio of the compound's atomic weight divided by the atomic weight of nitrogen contained in the compound. The following example illustrates the concept.

EXAMPLE 4.23	**INORGANIC NITROGEN COMPOUNDS EXPRESSED AS NITROGEN**

Ammonia is converted to nitrite and then nitrate by the process of nitrification. If 10 mg/L of ammonia is converted, determine the concentration of ammonia as N and the concentrations of the nitrite and nitrate, both in themselves and as N. The reactions are

$$NH_3 + 1\tfrac{1}{2}O_2 \rightarrow H^+ + NO_2^-$$
$$\quad (17) \qquad\qquad\qquad (46)$$

$$NO_2^- + \tfrac{1}{2}O_2 \rightarrow NO_3^-$$
$$\quad 46 \qquad\qquad 62$$

SOLUTION: Ammonia can be expressed as N by multiplying by the ratio of molecular weights:

$$C = 10\,\frac{mg\,NH_3}{L} \times \frac{14\,mg\,NH_3\text{-}N}{17\,mg\,NH_3}$$

$$= 8.2\,\frac{mg\,NH_3\text{-}N}{L} \quad \text{or} \quad 8.2\,\frac{mg\,NH_3}{L} \quad (as\,N)$$

The nitrite produced from the reaction is

$$C = 10\,\frac{mg\,NH_3}{L} \times \frac{46\,mg\,NO_2^-}{17\,mg\,NH_3}$$

If we express the nitrite as N, we obtain

$$C = 27 \frac{\text{mg NO}_2^-}{\text{L}}$$

$$= 27 \frac{\text{mg NO}_2^-}{\text{L}} \times \frac{14 \text{ mg N}}{46 \text{ mg NO}_2^-}$$

$$= 8.2 \frac{\text{mg NO}_2^- \text{-N}}{\text{L}} \quad \text{or} \quad 8.2 \frac{\text{mg NO}_2^-}{\text{L}} \quad (\text{as N})$$

Notice that the concentration of nitrite produced is the same as the ammonia measured as N. Now let us calculate the amount of nitrate produced from the 27 mg NO$_2^-$/L nitrite:

$$C = 27 \frac{\text{mg NO}_2^-}{\text{L}} \times \frac{62 \text{ mg NO}_3^-}{46 \text{ mg NO}_2^-}$$

$$= 36.4 \frac{\text{mg NO}_3^-}{\text{L}}$$

and the concentration of nitrate as N:

$$C = 36.4 \frac{\text{mg NO}_3^-}{\text{L}} \times \frac{14 \text{ mg N}}{62 \text{ mg NO}_3^-}$$

$$= 8.2 \frac{\text{mg NO}_3^- \text{-N}}{\text{L}} \quad \text{or} \quad 8.2 \frac{\text{mg NO}_3^-}{\text{L}} \quad (\text{as N})$$

So you can see that if we express the concentration of ammonia as N, the concentration of nitrite or nitrate produced from it is the same value as if expressed as N.

ADSORPTION

Adsorption is a surface phenomenon in which a solute (soluble material) concentrates or collects at a surface. This contrasts with absorption, in which a substance penetrates the material. Adsorption can occur between any two surfaces—a liquid and a solid, a gas and a solid, a liquid with another liquid, or a gas and a solid. Such processes abound in nature. Organics and metals will adsorb to soils. Organics themselves will adsorb to other organics and sediments on ocean, lake, or river floors. Several adsorption processes devised by humans—including activated carbon adsorption for the removal of contaminants from air and water, and ion exchange for removal of contaminants from water—have applications in environmental engineering. And the natural processs are often of interest to environmental engineers and scientists.

Clay materials contained in river sediment and soils, for example, strongly adsorb many organics and metal ions, a beneficial case being adsorption of pesticides applied to crops. Agricultural pesticides are often bound strongly to the clays and organics in the upper soil strata, preventing and in some instances drastically reducing contamination of ground water. The adsorbed pesticides are then degraded by myriad soil bacteria.

In adsorption, the solute comes out of the fluid phase and attaches to the adsorbent. The driving force for accumulation of the material at the interface between substances can be twofold. First, the material may not be very soluble in water and may therefore be more stable on the surface of an adsorbent. Second, the solute may have an affinity for the adsorbent. From this we can deduce that materials that are highly insoluble in water are better candidates for adsorption than highly soluble materials. Also, if we provide a material for which the solute has a high affinity, we also strengthen the force of adsorption. Forces binding the solute on the adsorbent surface can be chemical, physical (van der Waals forces), or electrostatic attraction.

Carbon Adsorption

Activated carbon is a material produced from coal, wood, and bone by the application of heat in the absence of air. The resulting material has a sizeable surface area for adsorption. Activated carbon can be used to remove toxic organics from both water and air. It is commonly used in aquarium filters. It is also used in some potable water plants to remove organics before or after chlorination, in industrial wastewater treatment to remove a variety of organics, and occasionally to remove residual organics following conventional wastewater treatment. And it is used in air pollution control to remove organic contaminants from contaminated air before discharge.

The amount of solute adsorbed per unit of adsorbent varies with the equilibrium or final concentration. So if activated carbon is used in a batch process, it reduces the concentration of the contaminant to some low value at equilibrium. The utilization of carbon is then high because the low equilibrium concentration means less contaminant is adsorbed per unit of carbon. The key to successful utilization of activated carbon as an adsorbent is to force the equilibrium to occur at as high a concentration as possible. This can be by using an adsorption column filled with activated carbon. As the contaminant passes through the column, the initial section will reach equilibrium at the higher influent concentration. If the column is relatively tall, most of the carbon can be equilibrated at the higher concentration, increasing the adsorbent utilization.

Several mathematical models (commonly called isotherms) have been developed to predict the mass of solute removed per mass of adsorbent used versus concentration. Two commonly used forms are the Freundlich isotherm and the Langmuir isotherm. Both will be presented here.

Freundlich Isotherm

This model, which depicts an empirical relationship, is expressed as

$$q_e = \frac{x}{m} = KC_e^{1/n} \qquad \qquad \textbf{4.52}$$

where

q_e = mass of solute adsorbed per mass of adsorbent used [mg adsorbed/mg carbon]

x = mass of solute adsorbed [mg or mol]

m = mass of adsorbent [mg]

C_e = equilibrium concentration of solute [mg/L or M]

K = experimental constant

n = experimental constant

The Freundlich equation can be linearized so that a plot of experimental data can be used to determine the parameters K and n. Taking the logarithm of both sides of the equation yields

$$\log\left(\frac{x}{m}\right) = \log K + \frac{1}{n}\log(C_e) \qquad \qquad \textbf{4.53}$$

If $\log(x/m)$ is plotted versus $\log(C_e)$, the data should fit a straight line. Otherwise, the Freundlich isotherm is not applicable. If the data generates a straight line, the y-intercept is $\log K$, and the slope of the line is $1/n$.

Data determining adsorption characteristics of a contaminant can be obtained by placing a known volume and concentration of the contaminant into several flasks with different amounts of adsorbent, usually activated carbon. The flasks are then mixed for several days until equilibrium is reached. The final concentration of contaminant in the liquid phase is then measured.

Langmuir Isotherm

Whereas the Freundlich isotherm is empirical, the Langmuir isotherm can be derived—by assuming that the solute forms only a monolayer of atoms on the adsorbate, that there is a finite adsorption area, that the process is reversible, and that an equilibrium exists. Although the final result is rather simple, the derivation is rather complex and lengthy. The math will be omitted here, but the result is

$$q_e = \frac{x}{m} = \frac{KQ^0 C_e}{1 + KC_e} \qquad\qquad 4.54$$

where

Q^0 = constant representing the mass of solute adsorbed per mass of adsorbent at saturation [mass/mass]

K = experimental constant [L/mg]

To linearize the Langmuir isotherm, we invert the equation, obtaining

$$\left(\frac{1}{(x/m)}\right) = \frac{1}{Q^0} + \frac{1}{KQ^0}\left(\frac{1}{C_e}\right) \qquad\qquad 4.55$$

If $(1/q_e)$ is plotted versus $(1/C_e)$, the data should fit a straight line. (You guessed it, if it does not, the Langmuir isotherm is not applicable! That is, one or more of the assumptions is not valid.) The y-intercept is $1/Q^0$. The slope is $1/(KQ^0)$. Another linearization can be obtained by multiplying both sides of Equation 4.55 by C_e.

EXAMPLE 4.24	CARBON ADSORPTION ISOTHERMS

A pharmaceutical manufacturer plans to install a new industrial production process. The waste stream for the process is expected to have a concentration of 7.5 mg/L methylene blue at a flow rate of 25 gpm. The company plans to use carbon adsorption to remove the methylene blue. The following adsorption data were obtained from the laboratory. The liquid volume in each flask was 200 mL. Determine if either or both isotherms are applicable (plot as a straight line in linearized form). Find both the Freundlich parameters K and $1/n$ and the Langmuir parameters K and Q^0 if applicable. Estimate the amount of activated carbon required to remove the contaminant each year if the carbon is removed by an adsorption column that reaches equilibrium with the carbon at the 7.5 mg/L concentration.

MASS OF ADSORBENT, mg	INITIAL CONCENTRATION, mg/L	FINAL CONCENTRATION, mg/L
98.6	25	0.04
58.1	25	0.11
26.3	25	0.49
15.7	25	1.2
8.8	25	3.2
2.9	25	10.2
0.8	25	19.7
Blank	25	25

SOLUTION: Solution A, for the Freundlich Isotherm

First calculate the amount of solute adsorbed per mass of adsorbent, x/m. As with the Langmuir isotherm, a spreadsheet can be set up to perform these calculations, graph the relationship, and determine the regression line. The accompanying table was created using a spreadsheet. Or you can use a hand-held calculator.

MASS OF CARBON, mg	C_0, mg/L	C_f, mg/L	$C_0 - C_f$, mg/L	MASS OF SOLUTE ADSORBED/MASS OF CARBON x/m,
98.6	25	0.04	24.96	0.051
58.1	25	0.11	24.89	0.086
26.3	25	0.49	24.51	0.186
15.7	25	1.2	23.8	0.303
8.8	25	3.2	21.8	0.495
2.9	25	10.2	14.8	1.02
0.8	25	19.7	5.3	1.32

Solute adsorption, x/m, is the product of the change in solute concentration and its volume (0.2 L) divided by the mass of carbon used. For the first row:

$$\frac{x}{m} = \frac{(C_0 - C_f)V_s}{m} = \frac{(25\,\text{mg solute/L} - 0.04\,\text{mg solute/L}) \times 0.2\,\text{L}}{98.6\,\text{mg carbon}}$$

$$= 0.051\,\frac{\text{mg solute adsorbed}}{\text{mg carbon}}$$

To determine the Freundlich parameters, you must find the intercept and slope of the C_f versus x/m plot. You can do this by graphing $\log(C_f)$ vs. $\log(x/m)$.

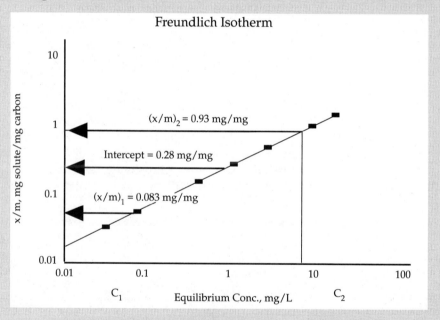

From the regression analysis using the spreadsheet, the slope of the line is 0.53, which is $1/n$. Therefore, $n = 1.89$. If the data are plotted by hand, the slope is computed in this way:

$$\text{Slope} = \frac{1}{n} = \frac{[\log(x/m)_2 - \log(x/m)_1]}{(\log C_2 - \log C_1)}$$

$$= \frac{\log(0.93\,\text{mg/mg}) - \log(0.083\,\text{mg/mg})}{\log(10\,\text{mg/L}) - \log(0.1\,\text{mg/L})}$$

$$= 0.52$$

Similarly, the intercept can be computed using the regression analysis in the spreadsheet program, or it can be determined visually from a manual plot of the data. (Remember that the intercept of a log–log plot occurs at $x = 1$, not zero!) From the spreadsheet, the intercept is $\log k = -0.557$, and k is thus 0.27. From the graph, the intercept is $k = 0.28$. Using the regression data, then, we see that the Freundlich isotherm is

$$\frac{x}{m} = KC^{1/n} = 0.27C^{0.53}$$

To determine the carbon required for one year, we must calculate x/m for 7.5 mg/L:

$$\frac{x}{m} = KC^{1/n} = 0.27(7.5)^{0.53}$$

$$= 0.79 \frac{\text{mg solute adsorbed}}{\text{mg carbon}}$$

Note: *It is also possible to read the value for (x/m) from the graph instead of calculating it. From the graph it would be approximately 0.80 mg solute adsorbed/mg carbon.*

To determine the mass of carbon required each year, you need to calculate the annual amount of contaminant discharged from the process. Use this formula:

$$C = \frac{m}{V}$$

where C is concentration, m is mass, and V is volume. This can be solved for the mass. The yearly volume is calculated by:

$$V = Qt = 15 \frac{\text{gal}}{\text{min}} \times 1 \text{ year} \times \frac{3.78\,\text{L}}{\text{gal}} \times \frac{365\,\text{days}}{\text{year}} \times \frac{24\,\text{hours}}{\text{day}} \times \frac{60\,\text{min}}{\text{hour}}$$

$$= 29.8 \times 10^6\,\text{L}$$

The mass is then

$$m = CV = \left(7.5\,\frac{\text{mg}}{\text{L}}\right) \times (29.8 \times 10^6\,\text{L}) \times \frac{\text{kg}}{10^6\,\text{mg}}$$

$$= 224\,\text{kg solute}$$

The required amount of carbon is then

$$\text{Mass carbon required} = \frac{\text{mass of solute removed}}{\left(\dfrac{x}{m}\right)}$$

$$= \frac{224\,\text{Kg solute adsorbed}}{\left(0.79\,\dfrac{\text{mg solute adsorbed}}{\text{mg carbon}}\right)}$$

$$= 284\,\text{Kg}$$

Solution B, for the Langmuir Isotherm:

As with Freundlich, the first task is to calculate the amount of solute adsorbed per mass of adsorbent, or x/m. And again, a spreadsheet can be set up to perform these calculations, graph the relationship, and determine the regression line. Or you can use a calculator. The following table is spreadsheet-generated.

m, mg	C_o, mg/L	C_f, mg/L	$C_o - C_f$, mg/L	x/m	$1/(x/m)$	$1/C_f$
98.6	25	0.04	24.96	0.0506	19.8	25.0
58.1	25	0.11	24.89	0.0857	11.7	9.09
26.3	25	0.49	24.51	0.186	5.37	2.04
15.7	25	1.2	23.8	0.303	3.30	0.833
8.8	25	3.2	21.8	0.495	2.02	0.312
2.9	25	10.2	14.8	1.02	0.980	0.0980
0.8	25	19.7	5.3	1.32	0.755	0.0508

It is clear from the accompanying figure that the graph is not a straight line. So these data do not follow Langmuir adsorption, and the isotherm cannot be used to solve this problem. Note, however, that at the lower three equilibrium concentrations (highest $1/C_f$) the numbers seem to be beginning to conform to Langmuir adsorption.

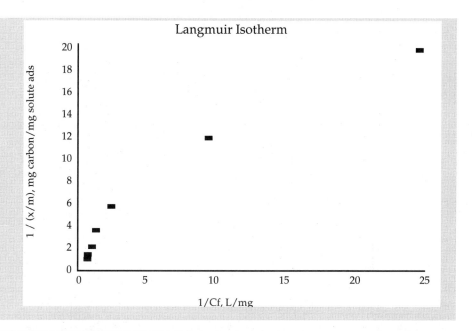

EXAMPLE 4.25

LANGMUIR ADSORPTION OF β-ENDOSULFAN

β-Endosulfan is an organic insecticide commonly used on vegetables and ornamental flowers. Carbon adsorption data for it is listed in the table. The initial concentration was 0.082 mg/L. Calculate the Langmuir constants. What q_e (mass adsorbed/mass of carbon) would be required to attain a contaminant concentration of 0.02 mg/L? The flask volume was 250 mL (.25 L).

CARBON DOSE, mg	C_f, mg/L
0	0.082
0.2	0.052
0.625	0.017
1.25	0.008
2.5	0.004

SOLUTION: The first step is to determine the change in concentration, q_e, $1/q_e$, and $1/C_e$. The following table presents this information.

CARBON DOSE, mg	C_f, mg/L	$C_0 - C_f$ mg/L	x/m mg/mg	$1/C_f$ L/mg	$1/(x/m)$ mg/mg
0	0.082	—	—	—	—
0.2	0.052	0.03	0.0375	19.2	26.7
0.625	0.017	0.065	0.026	58.8	38.5
1.25	0.008	0.074	0.0148	125	67.6
2.5	0.004	0.078	0.0078	250	128

A graph of the two left columns will appear as a straight line if the data conform to the Langmuir isotherm. A plot of the data is shown with a linear regression line.

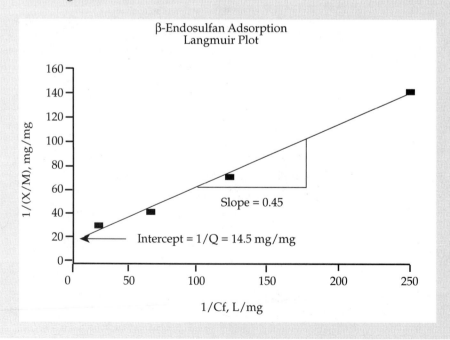

β-Endosulfan Adsorption
Langmuir Plot

The slope and intercept can be used to obtain the Langmuir constant s.

$$\frac{1}{Q^0} = 14.5 \frac{mg}{mg}$$

$$= 0.069 \frac{mg}{mg}$$

$$\frac{1}{KQ^0} = 0.45$$

$$K = \frac{1}{0.45 \times Q^0} = \frac{1}{0.45 \times 0.069}$$

$$= 32.2 \frac{L}{mg}$$

The value of q_e at an equilibrium concentration of 0.02 mg/L can now be calculated using the Langmuir equation.

Ion Exchange

Ion exchange is an adsorption process in which one ion is exchanged for another ion of like charge. Figure 4.3 depicts a divalent cation such as Cu^{2+} or Ca^{2+} being exchanged for a monovalent cation such as Na^+ or H^+. In ion exchange there is an exchange of charge equivalence. That is, for each X amount of charge adsorbed, the same amount of charge will be released or desorbed. Cation, anion, and mixed exchange resins are commercially available. Soils, some minerals, and organics are all natural exchange resins.

Ion exchange is used in many industrial processes, from the removal of cyanide $[CN^-]$ and toxic metals such as hexavalent chromium $[Cr^{6+}]$ to reduction of acids in orange juice. Water softening can also be accomplished using the ion exchange process.

Ion exchange resins can be regenerated. With the contaminant removed, they can then be reused. Regenerants in general use are salt (sodium chloride), sulfuric acid, and sodium hydroxide. During regeneration, the contaminant flow to the resin is stopped. A concentrated regenerant is then pumped through the ion exchange column. Figure 4.4 depicts a typical household water softener for the removal of calcium and magnesium from the water. During operation the water flows through the cation exchange column. The Ca^{2+} and Mg^{2+} ions are exchanged on the resin surface with Na^+ ions while carbonate and other anions pass through the column without adsorbing to the resin surface. After treatment, the water contains the same anions as before. Calcium and magnesium ions are

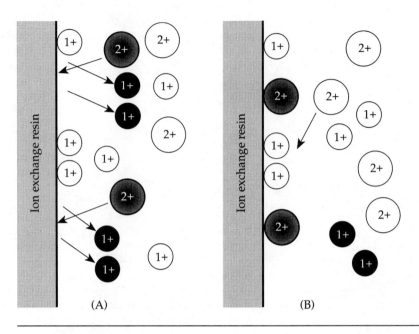

FIGURE 4.3   Ion exchange of a 2+ cation for a two 1+ cations.

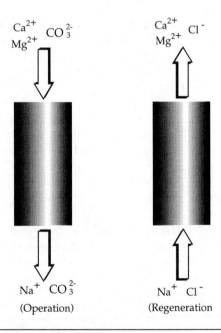

FIGURE 4.4   Operation and regeneration of a typical household water softener using sodium chloride.

retained on the resin, and sodium ions are released to the water on an equivalent charge basis—two sodium ions for each calcium or magnesium ion.

During regeneration, a strong solution of sodium chloride is passed through the column, forcing the calcium and magnesium ions off the resin. The regenerant water contains calcium, magnesium, excess sodium, and chloride ions.

Review Questions

1. An anaerobic digester is a biological reactor often used to treat waste-water sludges—liquids containing from 0.5 to 5.0% solids. If a digester has a bisulfide [HS^-] concentration of 30 mg/L and a pH of 7.7, what is the partial pressure of hydrogen sulfide in the digester gas? *Hint:* Calculate the $H_2S(aq)$ concentration and then use Henry's law.

3. An anaerobic digester has a methane concentration of $9.75 \times 10^{-4} M$ and a carbon dioxide concentration of $1.19 \times 10^{-2} M$. What are the partial pressures of the two gases?

3. Freight train cars usually carry a maximum net weight of about 100,000 lbs. If a train is carrying three cars of concentrated sulfuric acid (assume that it is 96% acid), how much lime [$Ca(OH)_2$] would be required to neutralize the acid if it is spilled in a derailment or wreck?

4. Calculate the ThOD of the following wastewaters:

   a. 45 mg/L of methanol, CH_3OH

   b. 10 mg/L of glucose, $C_6H_{12}O_6$

   c. a mixture containing 45 mg/L of methanol and 10 mg/L of glucose

   d. $10^{-4} M$ acetic acid, CH_3COOH

   e. 5 mg/L of trichloroethane, CCl_3CH_3. *Hint*: The chlorine goes to HCl.

5. Use Gibbs free energies to calculate the saturation concentration of dissolved oxygen at 25°C in water. Compare this to the value found in the appendix.

6. Use Gibbs free energies to calculate the Henry's law constant for oxygen in water. Compare this value to that listed in Table 4.1.

7. Table 4.4 lists several acid reaction constants including that for carbonic acid [$H_2CO_3^*$]. Using the values for Gibbs free energy, determine the K_a for "real" [H_2CO_3]. Is it stronger than [$H_2CO_3^*$]?

8. Using Gibbs free energies, determine the fraction of [$H_2CO_3^*$] that is actually $CO_2(aq)$.

9. From Problems 7 and 8, explain why "composite" K_a is lower than the "real" acid K_a.

10. Determine the pH when the following compounds are dissolved in water in the indicated amounts. Assume that the compounds dissolve and dissociate completely.

   a. $10^{-5} M$ sodium hydroxide [NaOH]

   b. $10^{-4} M$ sodium acetate [CH_3COONa]

   c. $10^{-6} M$ magnesium hydroxide [$Mg(OH)_2$]

   d. $10^{-6} M$ calcium hydroxide [$Ca(OH)_2$]

11. Determine the pC of these species:

   a. 2.3×10^{-5} mol/L of Ca^{2+} ions

   b. $10^{-7} M$ OH^-

   c. $5 \times 10^{-4} M$ Na^+

   d. $3 \times 10^{-5} M$ K

   e. 150 mg/L Fe^{3+}

12. Write the electroneutrality equation for the following compounds dissolving into water (assume complete dissociation):

   a. $10^{-5} M$ sulfuric acid, H_2SO_4

   b. $10^{-4} M$ sodium chloride, NaCl, and $3 \times 10^{-5} M$ sodium acetate, CH_3COONa

   c. $10^{-5} M$ acetic acid, HCH_3COO, and $4 \times 10^{-5} M$ sodium acetate, CH_3COONa

   d. $4 \times 10^{-6} M$ calcium hypochlorite, $Ca(OCl)_2$

   e. $10^{-4} M$ sodium carbonate, Na_2CO_3, and $10^{-4} M$ sodium bicarbonate, $NaHCO_3$

   f. $2 \times 10^{-5} M$ sodium nitrate, $NaNO_3$, and $5 \times 10^{-5} M$ ammonium nitrate, NH_4NO_3

13. What is the pH of a solution saturated with lime [$Ca(OH)_2$]? You may assume that calcium does not complex with the water.

14. Determine the pH of rainwater unaffected by acidic air pollution. *Hint:* The partial pressure of CO_2 in the atmosphere is $10^{-3.5}$ atm. Assume that this is in equilibrium with distilled water. Surprised?

15. List the possible species and write the necessary equations (equilibrium, mass balance, and electroneutrality) to solve the following chemical systems. (Do not solve the equation set unless directed to do so by your instructor.

   a. $10^{-5} M$ sodium acetate $+ 10^{-6} M$ acetic acid

b. $10^{-5} M$ sodium dihydrogen phosphate $+ 10^{-6} M$ disodium hydrogen phosphate

c. $10^{-4} M$ hydrochloric acid

d. $10^{-4} M$ acetic acid

16. Use the Henderson–Hasselbach relationship to graph the percent ionization of hypochlorous acid from a pH of 0 to 14. Make pH the independent variable (x-axis) and percent ionization from 0 to 100% on the dependent axis (y-axis). Use a spreadsheet to graph the relationship if your instructor so directs.

17. Carbonic acid $[H_2CO_3]$ partially ionizes to bicarbonate $[HCO_3^-]$ and carbonate $[CO_3^{2-}]$. At what pH does the bicarbonate concentration equal the carbonate concentration?

18. Lindane, a common insecticide, is toxic to humans. Managers at a manufacturing plant wish to remove Lindane in the wastewater from its operation by using carbon adsorption. Use the following data to determine the Freundlich and/or Langmuir parameters for acrylonitrile. Are both isotherms applicable? Which provides the best fit? How much carbon would be required to treat 5000 gallons at a concentration of 25 mg/L? *Optional*: Plot the data using a spreadsheet and obtain the parameters using linear regression.

CARBON DOSE, mg	EQUILIBRIUM CONCENTRATION, mg/L
0	1.018
0.2	0.788
0.5	0.586
1.0	0.255
2.0	0.120
4.0	0.032
10.0	0.006

Note: The test volume was 200 mL.

19. Acrylonitrile is used as a feedstock in the manufacture of acrylic fibers. It is also used in the plastics, adhesives, and pharmaceutical industries. Unfortunately, this substance is toxic, explosive, and flammable. It is possible to use carbon adsorption to mitigate its effects and diminish the dangers in its use. Use the following data to determine the

Freundlich parameters for acrylonitrile. How much carbon would be required to treat 5000 gallons at a concentration of 25 mg/L? *Optional*: Plot the data using a spreadsheet and obtain the parameters using a linear regression program.

CARBON DOSE, mg/L	EQUILIBRIUM CONCENTRATION, mg/L
0	48.7
1,000	40.85
2,000	31.77
5,000	17.86
10,000	8.68
20,000	2.98

20. A waste stream contains potassium dichromate $[K_2Cr_2O_7]$. Sketch and label an ion exchange system that could separate and selectively remove the dichromate but not the potassium. *Hint*: The potassium will ionize in solution, becoming $[K^+]$ and $[Cr_2O_7^{2-}]$.

21. Cyanide is widely used in electroplating. Common contaminants in the water used to rinse zinc-plated parts are sodium, zinc, and copper cations and cyanide anions. What type of ion exchange system could be used to recover the zinc, copper, and cyanide from the rinsewater? Sketch the setup. *Hint*: Use at least two separate ion exchange units.

22. What type of regenerants could be used for the cyanide-based plating solution in Problem 22?

23. A water sample has 45 mg Ca^{2+}/L, 25 mg Mg^{2+}/L, and 0.5 mg Sr^{2+}/L. What is the total concentration of divalent cations, expressed as calcium carbonate?

24. A water contains 25 mg L/NH_4^+, 2 mg L/NO_2^-, and 10 mg L/NO_3^-. Express each concentration as mg L/N. If the ammonium and nitrite are converted to nitrate, what is the nitrate concentration as N?

References

1. W. Stumm and J. J. Morgan, *Aquatic Chemistry, 2nd ed.* (New York: Wiley, 1981).

2. V. L. Snoeyink and D. Jenkins, *Water Chemistry* (New York: Wiley, 1980).

3. L. G. Sillen and A. E. Martell, *Stability Constants*, Special Publication No. 17, Chemical Society of London (1964).

4. L. G. Sillen and A. E. Martell, *Stability Constants Supplement No. 1*, Special Bulletin No. 25, Chemical Society of London (1971).

5. D. O. Wagman et al., "The NBS Tables of Chemical Thermodynamic Properties," *Journal of Physical and Chemical Reference Data*, vol. 11, supplement no. 2 (1982).

Additional References

L. O. Benefield, J. F. Judkins, and B. L. Weand, *Process Chemistry for Water and Wastewater Treatment* (New York: Prentice-Hall, 1982). A more extensive but simplified approach to environmental chemistry.

C. N. Sawyer and P. L. McCarty, *Chemistry for Environmental Engineering*, 3rd ed. (New York: McGraw-Hill, 1978). A rather old, but good textbook on environmental chemistry.

T. D. Reynolds, *Unit Operations and Processes in Environmental Engineering* (Boston: PWS-Kent, 1982). Contains extended explanations of adsorption and ion exchange.

Robert C. Weast, ed., *CRC Handbook of Chemistry and Physics*, 62nd ed. (Boca Raton, Florida: CRC Press, 1981).

5 Organic Chemistry

"OF ALL THE CHEMICALS THAT HAVE BEEN TOSSED INTO THE CALDRON OF PUBLIC ANXIETY, 2,3,7,8-TETRACHLORODIBENZO-p-DIOXIN, TCDD, HAS ACHIEVED THE MOST NOTORIETY AND EVOKED THE GREATEST FEARS FOR THE LONGEST PERIOD OF TIME. . . . IT IS ONLY RECENTLY THAT A CLEARER PICTURE OF THIS CREATURE [TCDD] HAS EMERGED, AND WHILE IT MAY NOT BE THE MAMMOTH ONCE BELIEVED, IT IS NO MOUSE EITHER."

DAVID J. HANSON

C & E News, August 12, 1992

Organic chemicals stored at a groundwater contamination site in Dayton, Ohio. A vacuum-extraction well can be seen in the center of the photo.

INTRODUCTION

Organic chemistry is the study of substances that contain carbon. The only inorganic carbon compounds are CO_2 and its relatives (carbonic acid, bicarbonates, carbonates), cyanides, and cyanates.

Originally, the vital force theory proclaimed that all **organic compounds** came from living organisms. However, in 1828 a German chemist, Friedrich Wöhler, formed urea, a known organic compound, from ammonium cyanate, an inorganic compound, thus disproving the theory. The following equation shows the reaction Wöhler was able to complete.

$$NH_4^+CNO^- \xrightarrow{\Delta} (NH_2)_2CO \qquad\qquad \textbf{5.1}$$

Production of Organic Chemicals

Ethylene is the organic chemical produced in the greatest amount in the United States. It is followed by propylene, urea, and ethylene dichloride [1]. The top five organic chemicals produced in the United States, and their yearly production, are listed in Table 5.1. These compounds account for a production of more than 350 pounds per person per year!

TABLE 5.1   The Top Five Organic Chemicals Produced in the United States in 1990

CHEMICAL	PRODUCTION, 1000 METRIC TONS
Ethylene	17,001
Propylene	10,034
Urea	7,171
Ethylene dichloride	6,033
Benzene	5,380

Properties of Organic Chemicals

Organic compounds have a number of common characteristics.

1. They are combustible. Complete combustion of hydrocarbons results in the formation of carbon dioxide and water. For a simple hydrocarbon fuel such as propane, the reaction is given by

$$C_3H_8 + 5O_2 \rightleftarrows 3CO_2 + 4H_2O \qquad\qquad \textbf{5.2}$$

This reaction yields 495.6 kcal/mol of heat. For more complex organics the end products may include other compounds.

2. Their melting and boiling points are lower than those for inorganic compounds. Table salt, sodium chloride, a common inorganic compound, has a molecular weight of 58.5 g, a melting point of 801°C and a boiling point of 1413°C. Butane, a hydrocarbon fuel, has a molecular weight of 58.12 g, a melting point of -138°C and a boiling point of -0.5°C.

3. Their solubility in water is limited. While table salt has a solubility of about 350 to 400 g/L of water, butane is only slightly soluble in water.

4. They undergo molecular reactions as opposed to ionic reactions. Butane, for example, combusts to carbon dioxide and water

$$C_4H_{10} + 6\tfrac{1}{2}O_2 \rightarrow 4CO_2 + 5H_2O \qquad \textbf{5.3}$$

whereas sodium chloride ionizes in water to form sodium and chloride ions:

$$NaCl \rightleftarrows Na^+ + Cl^- \qquad \textbf{5.4}$$

5. They can have very high molecular weights. Organic molecular weights vary from quite light, such as 16 for methane, to molecular weights in the 10,000s for some polymers.

6. They are **Isomers** (different compounds with the same chemical formula). For example, C_5H_{12} can be any one of three different compounds: *n*-pentane, isopentane, or neopentane. All have the same chemical formula and molecular weight. However, they have different structures, as shown in Figure 5.1. And the three have different properties, some of which appear in Table 5.2.

7. They form a substrate for microorganisms. Sources of organic compounds now include both nature and chemical synthesis, and many,

TABLE 5.2   Selected Properties of the Isomers of Pentane

NAME	M.P., °C	B.P., °C
n-pentane	-130	36.1
2-methylbutane, or isopentane	-159.9	27.8
2,2-dimethylpropane, or neopentane	-16.5	9.5

SOURCE: *CRC Handbook of Chemistry and Physics,* 62nd ed. (Boca Raton, Florida: CRC Press, 1981).

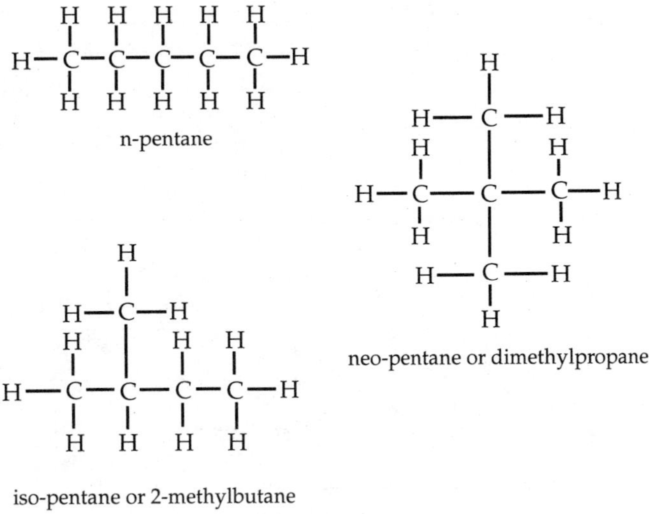

H H H H H
| | | | |
H—C—C—C—C—C—H
| | | | |
H H H H H

n-pentane

H
|
H— C —H
H | H
| |
H— C — C — C—H
| | |
H | H
H— C —H
|
H

neo-pentane or dimethylpropane

H
|
H—C—H
H | H H
| | |
H— C — C — C — C—H
| | | |
H H H H

iso-pentane or 2-methylbutane

FIGURE 5.1 Structure of the isomers of pentane.

although possibly not all, of them can be metabolized by microorganisms. In other words, organics are biodegradable.

THE CARBON ATOM

The sixth element on the periodic chart, carbon has an atomic number of 6. A carbon atom contains six neutrons, six protons, and six electrons, its molecular weight is 12.01115 g. Normally, carbon has a valence of ±4 and combines, forming tetrahedral structures. Bonding is usually covalent, involving shared electrons. Carbon readily bonds to itself, forming chain compounds. (*See* Figure 5.3 for octane, page 124.)

Bond Energy

The carbon bond is unique in chemistry in that it enables carbon molecules to form extended chains. This is the most important property distinguishing carbon compounds from others. The binding energy of a single carbon–carbon connection is approximately 100 kcal/mol.

Hydrogen Bonding

Hydrogen bonding occurs between oxygen and hydrogen atoms in the water molecule. To a great extent, the process is a function of **electronegativity** differences of the oxygen and hydrogen atoms. If water and

TABLE 5.3   Hydrogen Bonding Potential of Methane and Water

	MOLECULAR WEIGHT	MELTING POINT, °C	BOILING POINT, °C
Methane	16.04	−182	−164
Water	18.02	0	100

ELEMENT	ELECTRONEGATIVITY
Carbon	2.55
Hydrogen	2.20
Oxygen	3.44

methane are compared, for example, we see that their molecular weights are quite similar. Yet when we look at the melting and boiling points of methane and water, we see striking differences as shown in Table 5.3. Whereas the carbon and hydrogen in methane have similar electronegativities, this is not true of the water molecule. Also, the tetrahedral structure of methane prevents hydrogen atoms of one methane molecule from being attracted to the carbon molecules of another. With water, there is a pronounced difference. Hydrogen is much more electronegative than oxygen, and the oxygen molecule is electron rich, presenting two electron pairs on the side opposite from the two hydrogen atoms. So water is a strongly dipolar molecule. For this reason hydrogen atoms of one water molecule are drawn to the oxygen of another water molecule. This intermolecular attraction causes water to be more stable than a comparatively sized nonpolar molecule such as methane. It also provides water and other polar molecules with unique properties. Polar compounds tend to be readily soluble in other polar compounds. And nonpolar hydrocarbons tend to be readily soluble in other nonpolar compounds. As a result, nonpolar hydrocarbons are relatively insoluble in water but highly soluble in other nonpolar hydrocarbon solvents. Similarly, polar organic compounds such as alcohols, ethers, and organic acids are highly soluble in the polar solvent, water.

HYDROCARBONS

Hydrocarbons are compounds containing only hydrogen and carbon. They are important because as energy sources they have powered the industrial revolution of this century. And they continue to do so. The earth contains vast hydrocarbon reserves in the form of natural gas, crude oil, and coal — though these reserves are dwindling rapidly. Hydrocarbons are also the feedstock for most synthetic organic chemicals, which include plastics, pesticides, and various building materials. As shown in Figure 5.2, approximately 88% of the current world energy needs are supplied by hydrocarbon fuels [2].

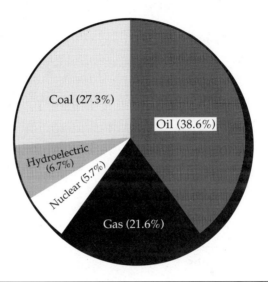

FIGURE 5.2 World energy sources in 1990.

Hydrocarbons are chemically unique. The electronegativity of hydrogen (2.2) and carbon (2.5) are similar, so the two form relatively nonpolar bonds. This is much different from water, which is highly polar. (Oxygen has an electronegativity of 3.44 versus hydrogen's 2.2.) Thus, hydrocarbons are excellent solvents for other nonpolar molecules but are very poor solvents, and they are insoluble in polar substances like water.

Additionally, hydrocarbons contain a tremendous energy per volume or mass compared to most other energy storage systems. When oxidized or burned completely, they form carbon dioxide and water. However, under conditions where insufficient oxygen is present, carbon monoxide [CO] and carbon [C] will be formed. In the air pollution field, these compounds are termed "products of incomplete combustion," or PICs. Many PICs, including carbon monoxide, are harmful or toxic.

Types of Hydrocarbons

Hydrocarbons are divided into four categories: alkanes, alkenes, alkynes, and aromatics. An example of each is shown in Figure 5.3. Alkanes are

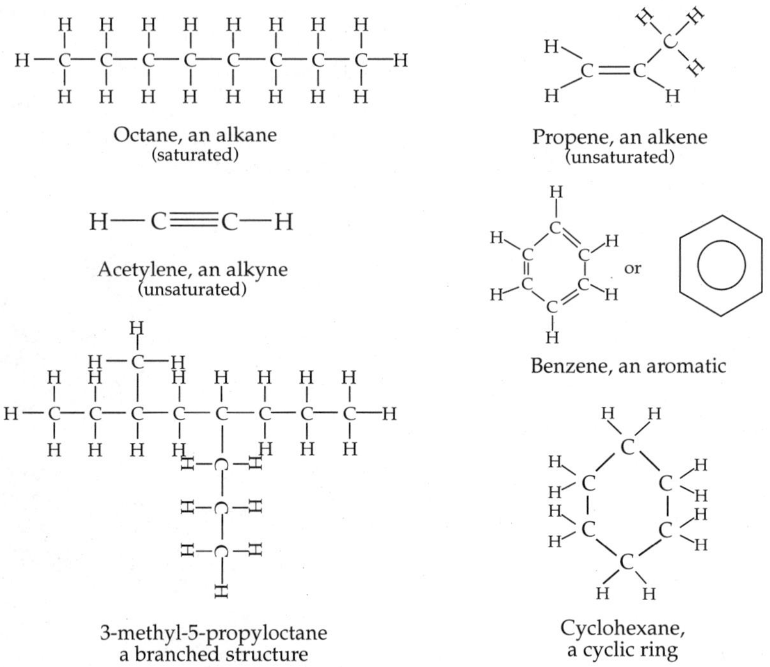

Octane, an alkane
(saturated)

Propene, an alkene
(unsaturated)

Acetylene, an alkyne
(unsaturated)

Benzene, an aromatic

3-methyl-5-propyloctane
a branched structure

Cyclohexane,
a cyclic ring

FIGURE 5.3 Categories of hydrocarbons.

hydrocarbons with single bonds between all carbon atoms. Alkenes are hydrocarbons with at least one double bond between two adjacent carbon atoms. Alkynes are hydrocarbons with at least one triple bond between two adjacent carbon atoms. These compounds may be straight chain, branched, or cyclic in structure. Branched and cyclic hydrocarbons are also shown in Figure 5.3.

Aromatic hydrocarbons are ringed carbon compounds that have effectively one and one-half pairs of electrons bonding adjacent carbon atoms. This unique structure provides stability and different chemical properties. Many aromatic compounds are formed from the basic six-carbon structure, the benzene ring.

A saturated hydrocarbon is one that has a hydrogen atom on each available carbon electron pair. There are no double or triple bonds. That is, the molecule is "saturated" with hydrogen. As indicated in Figure 5.3, octane is saturated, and propene and acetylene are unsaturated. (Although the benzene ring is also unsaturated, the term is not usually applied to aromatic compounds.) Polyunsaturated compounds have multiple double or triple bonds (as in polyunsaturated cooking coils). Figure 5.4 is an example of a polyunsaturated hydrocarbon.

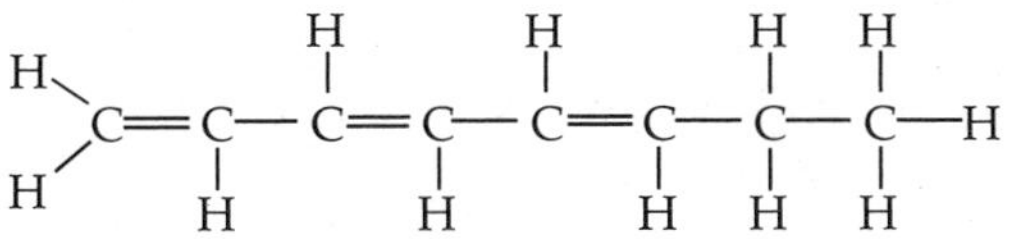

FIGURE 5.4 A polyunsaturated hydrocarbon.

Some common properties of the first ten normal (or *n-*) alkanes are shown in Table 5.4. Saturated hydrocarbons are more difficult to break down in the natural environment than unsaturated hydrocarbons.

Naming Straight-Chain Hydrocarbons

The naming procedure for organic compounds has been established by the International Union of Pure and Applied Chemists (IUPAC). The process may be difficult for more complex compounds, but it is easy to learn the names of many common straight- and branched-chain organics by memorizing a few basic rules and base names or prefixes. Table 5.4 shows the naming system for the first ten *n*-alkanes. All alkanes end with "-ane." All alkenes (carbon chains with a double bond) end with the suffix "-ene." All

TABLE 5.4 Properties of the First Ten *n*-Alkanes

NAME	AS A PREFIX	NUMBER OF CARBONS	APPROXIMATE MOLECULAR WEIGHT, g	MELTING POINT, °C	BOILING POINT, °C
Methane	Meth-	1	16.04	−182	−164
Ethane	Eth-	2	30.07	−183.3	−88.6
Propane	Prop-	3	44.11	−189.7	−42.1
Butane	But-	4	58.12	−138.4	−0.5
Pentane	Pent-	5	72.15	−130	36.1
Hexane	Hex-	6	86.18	−95	69
Heptane	Hept-	7	100.21	−90.6	98.4
Octane	Oct-	8	114.23	−56.8	125.7
Nonane	Non-	9	128.26	−51	150.8
Decane	Dec-	10	142.29	−29.7	174.1

SOURCE: *CRC Handbook of Chemistry and Physics,* 62nd ed. (Boca Raton, Florida: CRC Press, 1981).

triple bonded carbons, or alkynes, end with the suffix "-yne." Thus, by using the prefixes shown in the table, one can name 28 hydrocarbons. (Why is it not 30? The single carbon atom in methane cannot double or triple bond to itself.) The last part of the naming procedure of double and triple bonded hydrocarbons entails indicating the location of the double or triple bond when it can be in more than one location. This is done by a numeric prefix, indicating the number of the carbon atom with the double bond. This number is always the lowest possible number, as shown in the following example.

EXAMPLE 5.1	NAMING HYDROCARBONS

Name the following hydrocarbons using the rules previously described.

a.

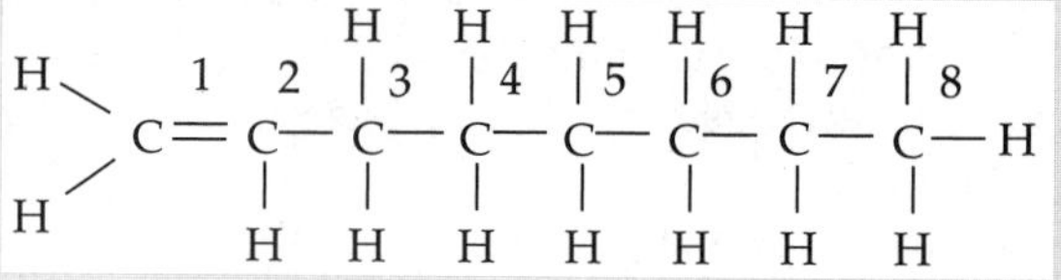

b.

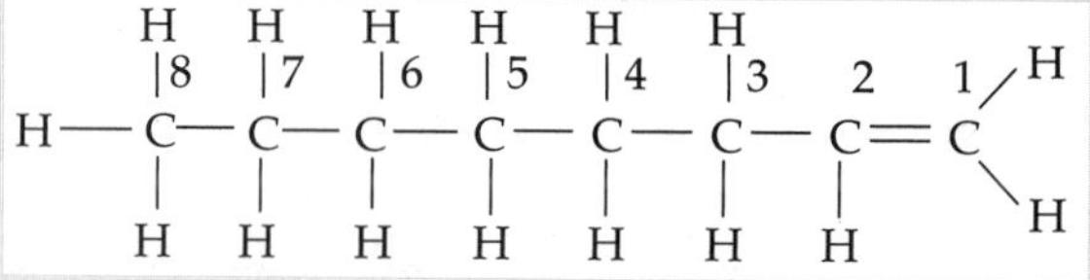

c.

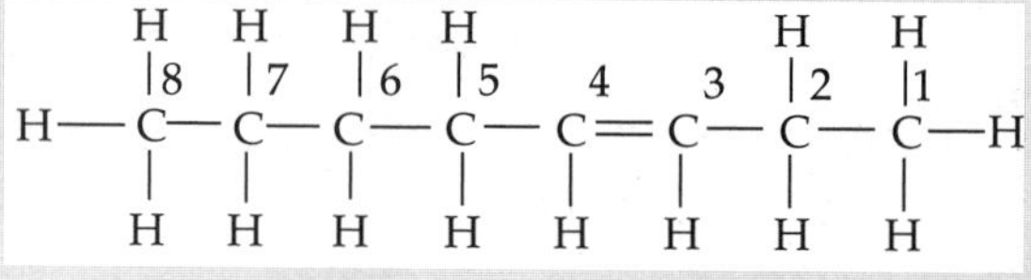

d.

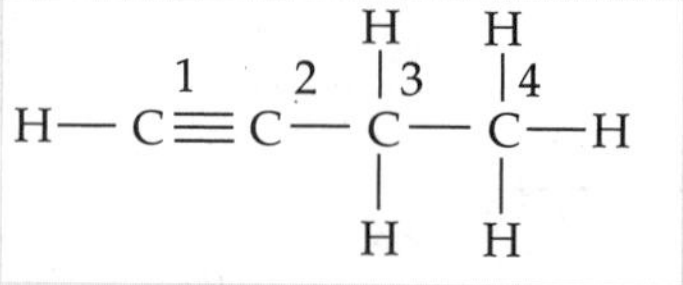

SOLUTION: a. This compound has eight carbons, so its base name is "oct-." It is an alkene because of the double bond, which is in the "1-" position. So its name is 1-octane.

b. This compound also has eight carbons, so its base name is also "oct-." It is also an alkene because of the double bond in the "1-" position.

This is not the "7-" or "8-" position because naming always uses the lowest numbers. Thus, its name is also 1-octene. The compound is identical to the previous one. It is a mirror image.

c. Again, the compound has eight carbon atoms, so the base name is "oct-." It has a double bond in the "3-" position (not the "5-" or "6-" positions). Therefore, its name is 3-octene.

d. This one has four carbon atoms, so its base name is "but-." It contains a triple bond in the "1-" position. Thus, the name is 1-butyne.

Naming of Branched Hydrocarbons

Branched hydrocarbons are those that have one or more additional chains attached to the main or longest chain. Branch chains, or substituent groups, are named by the length of the group and its location on the longest chain. If there is more than one substituent group, the groups are placed alphabetically in the prefix listing. If there are more than one of the same substituent group, its number is repeated and a di- or tri- is placed ahead of the substituent name. Branch chain naming is demonstrated in the following example.

EXAMPLE 5.2	NAMING BRANCH CHAIN HYDROCARBONS

Name the following branch chain hydrocarbons

e.

SOLUTION: **a.** The longest carbon chain is six carbons, so the base name is hexane. Since the branch group is one carbon, the prefix "meth-" is applicable. The branch chain is located on the "2-" carbon (not the "5-" carbon), so the name is 2-methylhexane.

b. The longest chain is again six carbons, yielding a base name of hexane. There are two single-carbon branches, resulting in a prefix of "dimethyl-." Both branch methyl groups are on the second or "2-" carbon. The name is then 2,2-dimethylhexane.

c. The base chain is six carbons long, giving us the base name of hexane. There is a two-carbon branch chain or "ethyl-" substituent group on the "3-" carbon. The name is thus 3-ethylhexane.

d. With a base chain six carbons long (not 5), the base name is hexane. There is a two-carbon branch chain or "ethyl-" substituent group on the "3-" carbon. The compound's name is then 3-ethylhexane — identical to the previous example.

e. The base chain is again six carbons long. However, there are three substituent groups: two methyl and one ethyl. The two methyl groups are on the "2-" carbon, and the ethyl group is on the "4-" carbon. This numbering method yields the lowest number, a "2-." (Numbering from the other direction produces a "3-" for the lowest number.) Note that the substituent groups are listed alphabetically. The compound is then 4-ethyl-2,2-dimethylhexane.

Aromatic Compounds

Aromatic hydrocarbons have a ring structure with multiple bonds. Benzene [C_6H_6], shown in Figure 5.5, is the simplest aromatic. The hydrogens are omitted in the drawing on the right side of the figure, but it is understood that they are a part of the compound. (Hydrogens are often omitted on structural drawings of organic compounds.) It has six carbons and six hydrogens. The term aromatic was originally applied because of the odor

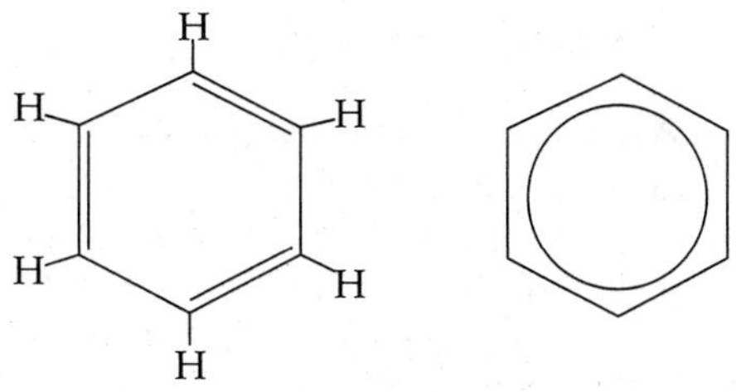

FIGURE 5.5 The structure of aromatic compounds.

of aromatic compounds.* However, it now has a chemical meaning as well in that these compounds have a unique ring structure. The carbon–carbon bond in an aromatic compound is neither a single nor a double bond. Instead, there are single bonds between each carbon atom, and three electron pairs (in a six-ring aromatic) become dispersed or delocalized throughout the ring structure. This leads to unusual ring stability. Replacement of the hydrogens by other active groups such as hydroxyl, halogen, sulfate, or nitrate occurs, but the ring structure usually remains intact. Biological degradation of these compounds in nature is usually quite slow.

Most, but not all, aromatic compounds have a six-ring structure. In addition, there are compounds in which several ring structures are present. For example, naphthalene (two benzene rings) and anthracene (three rings) are components in coal tar, shown in Figure 5.6. Again, note that the hydrogens are omitted.

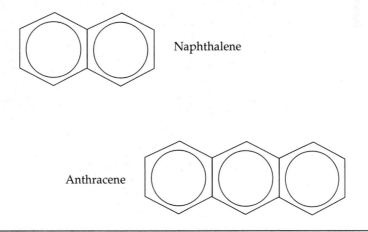

Naphthalene

Anthracene

FIGURE 5.6 Polycyclic benzenoid compounds.

---

*Many common spices contain aromatic compounds as their active ingredients. Cumin contains cumene; black pepper contains piperine; clove contains eugenol; cinnamon contains cinnamaldehyde. All are aromatic chemicals.

Naming Aromatic Compounds

When substituent groups are attached to a benzene ring, the base name is often changed. For example, when an "–OH" is attached to benzene, the common name is phenol, not hydroxybenzene. When a methyl group is attached, the common name is toluene. These names have been adopted by IUPAC. However, most IUPAC names use the substituent(s) as a prefix, followed by benzene. If more than one substituent is attached to the ring, numbers are used to designate the positions. As with other organic compounds, the numbers are arranged to obtain the lowest first number. An additional naming method for di-substituted benzene aromatics uses the designations *ortho*, *para*, and *meta*. *Ortho*, or an "*o-*" preceding the substituent groups, indicates the two substitutions are in the "1" and "2" positions. *Meta*, or "*m-*", indicates the substitutions occur in the "1" and "3" positions. And *para*, or "*p-*", indicates substitution at the "1" and "4" positions. Four examples are shown in Figure 5.7. Aromatic compounds are also named with the ring structure listed as a substituent group, phenyl. For a more extensive discussion of the naming of aromatics, refer to an organic chemistry textbook or the *CRC Handbook of Chemistry and Physics* [3].

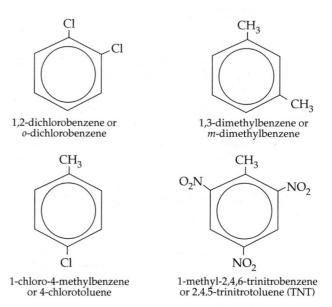

1,2-dichlorobenzene or
o-dichlorobenzene

1,3-dimethylbenzene or
m-dimethylbenzene

1-chloro-4-methylbenzene
or 4-chlorotoluene

1-methyl-2,4,6-trinitrobenzene
or 2,4,5-trinitrotoluene (TNT)

FIGURE 5.7 Naming aromatic compounds.

Biphenyls

Biphenyls are aromatic compounds with two **phenyl** or benzene rings attached by a single carbon–carbon bond as shown in Figure 5.8. Polychlorinated biphenyls (PCBs) were used extensively in industry until 1972. More than 500,000 tons were manufactured prior to that date [4]. Because of their widespread use and release, PCBs are almost ubiquitous in nature, and they degrade slowly, so they will remain in the environment for years to come. Many hazardous waste sites being cleaned up today contain PCBs.

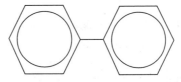

FIGURE 5.8   Biphenyl and polychlorinated biphenyls.

Hydrocarbon Fuels

As noted, hydrocarbon fuels comprise some 88% of the world's energy supplies at present. The three major fuels are petroleum or crude oil, coal, and natural gas. Petroleum is a complex mixture of organic carbons, primarily alkanes. It also contains small amounts of nitrogen, oxygen and sulfur compounds. The specific content of nitrogen, oxygen, and sulfur, as well as the specific distribution of hydrocarbons, varies from field to field. Current known world crude oil reserves are expected to last for another 60 years at current consumption rates [2].

Coal is a complex mixture. A large percentage of the carbon in coal is contained in linked aromatic rings. The proportion of these varies from a low of about 40% for subbituminous coals to almost 100% in anthracite coal. Small amounts of oxygen and nitrogen are present in coal as well [5]. The most troublesome component of coal is organic and inorganic sulfur. Organic sulfur is contained in the carbon structure; inorganic sulfur occurs primarily as iron pyrites. The amount of sulfur in coal varies with type and region. Because sulfur forms acids in coal combustion and because these acids return to earth as acid precipitation if they are not removed, sulfur content is a vital consideration in environmental engineering. Low-sulfur coals do not require equipment to remove the sulfur from coal-fired electric power plants. High-sulfur coals do. Western coals are low in sulfur. Illinois, Kentucky, and West Virginia coals are usually higher in sulfur. The United States has greater coal reserves than the Middle East does petroleum reserves. So if ways of minimizing the adverse environmental impacts of burning coal can be developed, coal can become a means of reducing U.S.

dependence on imported oil. Current world coal reserves will last for an estimated 225 years at current consumption levels [2].

Natural gas is primarily methane, but it contains small amounts of other short-chain hydrocarbons such as ethane. And it has almost no sulfur. Natural gas accounts for a smaller fraction of world energy use than either petroleum or coal, but it still supplies more energy than hydroelectric and nuclear energy combined (see Figure 5.2). Current world reserves of natural gas are thought to be sufficient to last for 60 years at present consumption [2].

OXYGENATED ORGANICS

Oxygenated organics (those containing oxygen) are important for several reasons. Many of their ions are sufficiently dipolar to allow hydrogen bonding with water, making them highly soluble. And many are more easily degraded than hydrocarbons themselves. Oxygenated organics also represent a large group of industrially and commercially important materials, including the alcohols, ethers, and phenols. Common household products in this category include sugars, flours, and alcohols.

Alcohols

Alcohols are important organic intermediates. The most common alcohol, ethanol or ethyl alcohol, is used as an organic intermediate and as a beverage. It can be produced by a variety of methods, the oldest being fermentation of high-sugar fruits and grains using microorganisms. Alcohols can also be gasoline additives. Their use reduces the amount of crude oil required to power gasoline engines and the air pollution associated with them. The 1991 Clean Air Act Amendments require addition of oxygen additives (alcohol contains an oxygen in the hydroxyl group) in several U.S. cities with carbon monoxide and ozone problems.* It has been estimated that an oxygen content of 2.7% in gasoline will reduce carbon monoxide emissions by 17% [6].

The functional group of an **alcohol** is the hydroxyl group [–OH]. Alcohols are classified according to the location of the hydroxyl group on the carbon chain. In primary alcohols the hydroxyl group is attached to a carbon that is bonded to only one other carbon. In secondary alcohols the hydroxyl group is on a carbon bonded to two other carbons. In a tertiary alcohol the hydroxyl group is connected to a carbon bonded to three other carbon atoms. This configuration is shown in Figure 5.9.

Many alcohols of commercial importance have common names. The

---

*The cities include Baltimore, Chicago, Hartford, Houston, Los Angeles, Milwaukee, New York, Philadelphia, and San Diego.

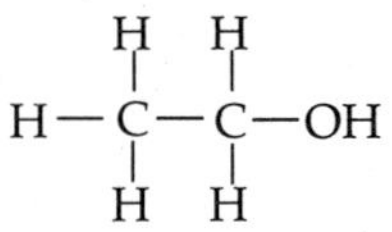

```
        H  H
        |  |
   H — C — C — OH          Ethanol, a primary alcohol
        |  |
        H  H


       H  OH  H
       |  |   |
   H — C— C — C — H         Iso-propanol, a secondary alcohol
       |  |   |
       H  H   H


             H
             |
         H — C — H
       H     |
       |     |
   H — C  —  C — OH         tert-butyl alcohol or
       |     |              2-methyl-2-propanol,
       H     |              a tertiary alcohol
         H — C — H
             |
             H
```

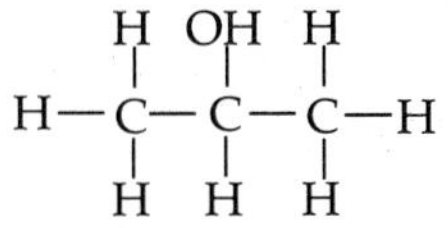

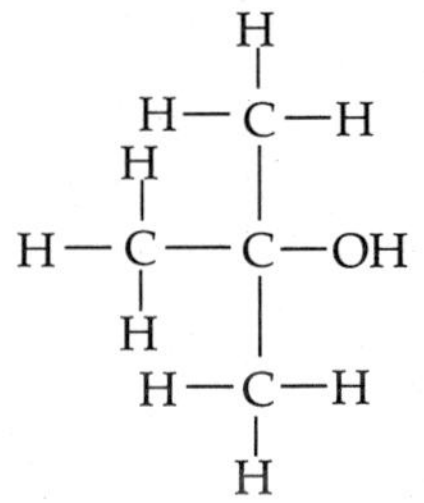

FIGURE 5.9   Classification of alcohols.

IUPAC system uses the ending or suffix "-ol" for alcohols. The location of the hydroxyl group is indicated by a numeric prefix where required. Thus, the isopropyl alcohol in Figure 5.9 would be 2-propanol, and *tert*-Butyl alcohol would be named 2-methyl-2-propanol.

Because of the electronegativity of the oxygen atom in the hydroxyl group, alcohol molecules are polar. However, as the molecular weight increases, the hydrocarbon (or nonpolar) portion of the molecule dominates and solubilities decline.

Although most, if not all alcohols, are toxic to humans, they are easily biodegraded by mixed microbial populations. Their end products are primarily carbon dioxide and water. Alcohol production facilities, including breweries and wineries, typically have large amounts of oxygen-demanding materials in their wastewater effluents. This water must be treated prior to discharge into lakes or streams.

Phenols

Phenol is an aromatic benzene ring with a hydroxyl substituted for one hydrogen. Phenol acts as a weak acid, releasing the proton on the hydroxyl. Phenol was once widely used as a disinfectant. Many disinfectants today are rated with phenol as the standard. Chloroxylenol, for example, is said to be 60 times as potent as phenol.

facturing, tanning, and photographic processing, and in the production of insecticides, herbicides and dyes. Chlorinated phenols are important pesticides.

Phenols are named using a procedure similar to benzene. The usual method is to list any groups by number and prefix. The hydroxyl is the 1 position, and other components number in sequence up to 6. As with other organics, the lowest number is taken when there is a choice. For example, it is 3-chlorophenol, not 5-chlorophenol, in Figure 5.10.

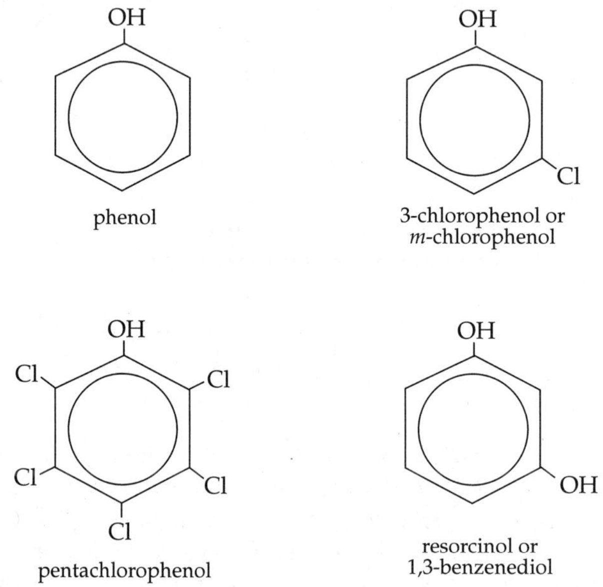

FIGURE 5.10   Phenol and phenol derivatives.

Thiols

Thiols or **mercaptains** are the sulfur analog of alcohols, the "–OH" being replaced with an "–SH." One noteworthy attribute of thiols is their unpleasant odors. The "rotten egg' smell of H_2S is one example. Another is 3-methyl-1-butanethiol, one of the "active ingredients" used by skunks for self-defense. Thiols are important in biochemistry and are present in living organisms and decaying matter.

Ethers

Ethers have two hydrocarbon groups bound by an interior oxygen molecule. The groups [R'-] and [R''-] in Figure 5.11 indicate generic hydrocarbon groups of unspecific length. Ethers are unable to hydrogen-bond to themselves because they do not have a hydrogen attached to an elec-

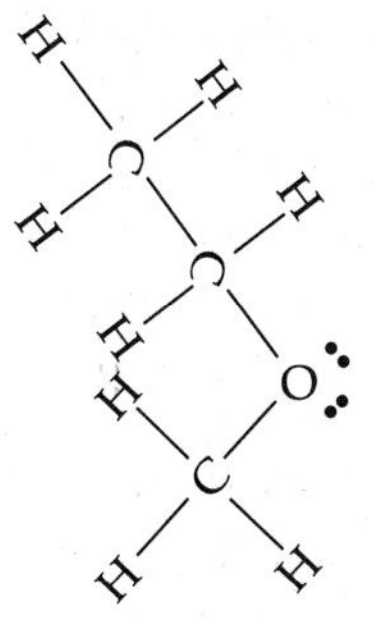

Methyl ethyl ether

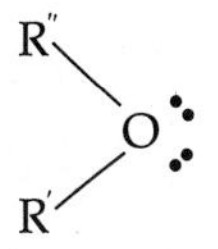

General ether form

FIGURE 5.11 The structure of ethers.

tronegative atom. However, they can hydrogen-bond to water or other dipolar compounds, thus increasing their solubilities. Ethers have solubilities similar to alcohols in water. Ethers are used as gasoline additives to reduce carbon monoxide emissions. One such additive is *tert*-butyl ether (also called MTBE).

Aldehydes and Ketones

Aldehydes and ketones contain the carbonyl functional group shown at the left in Figure 5.12. An aldehyde's functional group is at one end of the hydrocarbon chain, bonded to a carbon and a hydrogen. In ketones the functional group is located internally with hydrocarbon chains at each end.

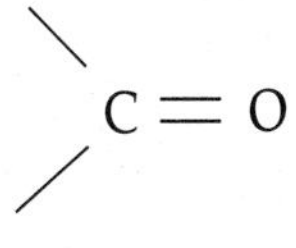

Carbonyl group

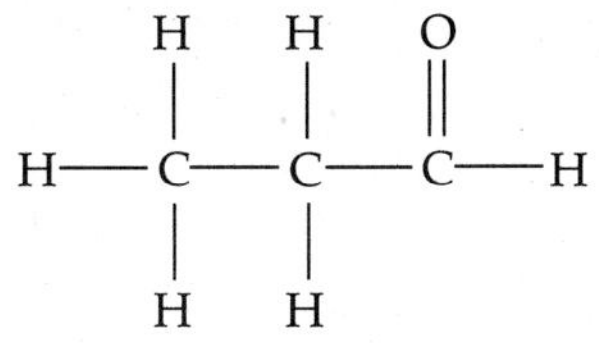

Propanal or Propionaldehyde

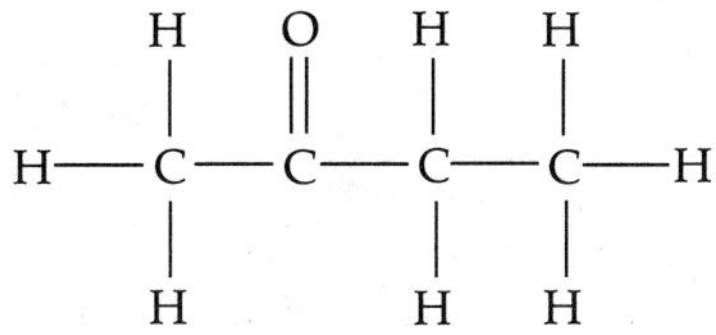

Ethyl methyl ketone, or Methyl ethyl ketone (MEK)

FIGURE 5.12 The carbonyl functional group.

Thus, the functional group bonds to two carbon atoms.

The IUPAC naming system for aldehydes uses an "-al" suffix. Ketones have prefixes indicating the two hydrocarbon groups followed by "ketone." Hydrocarbon group names are listed alphabetically. Ketones thus have three-word names. Two naming examples appear at the right in Figure 5.12. Propanal is the IUPAC name, and propionaldehyde is the common name. Ethyl methyl ketone is the IUPAC name but methyl ethyl ketone (or MEK) is the common reference. (Note that the common name does not conform to the IUPAC system.)

Aldehydes and ketones are both important in the synthesis of other chemical compounds and ketones are also used as industrial solvents. MEK, for example, is widely used.

Carboxylic Acids

Carboxylic acids are important chemical intermediaries in the preparation of several chemicals, including acetic acid (household vinegar), and they help in biological degradation. They contain the acyl group, which is a carbon double bonded to an oxygen. The carboxylic acid group is actually the acyl group with a hydroxyl attached. Both the acyl and carboxylic groups are shown at the left in Figure 5.13. At the right you will see two examples of carboxylic acids, acetic acid and butyric acid. Like other carboxylic acids, these have strong odors. Butyric acid is what makes rancid butter smell bad.

Because of the electronegativity of the oxygen on the acyl group, carboxylic acids are polar, and the proton on the hydroxyl group can be

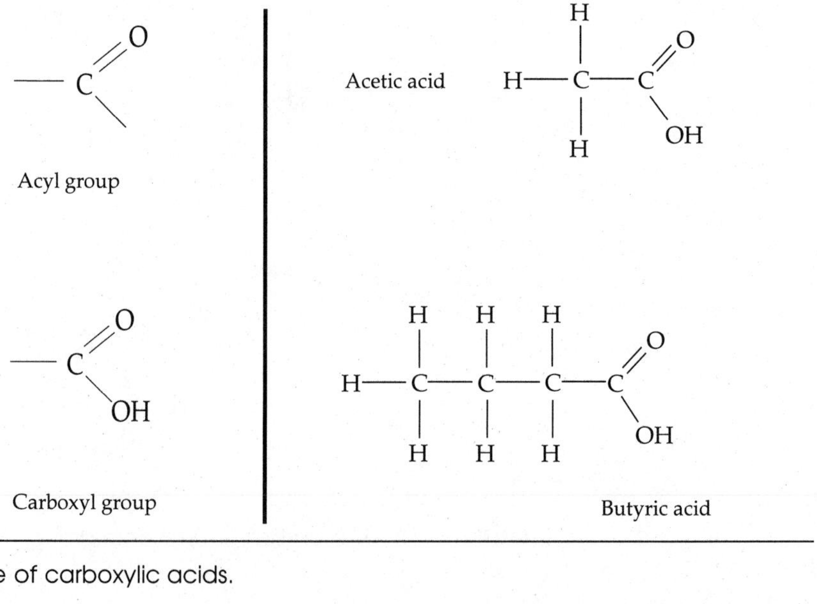

FIGURE 5.13 The structure of carboxylic acids.

carboxylic acids are polar, and the proton on the hydroxyl group can be donated, giving these compounds their acid property. The ionization of acetic acid is shown in this equation:

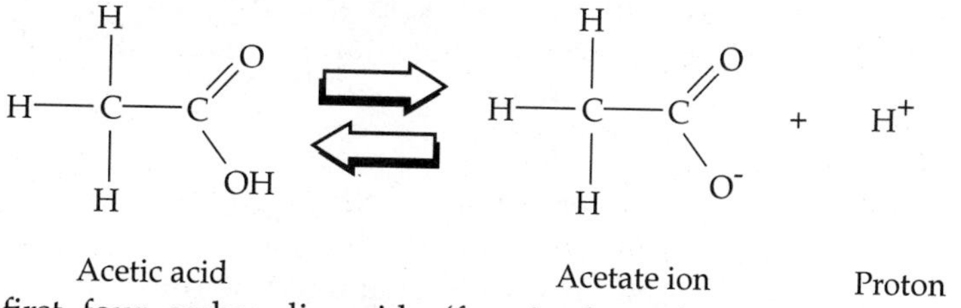

Acetic acid   Acetate ion   Proton   **5.5**

The first four carboxylic acids (formic through butyric) are completely miscible in water. The longer the hydrocarbon chain, the more like a hydrocarbon the acid acts. As a result, longer chain acids are less soluble. Decanoic acid (ten carbons) is almost totally insoluble.

Carbohydrates

Carbon dioxide and water combine in photosynthesis, using the chlorophyll in green plants and energy from the sun to form carbohydrates — compounds with an empirical formula of $C_a(H_2O)_b$. Carbohydrates are a major energy storage medium for plants and a significant energy source for many other life forms. The photosynthesis reaction is written as

$$aCO_2 + bH_2O + \text{Sunlight} \rightarrow C_a(H_2O)_b + aO_2 \qquad \textbf{5.6}$$

Organic catalysts mediate the process, and the reverse reaction produces energy from carbohydrates.

Carbohydrates include the basic sugars as well as cellulose, starch, and glycogen. The ending on most sugars is "-ose." The basic carbohydrate unit is the monosaccharide, a three- to six-carbon unit. More complex sugars are formed from these basic structures. Common table sugar, or sucrose, is a disaccharide composed of the monosaccharides fructose and glucose. Sucrose and its components are illustrated in Figure 5.14.

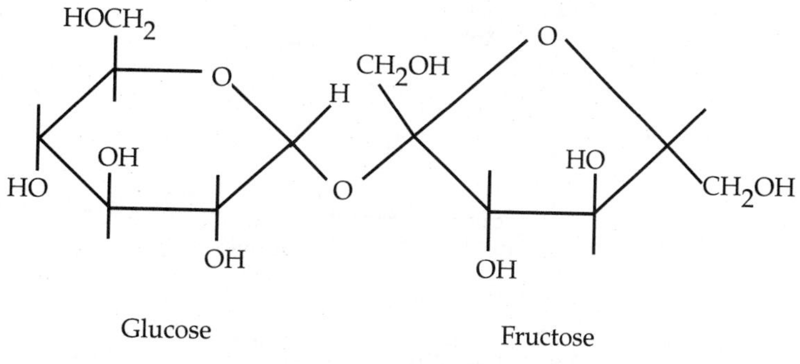

Glucose   Fructose

FIGURE 5.14 The disaccharide sucrose and its components, fructose and glucose.

NITROGENOUS COMPOUNDS

Nitrogenous organic compounds are important in both biological processes and industry. Two major groups of nitrogen-containing organics are amine and nitro compounds. Both have environmental applications.

Amines

Amines contain an "$-NH_2$" group or compounds in which the "N" is contained within the carbon chain. The nitrogen in an amine has three pairs of electrons for bonding and one free pair. (At lower pH values, bonding with a proton, H^+, to the fourth electron pair may occur.) As with carbon, the bonding of nitrogen is roughly tetrahedral. The three bonds and the one unbonded electron pair form angles of about 110°.

Primary amines have one carbon and two hydrogens attached to the nitrogen in the amine group. In secondary amines two carbons and one hydrogen attach to the nitrogen. And in tertiary amines the nitrogen connects with three carbons. An example of each type of amine is shown in Figure 5.15.

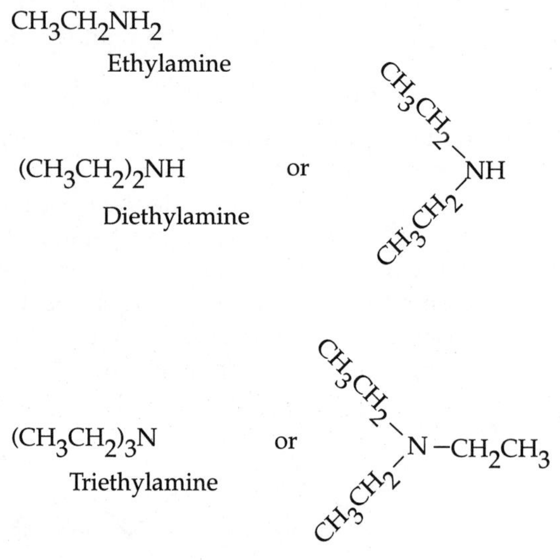

$$CH_3CH_2NH_2$$
Ethylamine

$$(CH_3CH_2)_2NH \qquad \text{or}$$
Diethylamine

$$CH_3CH_2 \diagdown NH$$
$$CH_3CH_2 \diagup$$

$$(CH_3CH_2)_3N \qquad \text{or}$$
Triethylamine

$$CH_3CH_2 \diagdown N-CH_2CH_3$$
$$CH_3CH_2 \diagup$$

FIGURE 5.15 Amine compounds.

Amino Acids

Amino acids are the building blocks of proteins, and proteins make up all life on earth. Amino acids are organic compounds with both a carboxylic acid group and an **amine** or "$-NH_2$" group.

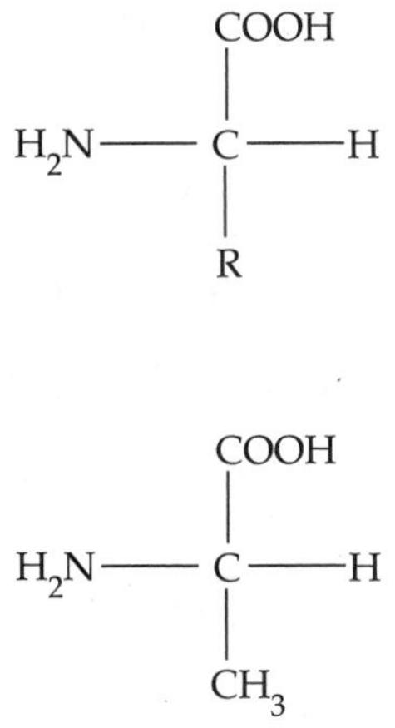

The amino acid structure

Alanine, a simple amino acid

FIGURE 5.16 The structure of an amino acid.

The structure of an amino acid is shown in Figure 5.16 along with an example, alanine. Proteins contain 22 amino acids. All living organisms can synthesize amino acids but some higher life forms cannot synthesize all 22. Humans can synthesize 14. The other 8 must be obtained from our diet. When proteins are degraded, the resulting compounds are simpler proteins and amino acids. When amino acids are degraded, ammonia results.

Much of the organic matter contained in domestic wastewater is composed of proteins. It therefore contains significant amounts of nitrogen. When the theoretical oxygen demand (discussed in the previous chapter) of a protein or other organic compound containing an amine is calculated, the nitrogen in the products is assumed to be ammonia. This is simply a convention, however. The nitrogen will actually be oxidized to nitrate, but that oxygen demand is calculated separately and termed nitrogenous oxygen demand or NOD. The following calculation for the ThOD of an amino acid will illustrate.

EXAMPLE 5.3 **ThOD OF AN AMINO ACID**

Calculate the ThOD of 50 mg/L of alanine, CH_3CHNH_2COOH (or $C_3H_7NO_2$).

SOLUTION: The end products are CO_2, H_2O, and NH_3. The first step is to write the basic equation, and balance it:

$$C_3H_7NO_2 + _O_2 \rightarrow _CO_2 + _H_2O + _NH_3$$

Or, in balanced form:

$$C_3H_7NO_2 + 3O_2 \rightarrow 3CO_2 + 2H_2O + NH_3$$
$$89.1 96.0$$

Thus, each 89.1 g of alanine requires 96 g of oxygen to oxidize it to carbon dioxide, water, and ammonia. The ThOD is then

$$ThOD = 50\ \frac{mg\ alanine}{L} \times \frac{96.0\ mg\ oxygen}{89.1\ mg\ alanine}$$

$$= 53.9\ mg\ oxygen/L$$

Nitrogenous Oxygen Demand

Nitrogenous oxygen demnd (NOD) is the amount of oxygen required to convert ammonia or organic nitrogen forms into nitrate $[NO_3^-]$. In calculation it is very similar to ThOD except the end product is nitrate. And, only the nitrogenous portion is considered. As an organic nitrogen compound is oxidized, it is assumed that the carbon goes to carbon dioxide, the hydrogen to water, and the nitrogen to ammonia. The NOD is then the oxygen required to oxidize the resulting ammonia to nitrate. In this manner, the carbonaceous and nitrogenous oxygen demands can be determined separately. The total oxygen demand is then the sum of the ThOD and NOD. NOD is a major portion of the oxygen demand of many wastewater discharges. The following example demonstrates NOD calculations.

EXAMPLE 5.4	CALCULATION OF NOD AND TOTAL OXYGEN DEMAND

Calculate the ThOD, NOD, and total oxygen demand for wastewater that contains

a. 25 mg/L serine, $CH_2OHCHNH_2COOH$, an amino acid.
b. 25 mg/L serine and 50 mg/L methanol, CH_3OH
c. 50 mg/L methanol and 30 mg/L ammonia, NH_3.

SOLUTION: a. The first step is to balance the oxidation of serine to end products of CO_2, H_2O, and NH_3.

$$C_3H_7O_3N + 2\tfrac{1}{2}O_2 \rightarrow 3CO_2 + 2H_2O + NH_3$$
$$105 \qquad\quad 80$$

$$ThOD_{ser} = 25\,\frac{mg\,Ser}{L} \times \frac{80\,mgO_2}{105\,mg\,Ser} = 19\,\frac{mg\,O_2}{L}$$

To determine the NOD, the ammonia from the previous reaction is oxidized to NO_3^-.

$$NH_3 + 2O_2 \rightarrow H^+ + NO_3^- + H_2O$$
$$64$$

Since we are determining the NOD of serine, not ammonia, we use the initial concentration of serine and the molecular weight of serine to determine the NOD.

$$NOD = 25\,\frac{mg\,Ser}{L} \times \frac{64\,mg\,O_2}{105\,mg\,Ser} = 15\,\frac{mg\,O_2}{L}$$

The total oxygen demand is then the sum of the ThOD and the NOD.

$$Total\ OD = ThOD + NOD = 34\,\frac{mg\,O_2}{L}$$

b. We already have the ThOD and NOD of serine from part (a). We must calculate the ThOD of methanol.

$$CH_3OH + 1\tfrac{1}{2}O_2 \rightarrow CO_2 + H_2O$$
$$32 \qquad\qquad 48$$

And the ThOD of methanol is then

$$ThOD_{MeOH} = 50\,\frac{mg\,MeOH}{L} \times \frac{48\,mg\,O_2}{32\,mg\,MeOH} = 75\,\frac{mg\,O_2}{L}$$

The ThOD of the two components is simply the sum of their respective ThODs, or

$$ThOD_{total} = ThOD_{Ser} + ThOD_{MeOH} = 19\,\frac{mg\,O_2}{L} + 75\,\frac{mg\,O_2}{L}$$

$$= 94\,\frac{mg\,O_2}{L}$$

Since methanol contains no nitrogen, it has no NOD. The total NOD is simply the NOD of serine, or 34 mg/L.

c. We calculated the ThOD of 50 mg/L methanol in part (b). We must now calculate the NOD of the ammonia.

$$NH_3 + 2O_2 \rightarrow H^+ + NO_3^- + H_2O$$

$$ 17 64$$

$$NOD = 30 \ \frac{mg \ NH_3}{L} \times \frac{64 \ mg \ O_2}{17 \ mg \ NH_3} = 113 \ \frac{mg \ O_2}{L}$$

Since ammonia has no ThOD, the ThOD$_{total}$ is 75 mg O$_2$/L, and the NOD$_{total}$ is 113 mg O$_2$/L.

Nitro Compounds

Nitro compounds contain the "–NO$_2$" group. They are important as explosives, dyes, antibiotics, and chemical intermediates. Two examples are nitrobenzene and nitroethane. Both are shown in Figure 5.17. Some nitro compounds are known or suspected carcinogens.

Many nitro compounds of interest are aromatic. They are named by adding nitro to the prefix of the major group. Where needed, a numerical prefix is used to indicate the location of the nitro group.

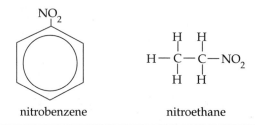

nitrobenzene   nitroethane

FIGURE 5.17 Nitrobenzene and nitroethane.

HALOGENATED ORGANICS

The elements in row VIIB of the periodic chart are halogens. They include fluorene, chlorine, bromine, iodine, and astatine. Chlorine is widely used in industry as well as in water and wastewater treatment plants. Fluorene, bromine, and iodine also have many industrial uses. Brominated and chlorinated organics are commonly used as pesticides. Chlorofluoro-carbons (CFCs) function as aerosol propellants and refrigerants, and they play a role in the manufacture of disposable styrofoam plastics. CFCs have been implicated in destruction of ozone in the upper atmosphere. Many, although not all, halogenated organics induce cancer and/or are known toxics.

The simplest halogenated organics are straight-chain hydrocarbons in which one or more hydrogens are replaced by halogens. Trichloromethane, an unwanted by-product of drinking water chlorination, is one example. Another is the common industrial solvent 1,1,1-trichloroethane. Both are shown in Figure 5.18.

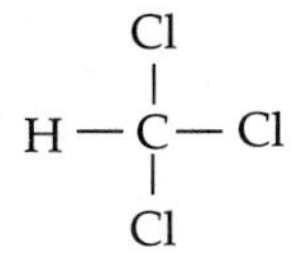

trichloromethane 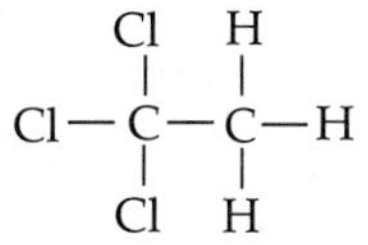   1,1,1-trichloroethane

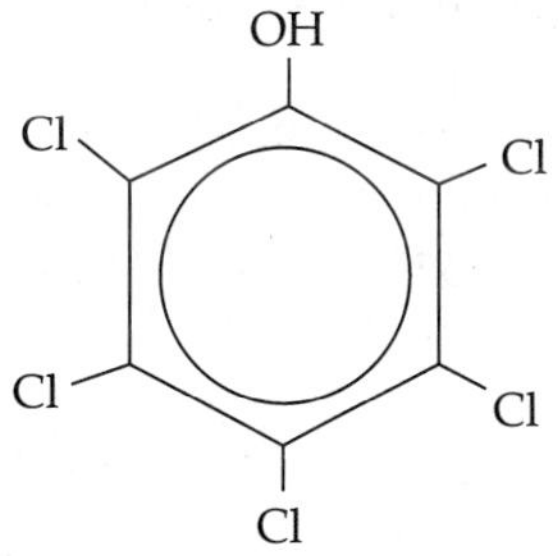

pentachlorophenol

FIGURE 5.18 Halogenated organics.

The structure of pentachlorophenol, a formulation used as a termite insecticide and wood preservative, is also shown in Figure 5.18. Pentachlorophenol has been associated with several adverse health effects.

Thus, halogenated organics are of great benefit to our way of life, yet they pose a serious threat to the environment. We are beginning to reduce the negative environmental effects of halogenated organics by restricting their use to more critical applications. The EPA is considering a ban on chlorine and chlorinated compounds because of their danger to humans and the environment.

PLASTICS

Plastics are organic polymers that did not exist 50 years ago. They are produced in large amounts each year. Production of the top five plastics in 1990 is shown in Table 5.5. This is more than 19 million metric tons per

TABLE 5.5   Production of the Top Five Plastics in 1990

PLASTIC	PRODUCTION, 1000 METRIC TONS
Low-density polyethylene	5,071
Polyvinyl chloride	4,123
High-density polyethylene	3,778
Polystyrene	2,273
Polypropylene	3,774

year! A polymer is a multiunit organic compound. Not all polymers are plastics, however. Other organic polymers include starch, cellulose, and wool.

Plastics are produced by combining many organic units or monomers. One example, polyethylene, is made of many ethylene units. Similarly, polypropylene is made of many propylene units (see Figure 5.19).

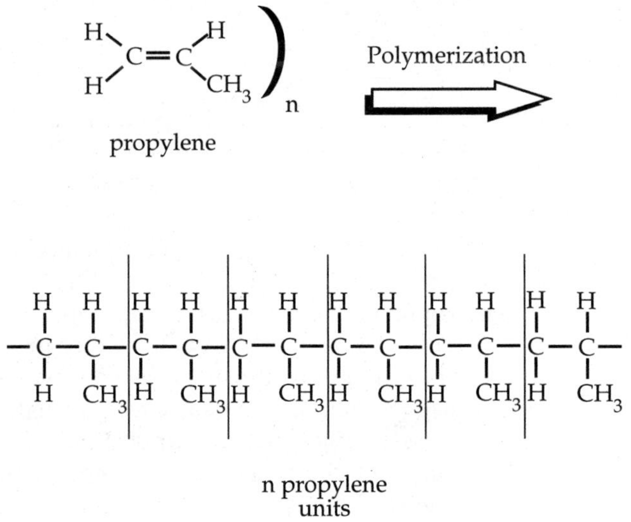

FIGURE 5.19   Polymerization of propylene.

Plastics are generally resistant to chemical attack. They also wear well, and with fiber reinforcing they can be quite rigid. Some are used for disposable products such as knives, forks, plates, cups, glasses, grocery bags, and trash bags. Others are used for exterior and interior parts of automobiles. The chemical resistance of plastic is an advantage in its use but a disadvantage after use, for plastics are generally difficult to biodegrade. In landfills they decompose very slowly. Recent attempts have been made to increase the biodegradability of some plastics by adding starch or other biodegradable materials to them. Thus far, however, few products made of these "biodegradable plastics" have been marketed. We should expect to see more of them and greater efforts to recycle plastics in the future.

CHEMICALS OF LIFE

Living organisms are made up of water and organic and inorganic chemicals. The organic chemicals determine what an organism is and how it functions. Most of the important biochemicals are composed of smaller chemical subunits that combine to form biological polymers, called biopolymers. The three major groups of biopolymers are proteins, polysaccharides, and nucleic acids. Each of these will be briefly introduced here.

Polysaccharides

High-molecular-weight carbohydrate polymers called polysaccharides are made up of units of carbohydrates such as glucose, fructose, or ribose. Three polysaccharides of particular importance are cellulose, starch, and glycogen. All are composed of units of glucose. However, glucose exists in two forms, α-glucose, and β-glucose. The only difference in them is the location of an "$-OH$" on the number 4 carbon atom. The three polysaccharides are distinguished by bonding at different locations within the basic glucose units, the extent of branching, and different molecular weights. Cellulose, the structural element in plants, is composed of linear chains (very little branching) of β-glucose units. Several chains hydrogen-bond to make larger structural units. A simple cellulose chain appears in Figure 5.20.

Starch is similar in overall structure to cellulose, but its bonding is slightly different. The difference, though small, makes a large difference in reactivity of starch. Many organisms can readily degrade starch. Few can degrade cellulose at significant rates. Whereas cellulose is the structural element of plants, starch is their major energy reserve. A single starch bond is shown in Figure 5.21.

The carbohydrate glycogen is used for energy storage in animals. Tens to hundreds of thousands of α-glucose units connect to make one glycogen unit within a cell. Studies indicate that glycogen has a molecular weight as high as 100 million [6]. Glycogen and starch differ in that glycogen is highly branched and starch is not. They are composed of the same form of glucose, however, and their main chains have the same type of bond. Yet at the branch points glycogen forms a different bond, as shown in Figure 5.22.

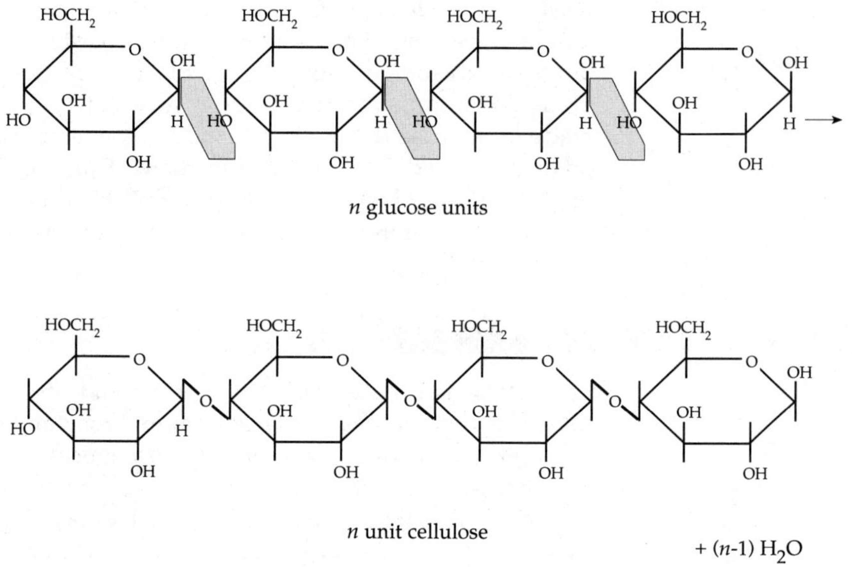

n glucose units

n unit cellulose

$+ (n\text{-}1)\ H_2O$

FIGURE 5.20 β-glucose units forming cellulose.

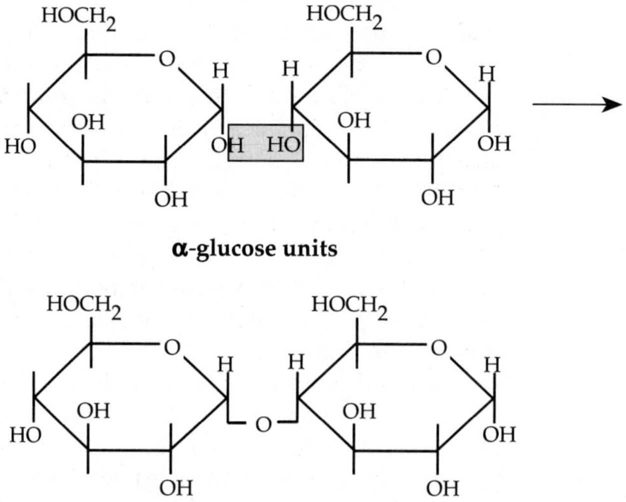

α-glucose units

FIGURE 5.21 α-glucose bonding in starch.

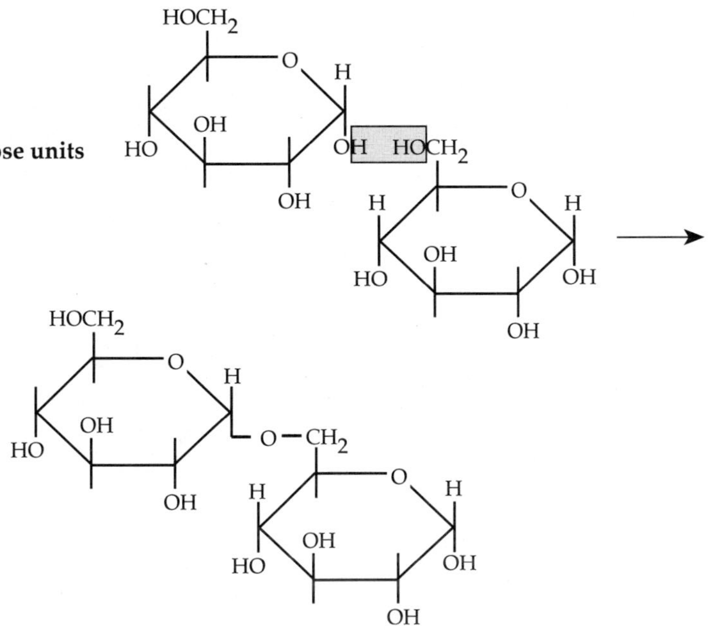

FIGURE 5.22   Branch bonding in glycogen.

Proteins

Proteins, composed of many amino acid units, are macromolecules — relatively large structures with molecular weights of several thousand upward. Proteins often are bound to nonprotein moieties. Two examples are glycoproteins, bound to carbohydrates, and lipoproteins, bound to greasy compounds called lipids. These are the most diverse of the three groups of biological polymers. Amino acids, introduced earlier, contain both an "–NH$_2$" and a "–COOH" group attached to the same carbon. To form proteins, amino acids combine into complex linkages. The carboxylic acid "–COOH" on one amino acid combines with the amine "–NH$_2$" group of another amino acid as shown in Figure 5.23. The resulting bond is a peptide link.

Many biological components, including enzymes, are proteins that act as catalysts and hormones that regulate biological activity. Proteins are also a major part of cell membranes and walls, blood in higher animals, silk, hair, nails, flagella and cilia in microorganisms, as well as many other biochemical materials.

Proteins are important in environmental engineering for a variety of reasons. They comprise a significant portion of living organisms. They mediate biological reactions. They are a major part of municipal wastewater. They are present in some industrial wastes as well as in the food fraction of municipal solid wastes.

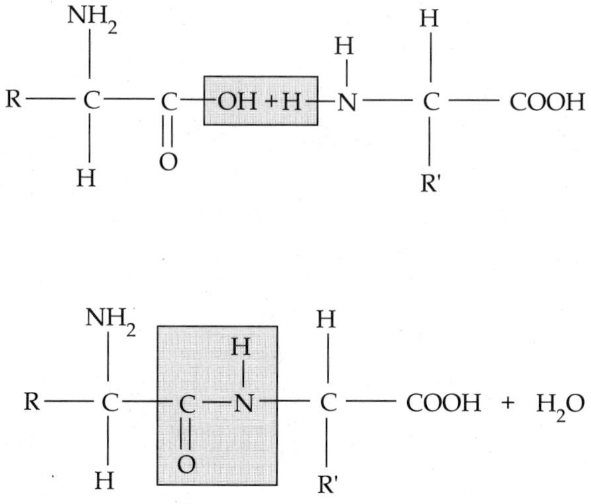

Peptide link

FIGURE 5.23 The peptide bond of amino acids.

Nucleic Acids

The blueprints of living organisms, nucleic acids contain the genetic codes. They are in everything from viruses to bacteria to mice to humans. Nucleic acids have three components: a pentose (5 carbon) sugar, a nitrogen base, and phosphate. Two sugars are used in nucleic acid. Ribose forms ribonucleic acid (RNA), and deoxyribose forms deoxyribonucleic acid (DNA). And five nitrogen bases are used. Both DNA and RNA use only four of the five. Cytosine, adenine, and guanine are used in both. Thymine is used only in DNA. Uracil is used only in RNA. All are cyclic organic nitrogens (but not aromatic). The five bases are illustrated in Table 5.6.

DNA is usually double stranded, although some viruses are known to have single stranded DNA. Similarly, RNA is typically single stranded, but again, some viruses are known to have double stranded RNA. The structure of a single DNA and a single RNA unit appear in Figure 5.24. DNA forms a double stranded helix, with paired complimentary bases forming hydrogen bonds. In DNA, thymine is always paired with adenine, and cytosine is always paired with guanine. This complementary pairing occurs for structural reasons. Adenine and guanine are larger than thymine and cytosine, and adenine and guanine cannot pair because there is not enough room between the double strands. Similarly, thymine and cytosine cannot pair because they do not extend far enough across the two strands to hydrogen-bond.

Although hydrogen bonds are weak compared to the **covalent bonds** holding the individual units together, there are many hydrogen bonds to

TABLE 5.6   Nucleic Acid Bases

BASE NAME	ACID	STRUCTURE
Cytosine	DNA & RNA	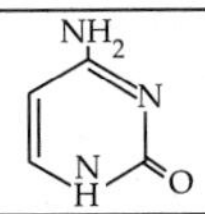
Thymine	DNA	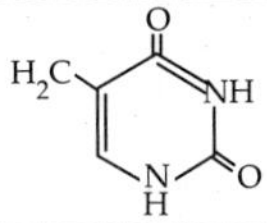
Uracil	RNA	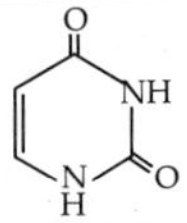
Adenine	DNA & RNA	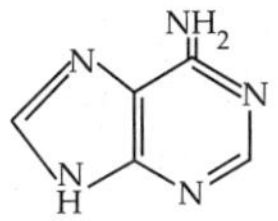
Guanine	DNA & RNA	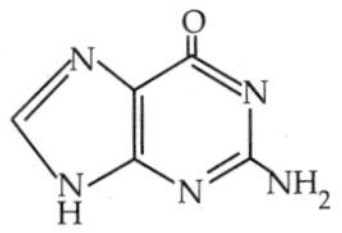

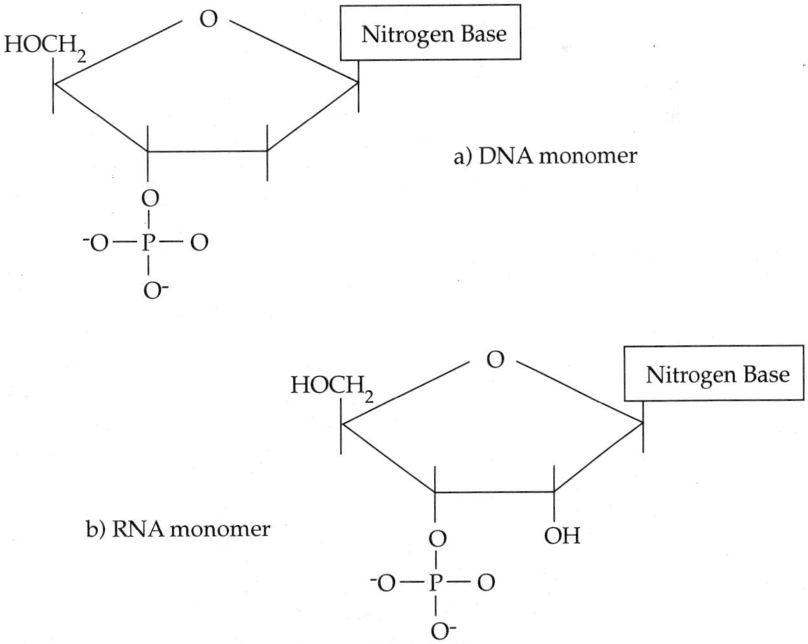

a) DNA monomer

b) RNA monomer

FIGURE 5.24   DNA and RNA monomer units.

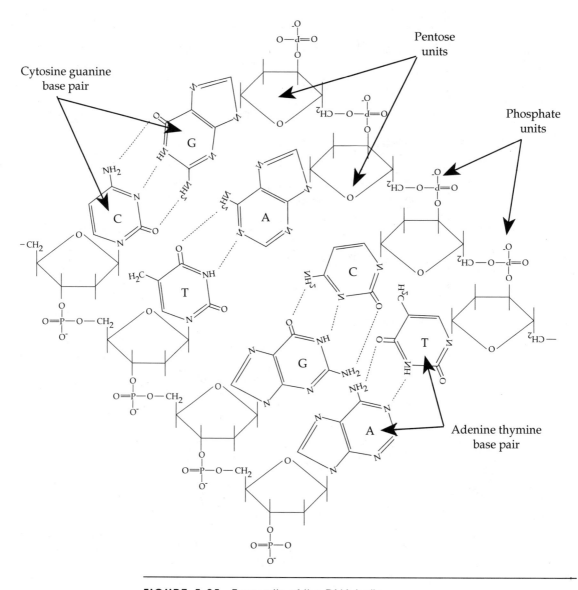

FIGURE 5.25   Four units of the DNA helix.

hold the strands together. Each A–T pair has two hydrogen bonds, and each C–G pair has three hydrogen bonds. Based on sheer numbers, their overall adhesion is high. Four units of a DNA strand are shown in Figure 5.25.

DNA and RNA are important for obvious reasons. They contain the genetic coding for any organism. And it is believed that some organic compounds can replace a nitrogen base in human DNA, causing health problems.

ORGANICS IN THE ENVIRONMENT

Organics have been in the environment since before the advent of life on earth. Most organics introduced into the environment by humankind were naturally occurring until the advent of synthetic organic chemistry in the 1940s and 1950s. Since that time we have introduced many new organics into the environment—organics with far-reaching effects.

Hydrocarbons

Hydrocarbons degrade slowly in the environment. However, there is evidence that soil microorganisms can break down members of this large group of compounds *in situ* (in place or in the soil) [7, 8]. Such microorganisms require a variety of nutrients for growth. Since hydrocarbons do not contain other nutrients such as nitrogen, oxygen, and sulfur, they are nutrient deficient. This is why cleanup crews for large crude oil or other hydrocarbon spills often add nutrients in the area of the spill. When the stricken *Exxon Valdez* disgorged oil in Prince William Sound on March 25, 1989, several million gallons of Alaskan crude washed ashore, contaminating almost 300 miles of rocky coast. Application of nitrogen : phosphorus-based fertilizers increased the degradation rate of the oil. Although environmental damage certainly occurred due to the spill, the effect is probably less than many initially expected because biodegradation was accelerated in this manner.

Halogenated Organics

Halogenated organics are almost everywhere. Many refrigerants and aerosol propellants are halogenated organics. The nonstick lining in cookware is a chlorofluorocarbon not known or suspected of causing health problems. Polychlorinated biphenyls (PCBs) were once commonly used in hydraulic applications and as an insulating oil in electric transformers, but they are toxic and suspected of being carcinogenic. The chlorination of drinking water to kill pathogenic organisms creates very small concentrations of halogenated organics known to be a health hazard. Halogenated organics are common solvents used in home and industry. Many pesticides are halogenated organics. Most hazardous waste disposal sites that are of environmental concern contain halogenated organics. So these ubiquitous compounds both improve our standard of living and cause environmental problems.

The Great Lakes of the United States and Canada exemplify the extent and persistence of halogenated organic contamination. Over the last two centuries almost every conceivable contaminant has been discharged into these lakes. By the 1960s the Great Lakes were highly polluted and supported little aquatic life. The problem was compounded by the fact that the lakes border two nations. Thus, any effective controls must be instituted by both governments. The first U.S./Canadian agreements to reduce pollution were signed in 1972. Other agreements were signed in 1978 and 1987. In particular, the countries have banned several halogenated pesticides, including DDT, PCBs, mirex, and heptachlor, and the level of these contaminants has decreased markedly during the past 20 years. In all but PCBs, contamination is now below targeted levels. The environmental quality of the Great Lakes has improved dramatically during the past 20 years [9].

The reduction in organics in these waters is the result of a combination of processes, including washout, chemical degradation, and biological degradation. Since the average detention time in the Great Lakes is more than 100 years (less than 1% of the water exits each year), if only washout occurred, the cleanup would take many years. However, other beneficial processes, including photodegradation and biochemical degradation, are also occurring in the lakes.

Most halogenated organics will degrade with or without microorganisms. However, the degradation rates are usually higher when microorganisms are present. Typical half-lives (time to reduce the concentration to one-half of the original) of a few halogenated organics are shown in Table 5.7 [10].

TABLE 5.7 Half-Lives of Halogenated Organics Without Microbial Degradation

COMPOUND	HALF-LIFE, YEARS
1,1,1-trichloroethane	0.5 to 2.5
Chloroethane	0.12
Trichloromethane	1.3
Trichloroethene	0.9 to 2.5
Dibromoethane	0.08

INFORMATION SOURCES

You can obtain additional information about organic chemicals from a variety of sources—including most any organic chemistry textbook. Several good ones are available.

Also, there is the *CRC Handbook of Chemistry and Physics,* latest edition, published by CRC Press [3]. This reference provides basic information on a wide range of inorganic and organic chemicals. The information in-

cludes structure, miscibility in water, and boiling and freezing temperatures. The book has a detailed description of the naming of organic
compounds as well.

The most recent edition of *The Merck Index, An Encyclopedia of Chemicals,
Drugs, and Biologicals,* published by Merck & Company in Rahway, New
Jersey [11] is also useful. It provides general chemical information, explains
uses of chemicals, gives toxicity information, and often includes additional
references. *The Merck Index* is an excellent reference for handling and safety
information on chemicals.

Review Questions

1. Name the following organic compounds:

a.

$$H-\underset{\underset{H}{|}}{\overset{\overset{H}{|}}{\underset{6}{C}}}-\underset{\underset{H}{|}}{\overset{\overset{H}{|}}{\underset{5}{C}}}-\underset{\underset{H}{|}}{\overset{\overset{H}{|}}{\underset{4}{C}}}-\underset{\underset{H}{|}}{\overset{\overset{H}{|}}{\underset{3}{C}}}-\underset{\underset{H}{|}}{\overset{\overset{H}{|}}{\underset{2}{C}}}-\underset{\underset{H}{|}}{\overset{\overset{H}{|}}{\underset{1}{C}}}-H$$

b.

$$\underset{H}{\overset{H}{>}}C\underset{1}{=}\underset{}{\overset{}{C}}\underset{2}{-}\underset{\underset{H}{|}}{\overset{\overset{H}{|}}{\underset{3}{C}}}-\underset{\underset{H}{|}}{\overset{\overset{H}{|}}{\underset{4}{C}}}-\underset{\underset{H}{|}}{\overset{\overset{H}{|}}{\underset{5}{C}}}-H$$

c.

$$H-\underset{\underset{H}{|}}{\overset{\overset{H}{|}}{C}}-\underset{\underset{H}{|}}{\overset{\overset{H}{|}}{C}}-\underset{}{\overset{}{C}}=\underset{}{\overset{}{C}}-\underset{\underset{H}{|}}{\overset{\overset{H}{|}}{C}}-\underset{\underset{H}{|}}{\overset{\overset{H}{|}}{C}}-H$$

d. $H-C\equiv C-H$

e.

$$H-\underset{\underset{H}{|}}{\overset{\overset{H}{|}}{C}}-\underset{\underset{H}{|}}{\overset{\overset{H}{|}}{C}}-\underset{\underset{OH}{\diagdown}}{\overset{\diagup O}{C}}$$

f.

$$H-\underset{\underset{H}{|}}{\overset{\overset{H}{|}}{C}}-\underset{\underset{H}{|}}{\overset{\overset{H}{|}}{C}}-\underset{}{\overset{\overset{O}{||}}{C}}-\underset{\underset{H}{|}}{\overset{\overset{H}{|}}{C}}-\underset{\underset{H}{|}}{\overset{\overset{H}{|}}{C}}-H$$

g.

$$\begin{array}{c} -\underset{|}{\overset{|}{C}}- \\ -\underset{5}{\overset{|}{C}}-\underset{4}{\overset{|}{C}}-\underset{3}{\overset{|}{C}}-\underset{2}{\overset{|}{C}}-\underset{1}{\overset{|}{C}}- \\ -\underset{|}{\overset{|}{C}}- \end{array}$$

h.

$$\begin{array}{c} -\underset{|}{\overset{|}{C}}- \\ -\underset{6}{\overset{|}{C}}-\underset{5}{\overset{|}{C}}-\underset{4}{\overset{|}{C}}-\underset{3}{\overset{|}{C}}-\underset{2}{\overset{|}{C}}-\underset{1}{\overset{|}{C}}- \end{array}$$

2. Use the *Merck Index* [11] to determine the use, toxicity, alternative
names, and general characteristics (boiling point, flash point, odor, etc.)
of the following compounds. Since many compounds have common as
well as formal names, the publication has a cross-index of names. Refer
to the cross-index to find the index number of the chemicals. The *Merck*

Index should be in library reference sections. If not, ask a science or engineering librarian for assistance.

a. methyl ethyl ketone

b. methyl alcohol

c. dimethyl ether

d. isopropyl alcohol

3. Name the following organic compounds:

a.

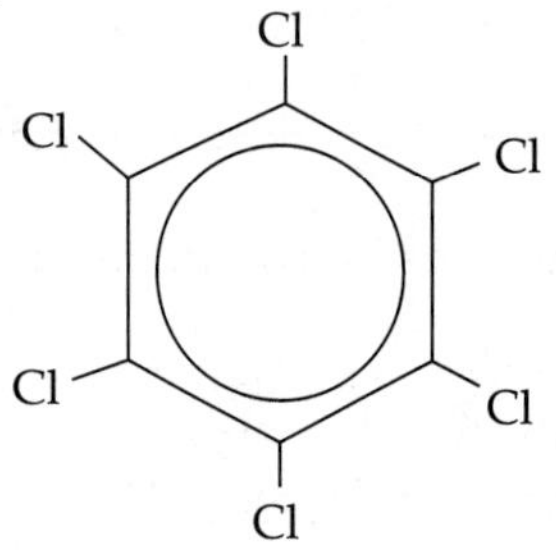

b.

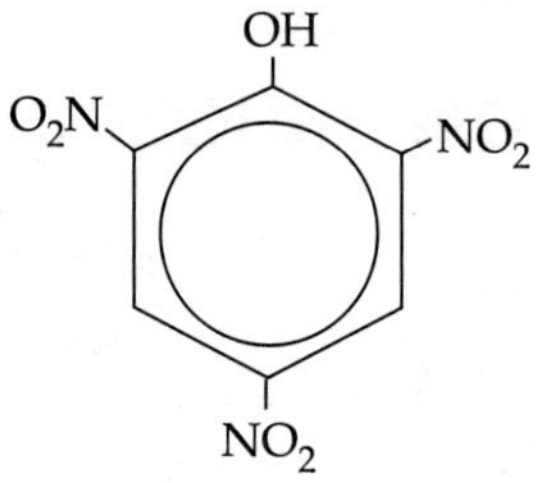

c.

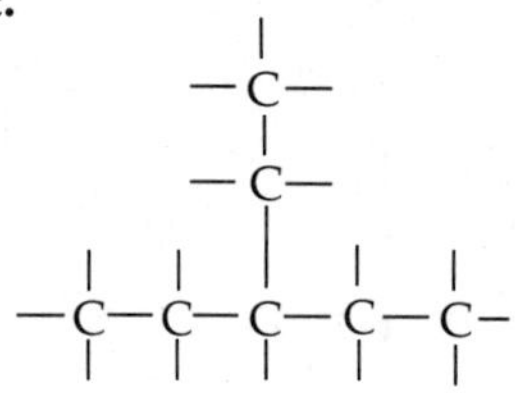

d.

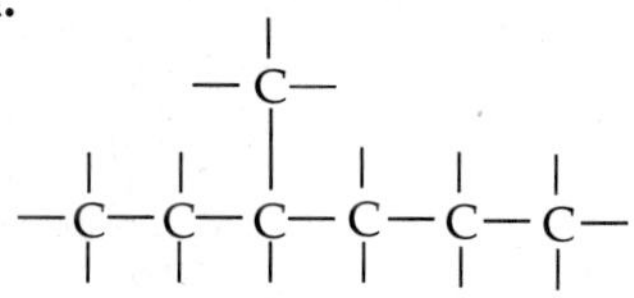

e.

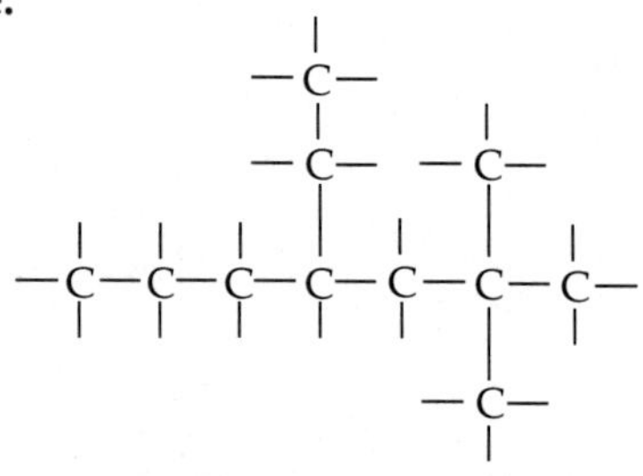

f.

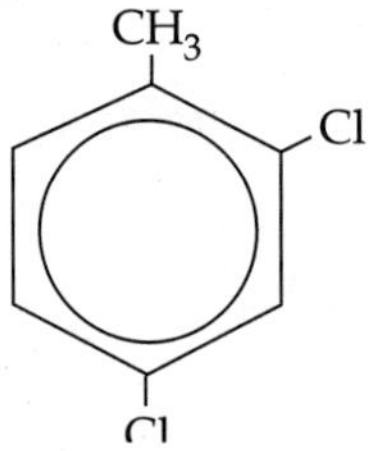

4. Write the structure of the following organic compounds:

a. Isopropyl alcohol

b. Butane

c. Butanol

d. *n*-butyl mercaptan

e. Cyclohexane

f. *n*-propanol

5. Why would you expect acetic acid or ethanol to be more soluble in water than in octane?

6. Using stoichiometry, calculate the amount of oxygen, mg O_2/mg, organic (ThOD) that is required to oxidize the following compounds to CO_2 and H_2O. Note that each is more oxidized than the previous compound. What is the relationship between the number of oxygens and the amount of oxygen required per mass of organic?

   a. Ethane $[C_2H_6]$; ethanol $[CH_3CH_2OH]$; acetic acid $[CH_3COOH]$

   b. Propane $[C_3H_8]$; n-propanol $[CH_3CH_2CH_2OH]$; propanoic acid $[CH_3CH_2COOH]$

7. Calculate the ThOD of the following:

   a. 100 mg/L of glycine, H_2NCH_2COOH

   b. 25 mg/L of ethanol, CH_3CH_2OH

   c. $10^{-5} M$ valine, $(CH_3)_2CHNH_2CHCOOH$

   d. 50 mg/L of glycine, H_2NCH_2COOH plus 25 mg/L of ethanol, CH_3CH_2OH

8. Calculate the NOD of the solutions in Problem 7.

9. Draw the structure of a single strand of RNA with the following bases: adenine, guanine, uracil, and cytosine (AGUC).

10. Draw the structure of a double strand of DNA with the bases listed in Problem 9.

References

1. E. V. Anderson, "Mexico's Chemical Industry Gears Up for North American Free Trade," *Chemical & Engineering News* (September 9, 1991), p. 11.

2. *BP Statistical Review of World Energy* (Cleveland: eThe British Petroleum Company, 1991).

3. *CRC Handbook of Chemistry and Physics*, 62nd ed. (Boca Raton, Florida: CRC Press, 1981).

4. T. W. G. Solomons, *Organic Chemistry*, 2nd ed. (New York: Wiley, 1980).

5. D. D. Whitehurst, "A Primer on the Chemistry and Constitution of Coal," in *Organic Chemistry of Coal*, John W. Larsen, ed., ACS Symposium Series no. 71, 1978.

6. D. Hanson, "Air Pollution Cleanup: Pact Set for Reformulating Gasolines," *Chemical and Engineering News* (August 26, 1991), pp. 4–5.

7. X. Wang, X. Yu, and R. Bartha, "Effect of Bioremediation on Polycyclic Aromatic Hydrocarbon Residues in Soil," *Environmental Science and Technology,* vol. 24, no. 7 (1990), pp. 1086–89.

8. J. M. Thomas and C. H. Ward, "In Situ Biorestoration of Organic Contaminants in the Subsurface," *Environmental Science and Technology,* vol. 23, no. 7 (1989), pp. 760–766.

9. B. Hileman, "The Great Lakes Cleanup Effort: Much Progress but Persistent Contaminants Remain a Problem," *Chemical and Engineering News* (February 8, 1988), pp. 22–39.

10. T. M. Vogel, C. S. Criddle, and P. L. McCarty, "Transformations of Halogenated Aliphatic Compounds," *Environmental Science and Technology,* vol. 21, no. 8 (1987), pp. 722–736.

11. *The Merck Index, An Encyclopedia of Chemicals, Drugs, and Biologicals,* 11th ed., Susan Budavari, ed. (Rahway, New Jersey: Merck & Company, 1989).

6

Microbiology and Microbial Growth

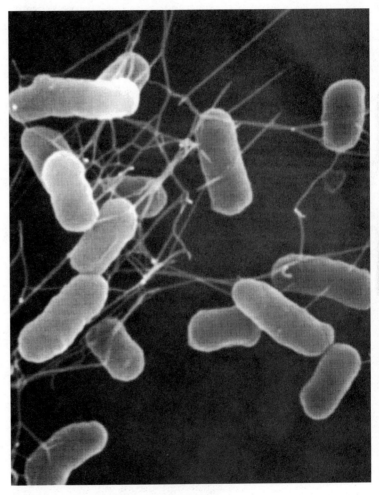

Scanning electron micrograph of
Xanthomonas sp. The bacteria are motile
by means of flagella, seen here
as rope-like strands. Micrograph courtesy
of John Bozzola, Southern Illinois
University at Carbondale.

INTRODUCTION

Microorganisms play a significant role in the natural environment and in engineered systems. They are present in all natural waters, serving beneficial purposes yet posing the threat of disease when contaminated water sources are used for human consumption. And microorganisms are a critical part of almost all municipal and many industrial wastewater treatment processes. So a clear understanding of their life cycles and chemistries is critical to environmental engineering.

The first known recorded observations of microorganisms were made by Dutch linen merchant and amateur scientist Antonie van Leeuwenhoek in 1673. Leeuwenhoek used a simple single-lens microscope for his observations. He sent detailed sketches of microorganisms with descriptive letters to the Royal Society of London, many of which were translated and published [1]. At this time no one realized that microorganisms are connected with disease. They were simply small creatures of interest.

Two centuries later, in 1876, a German physician named Robert Koch used the anthrax bacterium (*Bacillus anthracis*) to prove that microorganisms cause infection. Koch observed that *B. anthracis,* a particularly large bacterium, was always present in the blood of cattle with anthrax. But, he asked, did the organism cause the disease or did the disease produce the organism? Koch took blood from an infected animal and injected a small amount of it into a healthy animal. The healthy animal then contracted anthrax. He repeated this procedure 20 times, and newly inoculated animals always got sick. Koch even grew *B. anthracis* on culture plates and then inoculated healthy animals with the microorganisms. The animals contracted anthrax every time. Thus, Koch proved the germ theory of disease.

It may seem obvious to us that microorganisms can cause disease, but Koch's work was revolutionary. It led to many new discoveries in the field of microbiology, including the causes of deadly diseases of the time.

Microorganisms of interest to environmental engineers vary from viruses that pose general threats to human health to the specific waterborne protozoan *Giardia lamblia* (which causes a severe intestinal illness) to bacteria that metabolize organic wastes in wastewater treatment systems.

MICROORGANISMS

Microorganisms are present in almost every imaginable place on earth. They are in the soil, in the air, at the bottom of the ocean, and on and in other life forms, including humans. Many microorganisms serve a beneficial function: others, **pathogenic** or disease-causing **organisms**, are harmful. Microbes are divided into two basic groups: procaryotic organisms, which do not have cellular membranes, and eucaryotic organisms, which do. **Procaryotes** are single-cell organisms, including only bacteria. Eucaryotic microorganisms may be single-cell or multicellular. They include algae, fungi, protozoa, and rotifers. Viruses are considered a separate group. We will discuss each of these categories in turn.

Bacteria

Bacteria are unicellur organisms that do not have a nuclear membrane. They vary in size from $0.1\,\mu$m to more than $5\,\mu$m. Their shape varies from cyclindrical to spherical. The common *eschericia coli* is $0.5\,\mu$m in diameter and $2\,\mu$m long. Bacteria have a cell wall, cell membrane, cytoplasm, and DNA, as shown in Figure 6.1. The nuclear material, a single strand of DNA, contains the genetic information of the cell. And while the cell wall provides structural integrity for the bacterium, the membrane selectively allows passage of nutrients and waste into and out of the cell. The cytoplasm is composed of organic and inorganic molecules necessary for cell function. Within it are thousands of ribosomes, which translate the genetic code and produce cell proteins.

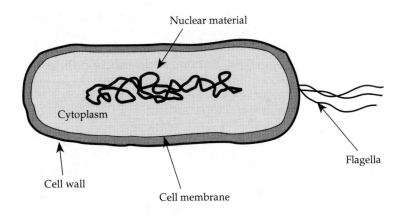

FIGURE 6.1 Structure of a bacterium.

Many bacteria have a means of motion, flagella. These are single strands of a hollow protein that the bacteria can rotate. The forward speed that bacteria can achieve with their flagella is about 20 to $80\,\mu$m/s, or about ten cell lengths per second [2]. In contrast, Olympic sprinters can attain speeds of approximately $10\,$m/s, or five body lengths per second.

Another property of certain bacteria is their ability to produce endospores (internally produced spores). The onset of endospore formation is usually triggered by nutrient exhaustion. Under these conditions the vegetative bacterial cell will form an endospore rather than divide. Thus, spore formation is not a method of reproduction; instead, it is a method of self-preservation brought on by negative environmental conditions. Spores are highly resistant to heat, chemicals, radiation, and drying. Conditions that would easily destroy a vegetative bacterial cell will not harm a spore. It can remain dormant for many years awaiting the return of favorable conditions in which to grow anew.

Bacteria can live in many places inhospitable to most other organisms. Certain forms, for example, can exist at the interface between water and fuel in hydrocarbon fuel tanks. Others thrive in the depths of the oceans where hot waters escape from fissures. Still others live in the mouth and teeth of mammals, including humans. And there are bacteria in and on our foods and in our water supplies.

Viruses

The simplest form of life known, **viruses** exist in two forms: infectious particles that contain DNA or RNA, a structural coat, and possibly some other chemicals; and as a part of a host cell's DNA or RNA. Shown in Figure 6.2 is the structure of a **virion**. Note that the virion particle contains nucleic acid and a capsule, or nucleocapsid, composed of capsomere units. Some viruses also have a coating on the nucleocapsid, termed an envelope. When a virus infects a host, the nucleocapsid attaches to the host cell wall, and the nucleic acid is then injected into the cell. The viral nucleic acid then inserts itself in the host's nucleic acid.

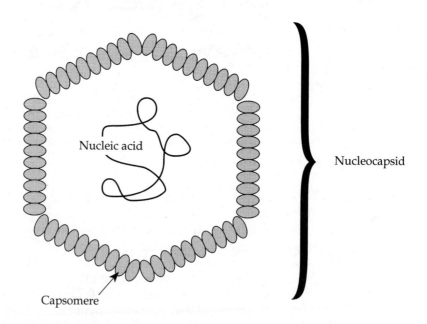

FIGURE 6.2 Structure of a virion.

Algae

Algae are chlorophyll-containing **eucaryotic organisms** that carry out oxygenic photosynthesis. Most algae are microscopic, but some, such as kelp, reach lengths of more than 100 feet. Photosynthetic activity occurs in membranous structures, chloroplasts, within the cell. Although algae utilize carbon dioxide and produce oxygen in the light, they concurrently respire, consuming oxygen and producing carbon dioxide. During periods of light, the net effect is production of oxygen. Algal photosynthetic production of oxygen far exceeds respiratory consumption of oxygen in the presence of light. However, during periods of darkness, respiration continues, without the photosynthesis. At this time algae consume oxygen.

Algae are important in environmental engineering for many reasons. They are at least partially responsible for taste and odor problems in public drinking water supplies originating in surfacewater, particularly lakes. In addition, they generate much of the suspended matter present in the effluent from lagoon wastewater treatment systems. Algae also grow where wastewater or farm runoff adds too much nitrogen and phosphorus to surfacewater. These nutrients frequently foster excessive algal production, termed algal blooms. Yet overall, algae are important in the environment as primary producers of biomass. Marine phytoplankton, algae, are responsible for approximately 90% of the photosynthesis on earth [3].

Protozoa

Protozoa are noteworthy in environmental engineering primarily for their infectious nature and their role in the mixed microbial cultures of wastewater treatment system. Many protozoa are parasitic to humans. Others are beneficial in that they consume large amounts of bacteria, helping to clarify wastewater. Protozoa are much larger than bacteria, and because they are predatory, most are motile. There are several forms. Flagellated protozoa have one or more flagella for motility. Ciliated protozoa have cilia covering the cell membrane. These, too, provide motility. Another common form of protozoa are the amoebas, a group characterized by a flowing movement.

Rotifers

Multicellular microorganisms called rotifers prey on smaller microorganisms, so they are of particular interest in wastewater treatment systems. They act as "clarifiers" for such systems by removing large quantities of dispersed bacteria. Their presence in water indicates a low level of pollution. A common rotifier is illustrated in Figure 6.3.

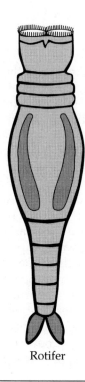

Rotifer

FIGURE 6.3   A common rotifer.

Metabolic Classification

Microorganisms can be classified by metabolic processes such as their energy source, oxygen source, and carbon source for synthesis.

Carbon Source.   All known life forms require carbon. **Heterotrophic** organisms obtain carbon from other organic matter or proteins. However, several classes of microorganisms can obtain carbon for synthesis from inorganic carbon such as carbon dioxide and its dissolved species (the carbonates). These organisms, a group that includes plants and algae, are called **autotrophic**.

Oxygen Source.   Some microorganisms require molecular oxygen [O_2]. For others it is toxic. Aerobic organisms, or **aerobes**, are those that require molecular oxygen. They include many lower organisms, such as bacteria and protozoa, but also all known higher life forms. **Anaerobic** organisms do not require molecular oxygen. In fact, molecular oxygen is toxic to them. This group includes several species of bacteria and some protozoa. These organisms, as well as all known life forms, require oxygen, but they obtain their oxygen from inorganic ions such as nitrate or sulfate or from protein. **Facultative organisms** prefer or preferentially use molecular oxygen, but will use other pathways for energy and synthesis if molecular oxygen is not available.

Energy Source Energy is produced in cells by either chemical or photo oxidation. A **chemotroph** obtains energy from the metabolism of chemicals, either organic or inorganic. A **phototroph** obtains energy from light, using photooxidation.

Microorganisms in the Environment

Microorganisms serve important and useful purposes in the environment. Some species of bacteria can convert atmospheric nitrogen to ammonia (nitrogen fixation). This is important for some symbiotic bacteria associated with major crops such as soybeans. Microorganisms are the principal decomposers of waste material, whether anthropogenic or natural. In landfills, microorganisms degrade organic wastes, producing simpler organics and eventually carbon dioxide, methane, and water. In these processes the minerals contained in the wastes return to the environment.

Microorganisms naturally present in water bodies degrade wastes discharged to the water. In so doing, however, they consume valuable oxygen essential for other aquatic organisms such as fish and shellfish. Where excessive amounts of wastes are present, this process can and does deplete the oxygen, resulting in very low levels of dissolved oxygen, or anaerobic conditions.

In other situations, microorganisms represent a threat to human health or agricultural products. If not removed, killed, or inactivated in surface-water sources used for human consumption, microorganisms can represent a significant health risk. Much of drinking water treatment is occupied with reducing or eliminating **pathogenic organisms**. The United States, Canada, Japan and Western Europe are the only major areas on earth where people can safely assume that public water supplies are practically devoid of health risk from microbial contamination. Pathogenic microorganisms also degrade or decompose many agricultural products, causing huge financial losses.

Microorganisms in Engineered Systems

Microorganisms are used in engineered systems. Specific strains of bacteria and yeasts (forms of fungi) help produce medicines and alcoholic beverages. These creatures have special characteristics. Most microorganisms of use in environment engineering, however, are not so narrowly defined. They will simply grow and predominate in a given system. We will see later, though, that some very special strains of microorganisms are being developed, particularly to treat hazardous or toxic substances.

Microorganisms are widely used for wastewater treatment. Such systems can be divided into two groups based on whether the bacteria attach to engineered surfaces provided for that purpose, or remain suspended in the liquid being treated. Attached bacteria have evolved a complex chemical attachment system. They develop a slimy substance on their cell walls. The chemical structures are varied but consist mostly of polysaccharides. A more technical term for this layer of slime is a capsule or glycocalyx [4]. In some cases, the glycocalyx* is used by bacteria as a means of attachment to a host. Have you ever walked in a stream that had slippery rocks? This covering is a biofilm. (The term does sound more appealing than slime layer.) In attached biofilm treatment systems such slime enables bacteria to attach to media. Microorganisms in the biofilm then assimilate the wastes, cleaning the water in the process. A biofilm layer is shown in Figure 6.4.

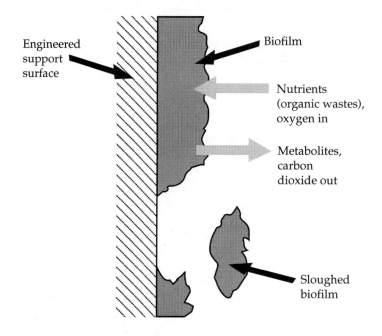

FIGURE 6.4 Attached biofilm layer.

Two biofilm treatment processes are the trickling filter and the rotating biological contactor. They are shown schematically in Figure 6.5.

---

* In addition to attachment, the slime layer serves an important function during infection. In some pathogenic bacteria, the mass of glycocalyx appears to act as a disguise, preventing a host's immune system from identifying and destroying the bacteria.

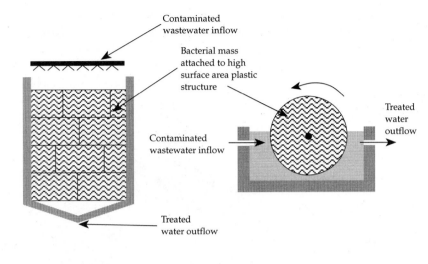

Contaminated
wastewater inflow

Bacterial mass
attached to high
surface area plastic
structure

Contaminated
wastewater inflow

Treated
water
outflow

Treated
water outflow

a) Trickling filter   b) Rotating biological contactor

FIGURE 6.5 Attached biofilm treatment processes.

In suspended growth systems, the bacteria are suspended in the water in large tanks (or reactors). The bacteria move through the process with the water. However, most bacteria do not grow at a sufficient rate to treat the wastes adequately. They must be separated from the treated water and returned to the beginning of the treatment tank. In large tanks this is accomplished by settling. The clear liquid rises to the top and flows to the next process or the stream. The microbial mass, which has a specific gravity slightly greater than water, goes to the bottom of the tank. This microbial mass, still mostly water, is pumped from the bottom and mixed with the new wastewater entering the plant. The glycocalyx that enables bacteria to adhere to surfaces causes the suspended bacteria to stick to each other, forming large clumps, or bacterial floc, which settles faster than individual bacteria.

On the other side of the coin, bacteria coated with glycocalyx get attached in places where they are not wanted. Biofilms are a significant problem, for example, in water treatment plants and distribution systems, cooling towers, and in some industrial water applications [5].

Microorganisms are also being employed to destroy toxic and hazardous wastes that have leaked onto or into the soil. In many cases this requires the injection of essential nutrients into the soil to allow these organisms to thrive and grow.

METABOLISM

Microbial **metabolism** is the sum of the processes sustaining the organism, including both production of new cellular materials and degradation of other materials to provide energy. Metabolism includes production of new cellular materials, or **anabolism**, and the degradation of substrate to produce energy to maintain the cell, or **catabolism**. Cellular metabolism is depicted in Figure 6.6.

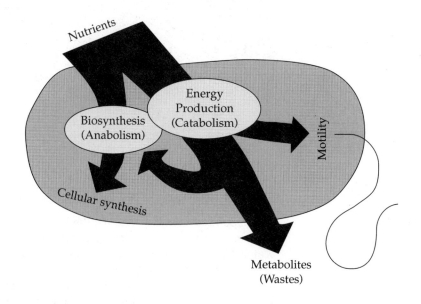

FIGURE 6.6 Cellular metabolic processes.

Energy Chemicals

As a cell metabolizes substrate, the energy from the substrate must be stored and transported to various parts of the cell. Several chemicals are involved in this process. However, the primary chemical for this storage is adenosine triphosphate (ATP). The oxygen-to-phosphate bond holding each phosphate group together in ATP has a high energy level compared to most other bonds. The high energy bond is indicated by the " ~ " in Figure 6.7. In most cases, only the last (circled) phosphate group is lost, producing the lower energy adenosine diphosphate (ADP). However, in some cases, a second phosphate group can also be utilized to produce energy. When only one phosphate group is left, the compound is adenosine monophosphate (AMP). Since energy production equates to ATP production, or addition of a phosphate to an ADP molecule, the process is often termed **phosphorylation**. Where oxygen is involved, it is **oxidative phosphorylation**. Where only substrate is used, the process is **substrate level phosphorylation**. And where solar energy is used, it is **photophosphorylation**.

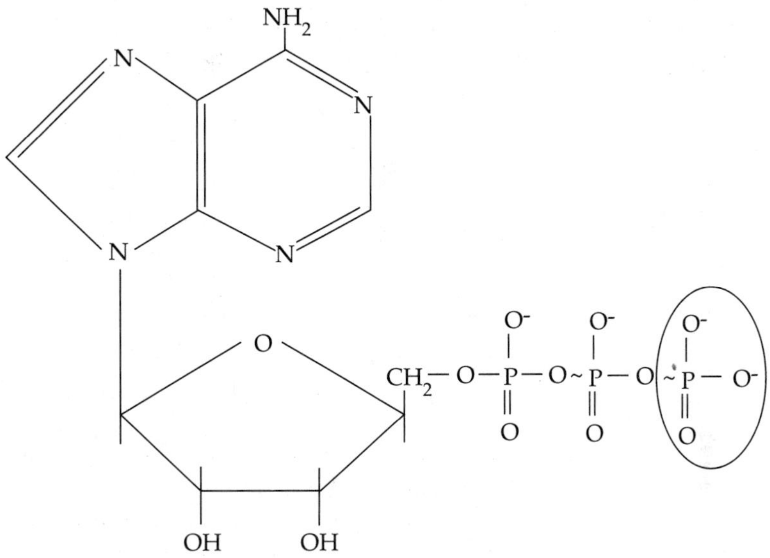

FIGURE 6.7 The structure of adenosine triphosphate.

A second energy storage chemical is nicotinamide adenine dinucleotide (NAD+). As we will see, this is important in the oxidation of many biological energy sources. During the oxidation process, a hydrogen pair, with its associated pair of electrons, is transferred to the NAD^+, producing NADH. The NADH then diffuses away from the enzyme mediating the oxidation. In aerobic systems the NADH is used to produce three high energy ATP's. The structure of NAD^+, and the added hydrogens making it NADH, are shown in Figure 6.8. In **fermentation** (no molecular oxygen) the NADH is regenerated with an organic electron acceptor, no ATP production occurs, and the energy is lost to the organism. In the next section the differences between fermentive and respiratory metabolism will be explored.

Energy Production

Organisms obtain their energy either by photo oxidation or by chemical oxidation. These processes are enzymatically catalyzed at each step. That is, a separate enzyme is required for each step. The energy produced is stored in chemicals, primarily ATP, for use in synthesis for repair, growth, and reproduction, and for use in motility. **Catabolism** (literally tearing apart) is the term for energy production from organic compounds. In the chemical oxidation of substrate, there is a substantial difference in the

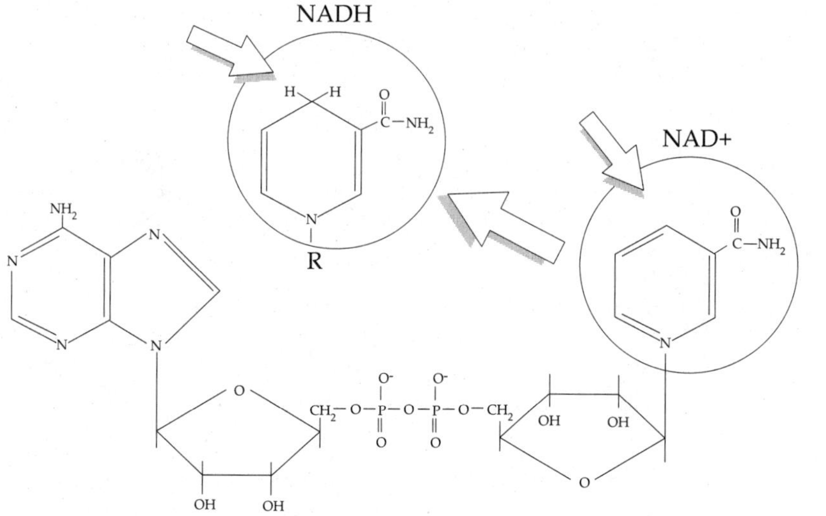

FIGURE 6.8 Nicotinamide adenine dinucleotide, NAD+ and NADH.

energy derived, depending on whether molecular oxygen is available as an electron acceptor. Where molecular oxygen is available, the energy yield is much higher. Some microorganisms can use oxygen if it is available but can continue to produce energy if it is not. These are referred to as **facultative organisms**. Organisms that absolutely require molecular oxygen are called **aerobic**. Those that do not use molecular oxygen — and in fact for them it is usually toxic — are known as **anaerobic** organisms.

In both cases the energy produced is usually stored in high-energy phosphate compounds, adenosine triphosphate being the most common. Organisms utilizing molecular oxygen for this process produce far more ATP per unit of substrate. When molecular oxygen is utilized, the process is termed **respiration**. When oxygen is not used, it is fermentation. Thus, **fermentation** is the production of energy without the benefit of oxygen as a terminal electron acceptor. Recall from basic chemistry, or Chapter 4, that oxidation is loss of electrons. Since these electrons cannot simply exist alone, an electron acceptor must be available. For respiratory systems this is molecular oxygen. For fermentation systems it is another organic compound. Glucose is a universal substrate for organisms. Many more complex compounds are converted to glucose to use in biological systems. Recall that starch, cellulose, and glycogen are glucose polymers. Using glucose as an example of both fermentation and respiration, we will see that the difference in energy production is significant.

Fermentation of glucose can result in several different end products. In alcohol fermentation by yeast, each mole of glucose produces 2 mol of ethanol. In another process, 1 mol of glucose produces 2 mol of lactic acid. And still other end products are possible. We will look at the lactic acid formation. For a cell, whether microorganism or organism, to produce ATP and energy in the fermentation of glucose, the glucose must be split apart. This is done by first adding two phosphate groups to the glucose molecule. This requires an investment of 2 mol of ATP for each mole of glucose. The process is shown in Figure 6.9. For the investment of 2 mol of ATP, the organism obtains 4 mol of ATP in return. Two moles of NADH are also produced, but without molecular oxygen to regenerate the NADH, its energy is lost. However, the organism does still need to regenerate the NADH since it is in limited supply. It accomplishes this by using other organic compounds, but without the gain of any additional ATP.

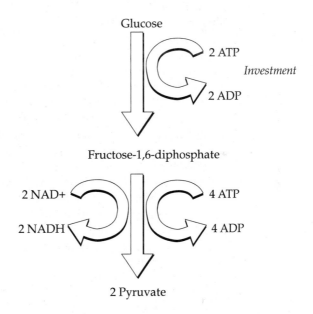

FIGURE 6.9   Fermentation of glucose, the Embden–Meyerhof–Parnas pathway.

Thus, in an anaerobic or fermentive environment, the metabolism of glucose produces 2 mol of pyruvate (or other similar end products) and 2 mol of ATP. Each mole of ATP provides 7.3 kcal of energy. So 1 mol of glucose provides 14.6 kcal of available energy under fermentative or anaerobic conditions.

Now, to compare the two processes, we turn to aerobic metabolism, or respiration. Respiration or oxidative phosphorylation, is energy metabolism using molecular oxygen. When glucose is oxidized aerobically, in the presence of molecular oxygen, the end products are carbon dioxide and water. In this biological process the energy release is stored in the form of ATP by the organism. The first steps in the process are the same as in fermentative metabolism. So a net production of 2 mol of ATP results from the metabolism of 1 mol of glucose. However, 2 mol of NADH are also produced. When molecular oxygen is available, each mole of NADH can produce 3 mol of ATP, or

$$\text{NADH} + \text{H}^+ + \ 3\text{PO}_4^{3-} + 3\text{ADP} + \tfrac{1}{2}\text{O}_2 \rightarrow \text{NAD}^+ + 3\text{ATP} + \text{H}_2\text{O} \quad \textbf{6.1}$$

Thus, with aerobic metabolism, the glucose yields six additional ATPs. When combined with the two from substrate metabolism, this adds to eight ATPs, or an available energy of 58 kcal — four times that from fermentation. And the pyruvate can be further metabolized to end products of carbon dioxide and water. This series of reactions produces an additional 30 ATPs. So 1 mol of glucose produces 38 mol of ATP in an aerobic system. The total available energy produced is then 277 kcal/mol of glucose. The fermentive metabolism produced only 5% of the available energy of the aerobic metabolism. We can see, then, that anaerobic organisms require much more substrate — about 20 times as much as aerobic organisms. This fact is useful in wastewater treatment. Anaerobic organisms must work harder!

Photophosphorylation is the production of ATP, and in some cases NADH, from solar energy. These complex processes will not be discussed in this text. Suffice it to say that the microorganisms utilize the sun's energy to provide energy for ATP production. In the next section we will see that ATP production, regardless of its origin, is used primarily for biosynthesis — cellular growth.

Biosynthesis

Biosynthesis, or anabolism, is the production of new cellular materials from other organic or inorganic intermediates. It requires the input of energy, usually from stored ATP. This process is the reverse of energy production or catabolism.

The primary constituents of most organisms, other than water, are macromolecules or organic polymers, which were discussed in Chapter 5. The monomers for these macromolecules, such as glucose and amino acids, are either obtained from the environment or synthesized from other compounds. Glucose is the primary constituent in cellulose, starch, and glycogen. Amino acids are the main components of the various proteins in organisms.

For biosynthesis to occur, the organism must have a variety of micro and macro nutrients present. Carbon, hydrogen, oxygen, nitrogen, and phosphorus are major ones, but several others are required as well. Table 6.1 selected charts nutrients needed for biosynthesis.

TABLE 6.1   Selected Nutrients Required for Biosynthesis

ELEMENT	PERCENTAGE OF DRY WEIGHT	COMMON SOURCES
Carbon	48	Carbon dioxide and organics
Oxygen	26	Dissolved molecular oxygen and water
Nitrogen	11	Ammonia, Nitrate, and amino acids
Hydrogen	5	Organics, water, and dissolved hydrogen gas
Phosphorus	<1	Phosphate and organophosphates
Potassium	<1	K^+
Sodium	<1	Na^+
Magnesium	<1	Mg^{2+}
Calcium	<1	Ca^{2+}
Sulfur	<1	SO_4^{2-}, HS^-, and sulfur containing amino acids
Iron	<1	Organic iron, Fe^{3+}

Microbial Genetics

The microbial DNA contains the organism's blueprint—the information necessary to construct the organism. During replication a new strand of DNA is constructed for each of the two existing strands, yielding two complete DNA molecules. (See Chapter 5, Organic Chemistry, for the structure of DNA and RNA.) In the DNA helix, each base pairs with a different base, creating a complementary set of information. So each DNA strand is said to have a complementary strand. The complementary pairing for the four DNA bases is shown in Table 6.2. During normal cell functions the messenger RNA, mRNA, is produced from a segment of DNA to carry information from the DNA to a ribosome for actual protein synthesis.

TABLE 6.2 Complementary DNA Base Pairings

DNA BASE	COMPLEMENTARY DNA BASE	COMPLEMENTARY RNA BASE
Adenine (A)	Thymine	Uracil (U)
Cytosine (C)	Guanine	Guanine
Guanine (G)	Cytosine	Cytosine
Thymine (T)	Adenine	Adenine

See Chapter 5, Organic Chemistry for base structures.

DNA Coding. Each three bases of a DNA strand contain the code for one amino acid. Since there are four possibilities for each, there are 4^3, or 64, possible contributions, but only 20 required amino acids. This means several amino acids have more than one coding sequence. There are also codes for the beginning and end of each protein. The sequence of the amino acid codes in the DNA correspond to the sequence in which the amino acids are attached in the protein. A section of DNA that codes for a particular protein is termed a gene.

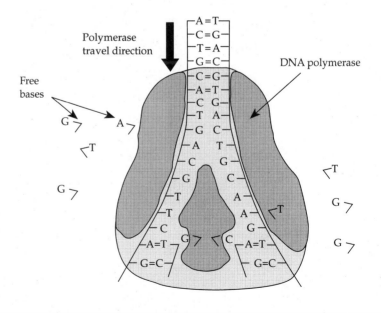

FIGURE 6.10 DNA replication.

Replication. During replication of an organism, the enzyme DNA polymerase splits a short section of the two DNA strands. It then begins to synthesize complementary DNA strands, base by base, for each of the two strands as shown in Figure 6.10. When the process is complete, the result is two complete molecules of DNA, each with one original strand and one new, complementary strand. The two new DNA molecules are identical to the original. All genetic information has then been replicated. As this process is occurring, the organism is also producing enough duplicates of other cellular components that the two organisms can function after cell division.

Protein Synthesis. To produce proteins, the DNA must also be replicated. However, in contrast to DNA replication, protein synthesis produces only a short strand of complementary RNA. This strand of mRNA, contains the information from a single gene (one protein unit or major subunit). The information contained in the mRNA is then used by a ribosome to produce a single protein or protein subunit. The process is diagrammed in Figure 6.11.

Engineering Applications. Microbes are being utilized in new environmental applications almost daily. The tremendous effort being made, on a

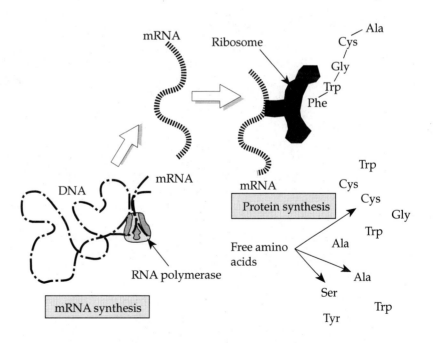

FIGURE 6.11 Messenger RNA and protein synthesis.

national scale, to clean up hazardous waste sites has resulted in new assignments for engineered organisms, and in new ways to apply or use naturally occurring organisms. There have been demonstrated successes in the use of microbes to treat groundwater and soil contamination in place. Where large amounts of soil and water have been affected, it is extremely expensive to pump water out of the ground and dig up the tainted soil. By adding required nutrients to the water or soil, or both, existing organisms have been shown to degrade contamination effectively in many cases. Special strains of oil-degrading organisms were a success in some areas affected by the *Exxon Valdez* spill as mentioned earlier.

And special techniques are altering the genetic code of organisms to add functions or create new ones. One of the first applications (non-environmental) of genetic engineering was to alter a bacterium to produce insulin. Today, engineers and scientists are developing ways to have microbes break down synthetic organics they could not normally degrade, or to increase the rate at which the synthetic organics decompose. These research programs should lead to improved methods of cleaning up some of our nation's hazardous waste sites. And they should help prevent hazardous waste from accumulating in the ground in the first place — in that toxic materials can receive microbial treatment rather than burial.

MICROBIAL DISEASES

Many diseases are caused by pathogenic microbes. Infectious diseases vary in intensity from the common cold to AIDS. In this section we will discuss general types of disease, the methods of disease transmission and how disease transmission relates to environmental engineering.

Pathogenic organisms infect hosts in many ways. Direct contact — including a handshake, kissing, or sex — can spread many diseases. While direct contact transmission is not normally of professional interest to environmental engineers, several methods of indirect contact are. One primary purpose of drinking water treatment is the reduction or elimination of waterborne disease transmission, and wastewater treatment is intended not only to reduce the impact of the discharge on receiving water quality, but also to reduce the likelihood of general disease transmission. Some landfilling activities are designed to minimize vector transmission of disease. Table 6.3 displays common methods of disease transmission and several representative diseases.

Waterborne Transmission

Although water's role in disease transmission may have been suspected earlier, it was not demonstrated or proved until 1855 when a cholera epidemic gripped London. John Snow, a physician, extensively studied the city's water supply and noticed its relationship to the disease. At that time

TABLE 6.3 Methods of Disease Transmission

METHOD OF TRANSMISSION	DISEASES
Direct contact	Syphilis, gonorrhea, poliomyelitis, chickenpox, and common cold
Indirect contact; waterborne or found in food	Typhoid fever, amoebic dysentery, and cholera
Indirect contact; airborne	Fungal diseases and histoplasmosis
Indirect contact; vectors	Plague and typhus, flea; encephalitis, malaria, and canine heartworm, mosquito; Lyme disease, deer tick

two independent, competing companies were supplying London's water. In many areas, only one company supplied it, but in some parts of the city the companies ran service lines in the same neighborhoods.

At the time of the epidemic, neither wastewater treatment nor water treatment existed. Wastewater was sent to the Thames River through underground sewers. A water company's only service was to pump the water to consumers. Both firms obtained water from the Thames. However, one's withdrawal pipe was upstream of London, before any sewage discharges. The other company withdrew its water opposite central London, after it had been contaminated with sewage.

Snow postulated that human wastes in the water carried the cause of cholera. (Recall that Robert Koch did not prove the germ theory of disease until 1876, some 20 years later.) Snow surveyed many homes in areas supplied by the two companies. The rate of cholera was almost ten times higher in homes supplied by the company obtaining water within London than in homes served by the company using uncontaminated water. This was true not only of the areas where the two companies were not in competition, but also in areas served by two water lines—where next door neighbors might be using water from different companies. Thus, without the benefit of the germ theory of disease, Snow determined that human wastes transmitted cholera.

Today, thanks to the work of environmental engineers, the transmission of infectious disease by water is almost nonexistent in developed countries. North America, Eastern Europe, and Japan have highly developed infrastructure to treat wastewaters before discharge and potable water prior to distribution. Unfortunately, in many developing nations, particularly Africa and Asia, a safe supply of drinking water is still not available.

Vector Transmission

A vector is an animal or insect that can carry and transmit an infectious disease agent from one host to another. Many mammals and insects act as vectors for human pathogens. Mosquitos can carry the malaria protozoan, *Plasmodium vivax*, a common disease in the tropics and subtropics. Before the development of DDT, malaria was a serious threat in the southern United States as well. Other examples of vector transmitters are the deer tick, which carries the spiral bacteria *Borellia burgdorferi*; the flea, which carries typhus; and the mosquito which carries canine heartworms. The vector-host cycle is shown in Figure 6.12.

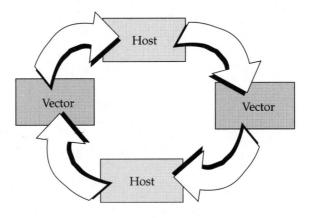

FIGURE 6.12   The cycle of a vector transmitted infectious disease.

Environmental engineers working in the solid-waste-field or in public health may be involved in vector control. Landfills, particularly improperly operated ones, may have problems with rats and mice, but these do not often represent a true public health threat as landfills are usually in remote areas. Vector control at landfills is accomplished in a variety of ways, but daily coverage of new solid wastes with at least six inches of soil is a primary method. Landfill operators also use poisons in some instances.

MICROBIAL GROWTH

Microorganisms use organic matter as substrate to produce energy for replication, motility, and repair of damaged cellular components. In waste-water treatment, microorganisms play an important role in destroying wastes and returning nutrients to the environment. In this section, we will look at ways to estimate microbial growth and substrate utilization.

When microorganisms are introduced into a closed system with an unlimited source of nutrients and substrate, there is a lag phase during which the organisms adapt to the new substrate and concentration. At this time little microbial growth occurs. (See Figure 6.13.) The organisms are manufacturing enzymes necessary to metabolize the newfound substrate. Once this initial phase is completed, the organisms will begin a period of rapid growth, their numbers increasing exponentially. At some point, their populations will reach a maximum. This can be the result of by-products or wastes building up within the system or of substrate concentration diminishing. In a closed system, however, there will be a maximum sustainable population. After some period of time, as waste products continue to accumulate and substrate is further consumed, a death phase will ensue, and the population will begin to decrease. The growth curve involved has been duplicated many times. Its shape and relationships are well established.

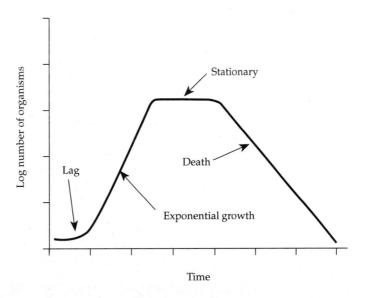

FIGURE 6.13   Batch growth curve for microorganisms.

Exponential Growth

Exponential growth occurs when both substrate and nutrients are in abundance, and when metabolites or waste products have not built up enough to dampen growth. During this period the only limitation on growth is the speed at which the organism can replicate. So the growth rate is proportional to the number of microorganisms, or

$$\frac{dX}{dt} = \mu X \qquad \qquad 6.2$$

where

X = concentration of microorganisms at time t

t = time

μ = proportionality constant or specific growth rate, [time^{-1}]

dX/dt = microbial growth rate, [mass/volume time]

The equation can be separated, yielding

$$\frac{dX}{X} = \mu dt \qquad \qquad 6.3$$

if the concentration of organisms is X_0 at time zero, then integration yields

$$\ln\left(\frac{X}{X_0}\right) = \mu t \qquad \qquad 6.4$$

or, in exponential form,

$$X = X_0 e^{\mu t} \qquad \qquad 6.5$$

This is the equation for the exponential growth phase shown in Figure 6.13.

EXAMPLE 6.1	EXPONENTIAL MICROBIAL GROWTH

A microbial system with an ample substrate and nutrient supply has an initial cell concentration, X_0, of 500 mg/L. The specific substrate utilization rate is 0.5/hr.

a. Estimate the cell concentration after 6 hours, assuming that log growth is maintained during the period.
b. Determine the time required for the microbial population to double during this phase.

SOLUTION: a. To determine the microbial concentration after 6 hours, we substitute into Equation 6.5, obtaining

$$X = X_0 e^{\mu t} = 500 \frac{mg}{L} \times e^{(0.5/hr \times 6hr)}$$

$$= 10,000 \frac{mg}{L}$$

Thus, in a period of six hours, the microorganisms increase 20-fold.

b. To determine the time for the concentration to double, we use the log form, Equation 6.4. Also, if the concentration doubles, then

$$\frac{X}{X_0} = 2,$$

or,

$$\ln \frac{X}{X_0} = \mu t$$

Solving for t, we obtain

$$t = \frac{\ln \left(\dfrac{X}{X_0} \right)}{\mu} = \frac{\ln 2}{0.5/hr}$$

$$t = 1.4 \, hr.$$

Therefore, the microbial population can double in only 1.4 hours. By comparison, the human population is now doubling about every 40 years.

The specific growth rate is not truly a constant. However, for the conditions noted earlier (high substrate concentration, abundant nutrients, and low waste concentration) it can be assumed constant. But as we will see in the next section, where nutrients or substrate are limited, μ varies with substrate concentration.

Substrate Limited Growth

In many engineered systems, as well as natural systems, the growth of microbes and rate of substrate utilization are limited by the amount of substrate. The situation is analogous to pizza consumption by a group of students confined to a room. At very low substrate levels (few pizza slices) the growth rate of the students (and rate of pizza consumption) is proportional to the amount of pizza available. That is, each newly appearing pizza slice is rapidly consumed. Double the number of pizza slices per time period, and the number of pizza slices consumed per time period will double. At very high substrate levels (many, many pizza slices) the growth rate of the students (and their pizza consumption) is independent of the number of pizzas. The students have more pizza available than they can possibly eat. At this point the consumption is actually proportional to the number of students.

Professor Monod [6] made this same observation. However, he used bacterial cultures, not pizza. (Poor Prof. Monod!). From extensive observations of microbial cultures, he postulated that the rate of bacterial growth, or the rate of substrate consumption, was as follows:

$$\mu = \frac{\mu_m S}{K_S + S} \tag{6.6}$$

where

μ_m = maximum specific growth rate [day^{-1}]

S = concentration of limiting substrate [mg/L]

K_S = Monod, or half-velocity constant [mg/L]

A typical Monod curve appears in Figure 6.14. It can be seen that the Monod, or half-velocity constant, corresponds to the concentration at which the specific growth rate is one-half the maximum.

If the Monod equation is substituted into the microbial growth equation, we obtain an expression for microbial growth at varying concentrations of substrates:

$$\frac{dX}{dt} = \frac{\mu_m S X}{K_S + S} \tag{6.7}$$

Both substrate concentration, S, and microbial concentration, X, vary with t. Equation 6.7 cannot be solved explicitly without simplifying assumptions, but it can be solved numerically using a computer. Solution provides the microbial growth rate as a function of substrate concentration. However, in general, environmental engineers are more interested in the amount of

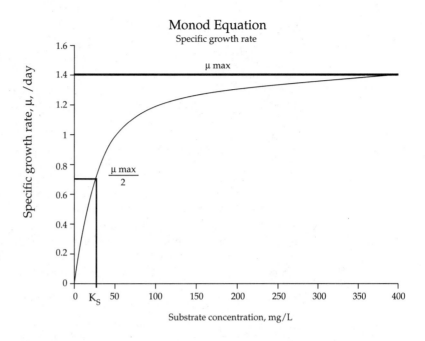

FIGURE 6.14 The Monod equation.

substrate consumed than the microbial growth rate, but the topics are related. A discussion of substrate utilization comes next.

Substrate Utilization

Substrate is utilized for several bacterial functions, including energy, cell maintenance, and cell growth or synthesis. Not all of these uses result in added cell mass, however. Only a fraction of the substrate is converted to cell mass. The fraction of substrate converted to microbial mass is the substrate yield, a quantity that can be expressed mathematically as

$$Y = \frac{\Delta X}{\Delta S}$$ **6.8**

where

Y = substrate yield, [mass of biomass/mass of substrate consumed]

In differential form this is

$$Y = \frac{dX}{dS} = \frac{\dfrac{dX}{dt}}{\dfrac{dS}{dt}}$$ **6.9**

and

$$\frac{dS}{dt} = \frac{1}{Y}\left(\frac{dX}{dt}\right)$$

6.10

Substituting for dX/dt from Equation 6.33, we arrive at

$$\frac{dS}{dt} = \frac{1}{Y}\left(\frac{\mu_m SX}{K_S + S}\right)$$

6.11

As with the expression for microbial growth, both the concentration of substrate, S, and the microbial concentration, X, vary with time. The equations for substrate utilization and microbial growth are not independent, however; they must be solved simultaneously, which further complicates the problem. Nevertheless, standard library subroutines in FORTRAN can be accessed for this purpose. Unfortunately, no general solution exists.

Review Questions

1. Define or describe the following:

   a. procaryotic
   b. eucaryotic
   c. heterotrophic
   d. autrophic
   e. phototrophic
   f. chemotrophic

2. Show that the half-velocity constant is, in fact, the substrate concentration at which the specific growth rate equals one-half of the maximum specific growth rate.

3. Describe the two viral forms.

4. Develop an expression for the doubling time of a microbial population based on μ, the specific growth rate.

5. Under ideal conditions, some bacteria can double in only 20 minutes. What is the specific growth rate in this case?

References

1. G. A. Wistreich and M. D. Lechtman, *Microbiology*, 3rd ed. (New York: Macmillan, 1980).

2. T. D. Brock and M. T. Madigan, *Biology of Microorganisms,* 6th ed. (New York: Prentice-Hall, 1991).

3. A. F. and E. T. Gaudy, *Microbiology for Environmental Scientists and Engineers* (New York: McGraw-Hill, 1980).

4. J. W. Costerton, G. G. Geesey, and K. J. Cheng, "How Bacteria Stick," *Scientific American*, vol. 238, no. 1 (1978), pp. 000–000 (An interesting article that is not difficult to read or understand.)

5. W. G. Characklis, "Fouling Biofilm Development: A Process Analysis," *Biotechnology and Bioengineering,* vol. 23 (1981), pp. 1923–60.

6. J. Monod, "The Growth of Bacterial Cultures," *Annual Review of Microbiology,* vol. 3 (1949), pp. 000–000.

7 Analysis of Treatment Processes

``NATURE PROVIDES HER CREATURES
WITH THE BEST POSSIBLE SYSTEMS;
THE ENGINEER HAS TO TRY TO
IMPLEMENT THEM IN DEVICES.''

ROBERT ZINTER
Optical Engineer

An empty aeration basin (biological reactor)
at the LeMay Wastewater Treatment Plant
in St. Louis, Missouri.

INTRODUCTION

This chapter presents concepts essential to the successful analysis of environmental problems, including air pollution control, water and waste-water treatment, and hazardous waste treatment. The topics covered are mass balances, chemical conversion, chemical reaction rates, and sedimentation principles.

BASIC FLUID PRINCIPLES

Since this is an introductory course, some students using this text will not have yet taken fluid mechanics, but a working knowledge of this science is needed in the design of many environmental treatment systems and in analysis of many natural systems. And at a minimum, volumetric flow rate and hydraulic retention time are involved in many applications discussed in this text. These two principles are introduced here. They will be employed in later sections as well.

Volumetric Flow Rate

As depicted in Figure 7.1, the volumetric flow rate is the volume of fluid passing an arbitrary plane in unit time, or

$$Q = Av \qquad\qquad \textbf{7.1}$$

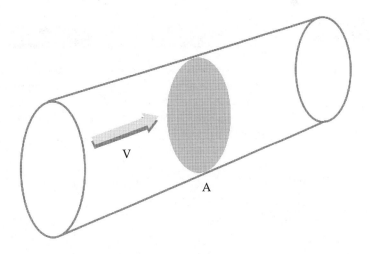

FIGURE 7.1 Volumetric flow rate.

where,

Q = volumetric flow rate [m³/day, ft³/s]

A = area across which the fluid passes [m², ft²]

v = fluid velocity [m/day, ft/s]

This relationship is useful in determining the required area for several processes in water and wastewater treatment and air pollution control. The relationship can also be used to determine the velocity of fluid moving in a pipe if its cross-sectional area and the volumetric flow rate are known.

Hydraulic Retention Time

The hydraulic retention time (HRT) is the average time a parcel of fluid is retained in a given tank or basin. It is the tank volume divided by the volumetric flow rate into the tank, or

$$\text{HRT} = \theta = \frac{V}{Q} \qquad\qquad 7.2$$

where

θ = hydraulic retention time [days]

V = volume [m³]

Q = volumetric flow rate [m³/day]

EXAMPLE 7.1	APPLICATION OF BASIC FLUID PRINCIPLES

As shown in the figure opposite, a wastewater basin has a diameter of 20 m and a depth of 3 m. The pipe feeding the basin is 40 cm in diameter. The velocity of the water in the pipe is 0.28 m/s. Determine the flow rate to the basin, the average vertical velocity in the basin, and the hydraulic retention time in the basin.

SOLUTION: To determine the volumetric flow rate, Q, we must first determine the pipe area:

$$A_p = \pi r^2 = (\pi)(20 \text{ cm})^2 \left(\frac{1 \text{ m}}{100 \text{ cm}} \right)^2$$

$$= 0.126 \text{ m}^2$$

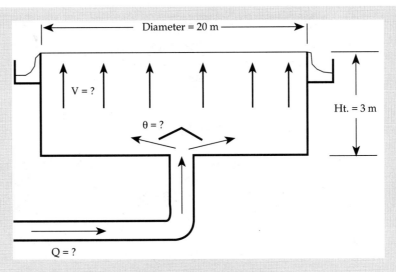

The flow rate, Q, can then be determined by

$$Q = Av = (0.126\,\text{m}^2)(0.28\,\text{m/s})$$
$$= 0.0352\,\text{m}^3/\text{s} = 3040\,\text{m}^3/\text{day}$$

The vertical velocity in the basin is obtained by first calculating the area of the basin, or

$$A_b = \pi r^2 = (\pi)(10\,\text{m})^2$$
$$= 314\,\text{m}^2$$

The velocity is then

$$v = \frac{Q}{A} = \frac{3040\,\text{m}^3/\text{day}}{314\,\text{m}^2}$$
$$= 9.68\,\text{m/day}$$

To calculate the HRT, we must first calculate the volume:

$$V = \pi r^2 h = \pi(10\,\text{m})^2(3\,\text{m})$$
$$= 942\,\text{m}^3$$

The HRT is then

$$\theta = \frac{V}{Q} = \frac{942\,\text{m}^3}{3040\,\text{m}^3/\text{day}}$$
$$= 0.31\,\text{day} = 7.4\,\text{hr}$$

CONVERSION

Conversion is the fraction of a species converted to product once it has entered a **system**. The conversion multiplied by 100 is the removal efficiency of a particular species in a process. For the following generalized reaction:

$$aA + bB \xrightarrow{k} pP + qQ \qquad\qquad \textbf{7.3}$$

the conversion of species A to product X, can be calculated as follows. (Note that, by convention, the conversion of A (the primary reactant) is not subscripted. However, the conversion of all other species is subscripted.)

$$X = \frac{(C_{A0} - C_A)}{C_{A0}} \qquad\qquad \textbf{7.4}$$

If the equation is solved for the effluent concentration, C_A, it yields:

$$C_A = C_{A0}(1 - X) \qquad\qquad \textbf{7.5}$$

It is important to be able to express the conversion of the other species in terms of the conversion of species A. When analyzing processes with multiple reactants and/or products, conversion can be utilized to eliminate the different effluent concentration variables. Each effluent concentration can be expressed in terms of its initial or influent concentration and the conversion of species A.

For each mole of A converted, there will be b/a moles of B converted. Or,

$$(n_{B0} - n_B) = \frac{b}{a}(n_{A0} - n_A) \qquad\qquad \textbf{7.6}$$

where,

$$n_{A0} = \text{mass of A at } t = 0 \text{ [mol]}$$
$$n_A = \text{mass of A at } t = t \text{ [mol]}$$
$$n_{B0} = \text{mass of B at } t = 0 \text{ [mol]}$$
$$n_B = \text{mass of B at } t = t \text{ [mol]}$$

If the volume of the reactor is assumed to remain constant, we can divide both sides of the expression by $C_{A0}V$. The expression then becomes

$$\frac{C_{B0} - C_B}{C_{A0}} = \frac{b}{a}\left(\frac{C_{A0} - C_A}{C_{A0}}\right) = \frac{b}{a}X \qquad\qquad \textbf{7.7}$$

This can then be solved for the concentration of B in terms of other known quantities:

$$C_B = C_{B0} - \frac{b}{a}C_{A0}X \qquad\qquad \textbf{7.8}$$

Similar relationships can be obtained for other reactants or products in chemical reactions. The following example illustrates the use of conversion.

EXAMPLE 7.2	CONVERSION OF REACTANTS

The reactor shown in Figure 7.2 has an inflow of 750 L/hr. The concentration of A in the influent is $0.3\,M$ and the concentration of B in the influent is $0.5\,M$. The conversion (of A) is 0.75. The reaction is

A + 2B → Products

Find the conversion of B, X_B, and the effluent concentration of A and B.

SOLUTION: The first step in the solution is to determine the effluent concentration of A. This can be obtained as follows:

$$C_A = C_{A0}(1 - X) = (0.3\,M)(1 - 0.75)$$
$$= 0.075\,M$$

For each mole of A converted to product, 2 mol of B are converted to product. Since we know the initial concentration of B, we can calculate its final concentration:

$$\text{mol/L of B converted} = 2(0.3\,M - 0.075\,M) = 0.45\,M$$
$$C_B = 0.5\,M - 0.45\,M = 0.05\,M$$

Alternatively, using Equation 7.8 yields

$$C_B = C_{B0} - \frac{b}{a}C_{A0}X = 0.5\,M - \left(\frac{2}{1}\right)(0.3\,M)(0.75)$$

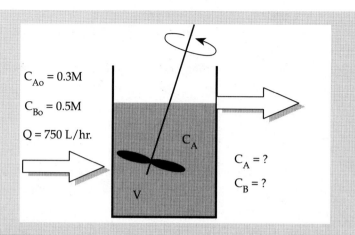

FIGURE 7.2   CSTR diagram for Example 7.2.

The conversion of B is then

$$C_B = 0.05\,M$$

$$X_B = \frac{C_{B0} - C_B}{C_{B0}} = \frac{0.5\,M - 0.05\,M}{0.5\,M}$$

$$= 0.9$$

MASS BALANCES

A **mass balance** is an organized statement of the inputs and outputs in an arbitrary but definable system. Mass balances are an important tool in analyzing flow systems in water and wastewater treatment, air pollution control, and stream pollution analysis, as well as other natural and engineered systems. As depicted in Figure 7.3, the net rate of mass accumulation within the system is equal to the rate of mass input to the system less the rate of mass output from the system less the net rate of mass conversion within the system. Or,

$$\begin{bmatrix} \text{Accumulation} \\ \text{rate} \end{bmatrix}_i = \begin{bmatrix} \text{Input} \\ \text{rate} \end{bmatrix}_i - \begin{bmatrix} \text{Output} \\ \text{rate} \end{bmatrix}_i - \begin{bmatrix} \text{Conversion} \\ \text{rate} \end{bmatrix}_i \qquad \textbf{7.9}$$

where the "i" subscript indicates passage across the system boundary. In

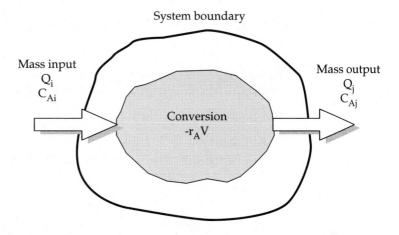

FIGURE 7.3 Mass balance.

equation form this is

$$\frac{dm_A}{dt} = \sum_{i=1}^{n} (C_{Ai}Q_i)_{in} - \sum_{j=1}^{n} (C_{Aj}Q_j)_{out} - r_A V \qquad 7.10$$

where,

m_A = mass of A [mass]

C_{Ai} = concentration of the ith component of species A entering the system [mass/volume]

C_{Aj} = concentration of the jth component of species A leaving the system [mass/volume]

Q_i = volumetric flow rate of the ith species entering the system [volume/time]

Q_j = volumetric flow rate of the jth species leaving the system [volume/time]

r_A = reaction rate of species A, [mass/volume·time]

V = volume of reactor

For systems at steady state with no accumulation, the time-dependent term goes to zero, and the equation reduces to

$$r_A V = \sum_{j=1}^{n} (C_{Aj}Q_j)_{in} - \sum_{i=1}^{n} (C_{Ai}Q_i)_{out} \qquad 7.11$$

Or, the rate of conversion of species A is the difference between the mass input rate and the mass output rate.

| EXAMPLE 7.3 | MASS BALANCE FOR A REACTOR |

Determine the net rate of mass conversion for the following systems:

a. There is no accumulation in the reactor.
b. Mass accumulates in the reactor at the rate of 0.05 mol/hr.

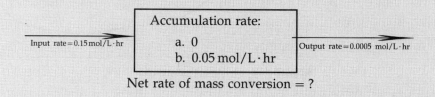

Net rate of mass conversion = ?

SOLUTION: **a.** By rearranging Equation 7.10, we know that the net rate of mass conversion is

$$\begin{bmatrix} \text{Conversion} \\ \text{rate} \end{bmatrix}_i = \begin{bmatrix} \text{Input} \\ \text{rate} \end{bmatrix}_i - \begin{bmatrix} \text{Output} \\ \text{rate} \end{bmatrix}_i - \begin{bmatrix} \text{Accumulation} \\ \text{rate} \end{bmatrix}_i$$

Since accumulation is zero,

$$\begin{bmatrix} \text{Conversion} \\ \text{rate} \end{bmatrix}_i = 0.15 \text{ mol/L} \cdot \text{hr} - 0.0005 \text{ mol/L} \cdot \text{hr}$$

$$= 0.1495 \text{ mol/L} \cdot \text{hr}$$

b. For the reactor with accumulation, the net rate of mass conversion is

$$\begin{bmatrix} \text{Conversion} \\ \text{rate} \end{bmatrix}_i = 0.15 \text{ mol/L} \cdot \text{hr} - 0.0005 \text{ mol/L} \cdot \text{hr}$$

$$- 0.05 \text{ mol/L} \cdot \text{hr}$$

$$= 0.0995 \text{ mol/L} \cdot \text{hr}$$

EXAMPLE 7.4	MASS BALANCE FOR A STREAM AND DISCHARGER

ACME Industries is located next to Spunkin Creek. The factory uses copper cyanide for plating both copper and brass. Estimate the maximum concentration of copper ACME can discharge in its effluent to meet the required maximum concentration, C_d, of 0.005 mg/L Cu^{2+} in the stream. The upstream copper concentration is below the detection limit, i.e., $C_u \simeq 0$ mg/L. Assume steady state conditions.

SOLUTION: We first use a mass balance on the flow into and out of the system. The flow after discharge can be calculated by a mass balance on the water entering and leaving the system (the concentration of water in water is unity and thus cancels):

$$Q_d = Q_u + Q_e$$

where the "e" subscript indicates effluent, the "u" subscript indicates up-stream, and the "d" subscript indicates downstream.

$$Q_d = 0.25 \, \text{m}^3/\text{s} + 0.08 \, \text{m}^3/\text{s} = 0.33 \, \text{m}^3/\text{s}$$

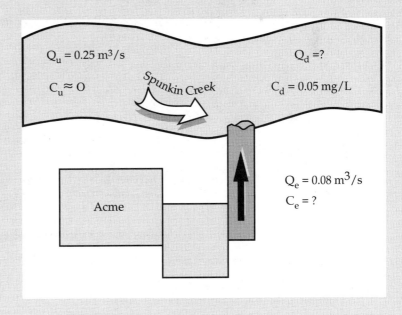

$Q_u = 0.25 \, \text{m}^3/\text{s}$

$C_u \approx 0$

Spunkin Creek

$Q_d = ?$

$C_d = 0.05$ mg/L

$Q_e = 0.08 \, \text{m}^3/\text{s}$

$C_e = ?$

Acme

The allowable concentration of copper in the effluent can then be determined by a mass balance on copper entering and leaving the system:

$$Q_u C_u + Q_e C_e = Q_d C_d$$

Solving for C_e (two equations and two unknowns), we get

$$C_e = \frac{Q_d C_d - Q_u C_u}{Q_e}$$

$$= \frac{(0.33\ \text{m}^3/\text{s})(0.005\ \text{mg/L}) - (0.25\ \text{m}^3/\text{s})(0\ \text{mg/L})}{0.08\ \text{m}^3/\text{s}}$$

$$= 0.021\ \text{mg/L}$$

EXAMPLE 7.5

MASS BALANCE ON A SECONDARY WASTEWATER TREATMENT FACILITY

Estimate the reaction rate, r_A, for the reactor shown in Figure 7.4. You may assume that there is no accumulation.

SOLUTION: The first step is to sum the flows into and out of the system, obtaining

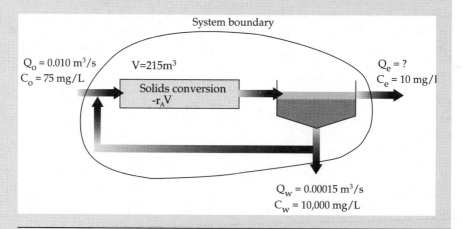

FIGURE 7.4 An activated sludge secondary wastewater treatment facility.

the effluent flow rate, Q_e:

$$Q_0 = Q_w + Q_e$$

Thus,

$$Q_e = Q_0 - Q_w = 0.010 \, \text{m}^3/\text{s} = 0.00015 \, \text{m}^3/\text{s}$$

and

$$Q_e = 0.00985 \, \text{m}^3/\text{s}$$

We can use Equation 7.12 to determine the reaction rate, r_A:

$$r_A = \frac{\sum\limits_{j=1}^{n} (C_{Aj} Q_j)_{\text{in}} - \sum\limits_{i=1}^{n} (C_{Ai} Q_i)_{\text{out}}}{V}$$

$$= \frac{[(10{,}000 \, \text{g/m}^3)(0.00015 \, \text{m}^3/\text{s}) + (10 \, \text{g/m}^3)(0.00985 \, \text{m}^3/\text{s})] - [(75 \, \text{g/m}^3)(0.010 \, \text{m}^3/\text{s})]}{215 \, \text{m}^3}$$

$$= 3.94 \times 10^{-3} \, \text{g/m}^3 \cdot \text{s} = 0.34 \, \text{kg/m}^3 \cdot \text{day}$$

Flux

Flux is the movement of a mass past a surface, plane, or boundary. It is useful in several environmental engineering applications, particularly in the analysis of sedimentation processes. Where the flux analysis is used for the solids in a sedimentation basin, it is often referred to as **solids flux**. Flux can be calculated by

$$SF_i = \frac{m_i}{A_i \cdot t} \qquad\qquad \textbf{7.12}$$

where,

SF_i = flux crossing the boundary i [kg/m$^2 \cdot$ hr]

m_i = mass crossing the boundary i in time t [kg]

A_i = area of boundary i [m]

t = time for the mass to cross the boundary i [hr]

This is depicted in Figure 7.5. If the right side of Equation 7.12 is multiplied by L/L, where L is the distance the appraching mass moves during time t, then the equation becomes:

$$SF_i = \frac{m_i}{A_i \cdot t} \times \frac{L}{L} = \frac{m_i}{A_i \cdot L} \times \frac{L}{t} \qquad\qquad \textbf{7.13}$$

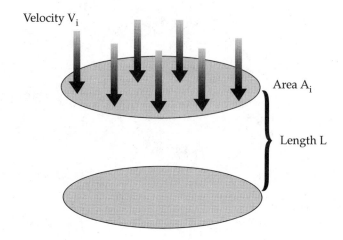

Velocity V_i

Area A_i

Length L

FIGURE 7.5 Definition of flux.

Thus,

$$SF_i = C_i V_i \qquad\qquad \textbf{7.14}$$

where,

C_i = the concentration of the material crossing the boundary i [kg/m³]

v_i = the velocity of the material crossing the boundary i [m/hr]

The following example demonstrates the application of flux.

EXAMPLE 7.6	MASS BALANCE AND FLUX FOR A SECONDARY CLARIFIER

A 25-m diameter secondary clarifier has an influent solids concentration of 2500 mg TSS/L. The flow rate into the clarifier is 17,500 m³/day. If the effluent solids are assumed to be zero, what return or recycle flow rate is required to attain a return solids concentration of 7500 mg TSS/L. Also, what is the solids flux across the boundary shown in Figure 7.6.

SOLUTION: We can perform a mass balance to determine the underflow or recycle solids concentration, X_u. Assuming no accumulation in the sedimentation tank,

Mass in = Mass out

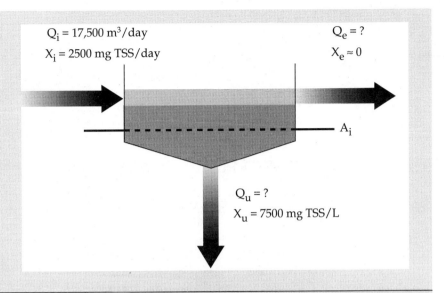

Figure with flows:

$Q_i = 17,500 \text{ m}^3/\text{day}$
$X_i = 2500 \text{ mg TSS/day}$

$Q_e = ?$
$X_e \approx 0$

A_i

$Q_u = ?$
$X_u = 7500 \text{ mg TSS/L}$

FIGURE 7.6 Clarifier definition for Example 7.6.

or

$$X_i Q_i = X_e Q_e + X_u Q_u$$

Since X_e is assumed to be zero,

$$Q_u = \frac{X_i Q_i}{Q_u} = \frac{(2500 \text{ mg TSS/L})(17,500 \text{ m}^3/\text{day})}{7500 \text{ mg TSS/L}}$$

$$= 5800 \text{ m}^3/\text{day}$$

To determine the flux across A_i, we need the mass moving across i per day, or

$$m_i = X_i V$$

where V is the volume applied per time. If we choose one day for t, then V is 17,500 m³. Thus, the mass is

$$m_i = (2500 \text{ mg TSS/L})(17,500 \text{ m}^3) \times \frac{\text{kg}}{10^6 \text{ mg}} \times \frac{10^3 \text{ L}}{\text{m}^3} = 44,000 \text{ kg}$$

And the flux is

$$SF_i = \frac{43{,}750\,\text{kg}}{\pi(25\,\text{m}/2)^2(1\ \text{day})}$$

$$= 89\,\text{kg/m}^2 \cdot \text{day} = 3.7\,\text{kg/m}^2 \cdot \text{hr}$$

REACTION KINETICS AND REACTOR DESIGN

Reaction kinetics is the study of the rate of chemical reactions and the factors that affect them. Such studies are frequently utilized in the design of chemical reactors, and many waste treatment processes are simply specialized chemical reactors. In addition, the concentrations of many chemicals of environmental importance are controlled by reaction rates. Examples of this are the levels of dissolved oxygen in streams, the production and transformation of air pollutants, and the reduction of water pollutants within wastewater treatment facilities.

Factors affecting chemical reaction rates are temperature, pressure (for gas phase reactions), chemical concentration, and catalyst or enzyme concentration. The following is a simplified introduction to chemical reaction rates. For a more detailed explanation of chemical kinetics and reactor design, students should consult references listed at the end of the chapter.

Chemical Reaction Rates

An **irreversible reaction** is one in which the reactant(s) proceed to product(s), but there is no backward reaction. That is, the products do not recombine or change to form reactants in any appreciable amount. An example is hydrogen and oxygen combining to form water in a combustion reaction. We do not observe water spontaneously separating into hydrogen and oxygen. In generalized form, irreversible reactions can be represented as

$$a\text{A} + b\text{B} \rightarrow \text{Products} \qquad\qquad \textbf{7.15}$$

A **reversible reaction** is one in which the reactant(s) proceed to product(s), but the product(s) react at an appreciable rate to reform reactant(s). Many biological reactions fit into this category. An example of a reversible reaction is the formation of adenosine triphosphate (ATP) and adenosine diphosphate (ADP). All living organisms use ATP (or a similar compound) to store energy. As the ATP is used, it is converted to ADP. The organism then uses food to reconvert the ADP to ATP. The generalized form of a reversible reaction is

$$aA + bB \rightleftarrows pP + qQ \qquad\qquad 7.16$$

An expression for the reaction rate can sometimes be written from the stoichiometric equation. When this is possible, the reaction is said to be an **elementary reaction**. The reaction rate for the generation of species A (in an irreversible reaction) is thus proportional to the concentration of each reactant raised to its stoichiometric coefficient. For the generalized reaction

$$aA + bB \xrightarrow{k} \text{Products} \qquad\qquad 7.17$$

the reaction rate thus becomes

$$r_A = kC_A^a C_B^b \qquad\qquad 7.18$$

where

C_A = concentration of reactant species A, [mol/L]

C_B = concentration of reactant species B, [mol/L]

a = stoichiometric coefficient of species A

b = stoichiometric coefficient of species B

k = rate constant [units are dependent on a and b]

The order of the reaction is the sum of the stoichiometric coefficients of the reactants. The overall order of the previous generalized reaction is $a + b$. The reaction is a order with respect to reactant A and b order with respect to B.

EXAMPLE 7.7	IRREVERSIBLE ELEMENTARY REACTION RATES

Write the reaction rate for the conversion of A to products for the following irreversible elementary reactions:

a. A $\xrightarrow{k}$ Products
b. A + 2B $\xrightarrow{k}$ Products

SOLUTION: **a.** For this reaction the stoichiometric coefficient of A is unity. Thus,

$$r_A = kC_A$$

The rate of conversion of reactant A to products is directly proportional to the concentration of A. The reaction is first order.

b. For this reaction, the stoichiometric coefficient of A is unity and of B is 2. The reaction rate is thus

$$r_A = kC_A C_B^2$$

The rate of conversion of species A to products is directly proportional to the concentration of A and directly proportional to the square of the concentration of species B. The reaction is first order with respect to A and second order with respect to B. The reaction is third order overall.

For a generalized reversible reaction,

$$aA + bB \overset{k_f}{\underset{k_b}{\rightleftharpoons}} pP + qQ \qquad\qquad\qquad \textbf{7.19}$$

the products can also recombine to form products, so the net reaction rate must take this into account. The rate at which reactant A is converted to products is then

$$r_A = k_f C_A^a C_B^b - k_b C_P^p C_Q^q \qquad\qquad\qquad \textbf{7.20}$$

where

k_f = forward rate constant [units depend on a and b]

k_b = backward rate constant [units depend on a and b]

C_P = concentration of product species P [mol/L]

C_Q = concentration of product species Q [mol/L]

p = stoichiometric coefficient of species P

q = stoichiometric coefficient of species Q

EXAMPLE 7.8 | **REVERSIBLE ELEMENTARY REACTION RATES**

Write the reaction rate for the conversion of A to products for the following reversible elementary reactions:

a. $A \overset{k_f}{\underset{k_b}{\rightleftharpoons}} P$ **b.** $2A + B \overset{k_f}{\underset{k_b}{\rightleftharpoons}} 2P$

SOLUTION: **a.** For this reaction the stoichiometric coefficients of both A and P are unity. Thus,

$$r_A = k_f C_A - k_b C_P$$

The rate of conversion of A to P is directly proportional to the concentration of B, but the rate at which P is converted back to A is directly proportional to the concentration of P.

b. The reaction rate is

$$r_A = k_f C_A^2 C_B - k_b C_P^2$$

Although few reactions are actually elementary, there are many instances where reactions may be modeled as elementary reactions with good accuracy. The concept of chemical reaction rates, when combined with reactor design, is a powerful tool for the environmental engineer.

Reactor Design

There are three basic reactor types: batch, continuous flow–completely mixed, and plug flow. Each is used, along with many variations, for the treatment of wastes and in modeling natural environmental systems.

Batch Reactors. Batch reactors are the simplest. As Figure 7.7 indicates, they consist of a tank in which the reactants are placed and then mixed

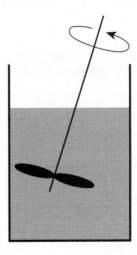

FIGURE 7.7 A batch reactor.

continuously for a period of time during which chemical reactions take place. The concentration of the reactant(s) decreases with time. We can use the mass balance equation, Equation 7.10, from the preceding section, to analyze the batch reactor:

$$\frac{dm_A}{dt} = \sum_{i=1}^{n} (C_{Aj}Q_i)_{in} - \sum_{j=1}^{n} (C_{Aj}Q_j)_{out} - r_A V \qquad \textbf{7.21}$$

There is no flow into or out of a batch reactor. Therefore, the flow terms are zero. This reduces the mass balance equation to

$$\frac{1}{V}\left(\frac{dm_A}{dt}\right) = -r_A \qquad \textbf{7.22}$$

For liquid phase reactions the volume does not change with time. Therefore, the volume term can be brought into the differential. Because m_A/V is the concentration, C_A, the equation becomes

$$\frac{dC_A}{dt} = -r_A \qquad \textbf{7.23}$$

To determine a relationship between the concentration of A and the time the reactants are in the batch reactor, we must equate the kinetic expression to the reactor mass balance, eliminating the term r_A. The resulting differential equation must then be solved to yield an equation in C_A and t. This mathematical process is demonstrated in the following example.

EXAMPLE 7.9	BATCH REACTOR APPLICATION

A wastewater contains contaminant A with an initial concentration of 1200 mg/L. It is to be treated in a batch reactor. The reaction of A to products is assumed to be first order. The rate constant, k, is 2.5/day. Determine the time required to convert 75% of A to products. Plot the conversion of A versus time for the first 10 days.

SOLUTION: We will use C_A for the concentration of contaminant A at time t, and C_{A0} for the concentration of contaminant A at $t = 0$. Since the conversion of A to products is first order, the resulting reaction is then

$$A \xrightarrow{k} Products$$

and the kinetic expression is

$$r_A = kC_A$$

Equating the rate equation with the batch reactor mass balance and eliminating r_A, we obtain

$$-\frac{dC_A}{dt} = kC_A$$

Using the initial condition that $t = 0$ at $C_A = C_{A0}$, we can solve this differential equation by separation of variables and integration to obtain

$$-kt = \ln\left(\frac{C_A}{C_{A0}}\right)$$

or

$$C_A = C_{A0}e^{-kt}$$

where C_A is the concentration of A at any time t. For the stated conditions we must solve for t, using the logarithmic form:

$$t = \frac{\ln\left(\dfrac{C_A}{C_{A0}}\right)}{-k}$$

However, we must first solve for the required effluent concentration C_A:

$$C_A = C_{A0}(1 - X) = (1200\,\text{mg/L})(1 - 0.75)$$

Where X is the fraction of A in the influent that is converted to product. Thus, if we know C_{A0}, we can solve for C_A:

$$C_A = 300\,\text{mg/L}$$

Substituting into the preceding equation for k, C_{A0}, and C_A, we can now solve for t:

$$t = \frac{\ln\left(\dfrac{300\,\text{mg/L}}{1200\,\text{mg/L}}\right)}{-2.5/\text{day}}$$

$$= 0.55\,\text{day}$$

To plot the conversion versus time curve for this reactor, we must solve for the conversion. To do so, we use the exponential form of the concentration versus time equation:

$$\frac{C_A}{C_{A0}} = e^{-kt} = (1 - X)$$

$$X = (1 - e^{-kt})$$

This equation can be plotted using a spreadsheet program, such as Quattro® Pro or Lotus® 1-2-3, or the hard way (by hand). The plot shown in Figure 7.8 was drawn using the graphics from a spreadsheet. The raw data are shown in Table 7.1 for the first 6 hours.

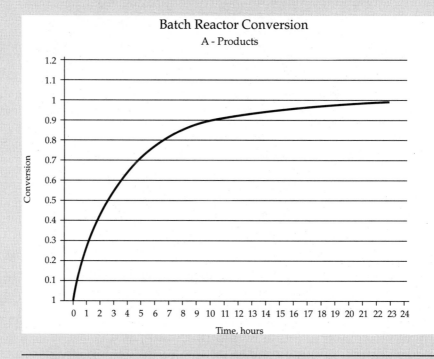

FIGURE 7.8 Batch reactor conversion.

Continuous Flow – Stirred Tank Reactors. CSTRs are flow reactors in which the contents are thoroughly mixed. As the influent enters the reactor, it is rapidly mixed throughout. Thus, the concentration in the reactor is the

TABLE 7.1   Batch Reactor
Conversion

TIME, hr.	CONVERSION
0	0
1	0.188
2	0.341
3	0.465
4	0.565
5	0.647
6	0.713

same as the effluent concentration. A diagram of a CSTR appears in Figure 7.9.

The CSTR mass balance can be derived from the general mass balance developed in the previous section. The general mass balance equation is

$$\frac{dm_A}{dt} = \sum_{i=1}^{n} (C_{Ai}Q_i)_{in} - \sum_{j=1}^{n} (C_{Aj}Q_j)_{out} - r_A V \qquad 7.24$$

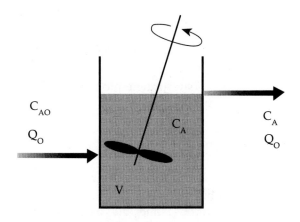

C_{AO}

Q_O

C_A

C_A

Q_O

V

FIGURE 7.9   Continuous flow-stirred tank reactor.

If we consider the CSTR at steady state, there is no accumulation of mass, and the time-dependent term goes to zero. If we also consider only one input and one output flow, the equation becomes

$$r_A V = C_{A0} Q_0 - C_A Q_0 \qquad\qquad\qquad \text{7.25}$$

where

Q_0 = volumetric flow rate into and out of the reactor [volume/time]

C_{A0} = reactant concentration entering the reactor [mass/volume]

C_A = reactant concentration in the reactor and in the effluent [mass/volume]

V = reactor volume [volume]

If we divide by the volume, V, we obtain

$$r_A = \frac{C_{A0} Q_0 - C_A Q_0}{V} \qquad\qquad\qquad \text{7.26}$$

However,

$$\theta = \frac{V}{Q} \qquad\qquad\qquad \text{7.27}$$

where

θ = hydraulic detention time [time]

If this is substituted into the mass balance equation, the CSTR mass balance becomes

$$r_A = \frac{C_{A0} - C_A}{\theta} \qquad\qquad\qquad \text{7.28}$$

To obtain the relationship between the concentration of a reactant in the effluent, C_A, and the detention time, θ, equate the applicable kinetic expression to the reactor mass balance, eliminating r_A. The resulting (algebraic) expression can then be solved for either detention time or effluent concentration.

| EXAMPLE 7.10 | DETENTION TIME IN A CSTR REACTOR |

For the wastewater in Example 7.9, estimate the detention time and reactor volume required if the flow rate to the CSTR reactor is $0.02 \, \text{m}^3/\text{s}$.

SOLUTION: From the previous example we know that the reaction is

$$A \xrightarrow{k} \text{Products}$$

and therefore the reaction rate is

$$r_A = kC_A$$

If we equate this equation to the CSTR mass balance, we obtain

$$\frac{C_{A0} - C_A}{\theta} = kC_A$$

Solving for θ yields

$$\theta = \frac{C_{A0} - C_A}{kC_A} = \frac{1200 \, \text{mg/L} - 300 \, \text{mg/L}}{(2.5/\text{day})(300 \, \text{mg/L})}$$

$$= 1.2 \, \text{days}$$

We can use the relationship for θ, Equation 7.35, to solve for the required reactor volume:

$$V = Q\theta = (0.02 \, \text{m}^3/\text{s})(1.2 \, \text{days})\left(\frac{86{,}400 \, \text{s}}{1 \, \text{day}}\right)$$

$$= 2070 \, \text{m}^3$$

| EXAMPLE 7.11 | CONVERSION IN A CSTR REACTOR |

A CSTR is used to treat a wastewater that has a concentration of $800 \, \text{mg/L}$. The hydraulic detention time of the wastewater in the reactor is 6 hours. The chemical reaction is an elementary irreversible second order reaction:

$$2A \xrightarrow{k} \text{Products}$$

The reaction constant, k, is $3.7 \, \text{L/mg} \cdot \text{day}$. Determine the conversion, X, for the process.

Note: *The conversion can be determined by solving first for the effluent concentration and then using that value to find the conversion. Or the relationship for the concentration can be substituted for the influent concentration and this solved directly for conversion. We will use both methods.*

SOLUTION A: The first step is to determine the reactor mass balance and the kinetic expressions. The reaction is second order irreversible, so the kinetic expression is

$$r_A = kC_A^2$$

The mass balance for a CSTR is

$$r_A = \frac{(C_{A0} - C_A)}{\theta}$$

When these are set equal, eliminating r_A, we obtain

$$\frac{C_{A0} - C_A}{C_{A0}} = kC_A^2$$

Rearranging, we obtain a quadratic equation with C_A as the unknown:

$$k\theta C_A^2 + C_A - C_{A0} = 0$$

If the values for k, θ, and C_{A0} are substituted into the relationship, the effluent concentration can be determined. It is $29 \, \text{mg/L}$. (Although $-30 \, \text{mg/L}$ is also a solution to the quadratic equation, it cannot be a solution to the problem. A negative concentration, and thus a negative mass, are physically impossible.)

The conversion can now be calculated:

$$X = \frac{C_{A0} - C_A}{C_{A0}} = \frac{800 \, \text{mg/L} - 29 \, \text{mg/L}}{800 \, \text{mg/L}}$$

$$= 0.96$$

SOLUTION B: The equation for the concentration of A (Equation 7.5) can be substituted into both the kinetic expression and the reactor mass balance. Equating those two expressions yields

$$\frac{C_{A0} - C_{A0}(1 - X)}{\theta} = k[C_{A0}(1 - X)]$$

$$\frac{C_{A0}X}{\theta} = k(C_{A0}^2 - 2kC_{A0}^2 X + k^2 C_{A0}^2 X^2)$$

$$k^3 \theta C_{A0}^2 X^2 - (C_{A0} + 2k^2 \theta C_{A0}^2)X + k\theta C_{A0}^2 = 0$$

This is also a quadratic equation that can, in this case, be solved directly for the conversion after values for k, θ, and C_{A0} are substituted. The conversion is the same as that in Solution A.

Plug Flow Reactors. In plug flow reactors (PFRs) the contents move along with minimal or no axial or longitudinal mixing. Thus, as shown in Figure 7.10, a thin "slice" of fluid entering the reactor has a uniform

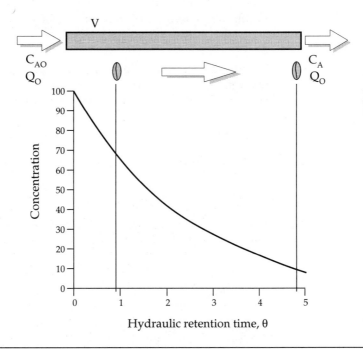

FIGURE 7.10 Schematic of a plug flow reactor showing the decrease in concentration as a "slice" of fluid progresses through the reactor.

concentration radially but does not mix with adjacent slices of fluid. As the slice makes its way through the reactor, the reactants are converted to products. The reaction in the slice is analogous to the reaction in a batch reactor. The difference is that the fluid in this case is actually flowing through the reactor. The hydraulic residence time, θ, is equivalent to the time, t, of the batch reaction. So the mass balance equation for the PFR is

$$r_A = -\frac{dC_A}{d\theta} \qquad\qquad 7.29$$

The analysis of a PFR is similar to that for a CSTR or batch reactor. As in the other two, a kinetic equation is equated to the mass balance equation, eliminating the reaction rate, r_A. The following example illustrates the concept.

EXAMPLE 7.12	HYDRAULIC RETENTION TIME IN A PFR REACTOR

For the wastewater described in the batch reactor exercise (Example 7.9), determine the hydraulic detention time, θ, and a reactor volume required if the flow rate to the PFR reactor is 0.02 m^3/day.

SOLUTION: We will use C_A for the concentration of contaminant A after a retention time of θ, and C_{A0} for the concentration of contaminant A at time $\theta = 0$. Since the conversion of A to products is first order, the reaction is

A $\overset{k}{\rightarrow}$ Products

and the rate equation is

$$r_A = kC_A$$

Equating the rate equation to the PFR reactor mass balance and eliminating r_A, we obtain

$$-\frac{dC_A}{d\theta} = kC_A$$

Using the initial condition that $\theta = 0$ at $C_A = C_{A0}$, we can solve the differential equation by separation of variables and integration, obtaining

$$-k\theta = \ln\left(\frac{C_A}{C_{A0}}\right)$$

or,

$$C_A = C_{A0}e^{-k\theta}$$

where C is the concentration of A at any time t. For the stated conditions we must solve for θ, using the logarithmic form:

$$\theta = \frac{\ln\left(\dfrac{C_A}{C_{A0}}\right)}{-k}$$

But we must first solve for the effluent concentration, C_A, using the initial concentration and the conversion. The concentration at time θ, C_A, is

$$C_A = C_{A0}(1 - X) = (1200\ \text{mg/L})(1 - 0.75)$$

$$= 300\ \text{mg/L}$$

Substituting into this equation for k, C_{A0}, and C_A, we obtain

$$\theta = \frac{\ln\left(\dfrac{300\ \text{mg/L}}{1200\ \text{mg/L}}\right)}{-2.5/\text{day}}$$

$$= 0.55\ \text{day}$$

Note that the hydraulic retention time θ is equivalent to the batch reaction time t.

The volume is then

$$V = Q\theta = (0.02\ \text{m}^3/\text{s})(0.55\text{d})\left(\frac{864005}{\text{d}}\right) = 950\ \text{m}^3$$

It is important to note that solving reactor design problems involves two steps: determination of the chemical rate equation or kinetic expression (that is, the rate of reaction of the species of interest) and writing of the reactor mass balance equation. Those two expressions can then be equated by elimination of the reaction rate, r_A. Solution of the resulting expression yields the concentration versus hydraulic retention time or detention time.

SEDIMENTATION PRINCIPLES

Sedimentation is the accumulation through gravity of particulate matter at the bottom of a fluid. This natural process is frequently used to separate contaminants from air, water, and wastewater. In water treatment colloidal particles are agglomerated by the addition of chemicals and then allowed to settle. In wastewater treatment the first major step is the settling of suspended solids. This is followed by biological treatment to convert the colloidal and soluble materials into a settleable biological mass, which is then removed by a second settling process. Sedimentation is also used in air pollution control to remove particles such as coal, wood, or metal dusts. An understanding of these processes is crucial in understanding water and wastewater treatment overall.

There are four types of settling: discrete, flocculant, hindered, and compression. Each is described in Table 7.2, and all occur in water and wastewater treatment. Discrete settling requires the lowest suspended solids concentration, followed by flocculant and hindered. Compression settling involves the highest concentration of suspended solids. The analysis for discrete settling is the simplest. A discrete particle's settling velocity can be described by an explicit equation. We will present the analysis of discrete settling only, although the other kinds of settling will be described.

TABLE 7.2   Water and Wastewater Treatment Settling Phenomena

SETTLING TYPE	DESCRIPTION	APPLICATIONS
Discrete	Individual particles settle independently, neither agglomerating nor interfering with the settling of the other particles present. This occurs in water with a low concentration of particles.	Grit chambers
Flocculant	Particle concentrations are high enough that agglomeration occurs. This reduces the number of particles and increases average particle mass. The heavier particles sink faster.	Primary clarifiers, upper zones of secondary clarifiers.
Hindred (zone)	Particle concentration is sufficient that particles interfere with the settling of other particles. Particles sink together, and the water is required to traverse the particle interstices.	Secondary clarifiers
Compression	In the lower reaches of clarifiers where particle concentrations are highest, particles can settle only by compressing the mass of particles below.	Lower zones of secondary clarifiers and in sludge thickening tanks.

Discrete Settling

Discrete settling, which occurs in grit chambers at wastewater treatment facilities, can be analyzed by calculating the settling velocity of the individual particles contained within the water. As shown in Figure 7.11, the forces acting on the particle are gravity in the downward direction, F_g, drag acting in the upward direction as the particle settles, F_d, and upward buoyancy due the water displaced by the particle, F_b. Equating the forces yields

$$F_g = F_d + F_b \qquad \qquad \textbf{7.30}$$

The gravitational force can be expressed as

$$F_g = m_p g \qquad \qquad \textbf{7.31}$$

where

$$g = \text{gravitational constant } [9.8 \, \text{m/s}^2]$$
$$m_p = \text{particle mass [kg]}$$

Using the density and volume of the particle, we obtain

$$F_g = \rho_p V_p g \qquad \qquad \textbf{7.32}$$

where

$$\rho_p = \text{density of the grit particle } [\text{kg/m}^3]$$
$$V_p = \text{particle volume } [\text{m}^3]$$

FIGURE 7.11 Forces acting on a discrete particle during settling.

The drag on the particle can be calculated by the drag equation from fluid mechanics:

$$F_d = \frac{1}{2} C_d A \rho_w v^2$$

7.33

where

C_d = drag coefficient, dimensionless

A = particle cross-sectional area [m^2]

ρ_w = density of water [kg/m^3]

v = velocity [m/s]

The buoyant force acting on the particle is

$$F_b = m_w g$$

7.34

where

m_w = mass of water displaced [kg]

Substituting the particle volume and density of water, we get

$$F_b = V_p \rho_w g$$

7.35

When these relationships are substituted into Equation 7.30, we obtain

$$\rho_p V_p g = \frac{1}{2} C_d A \rho_w v^2 + \rho_w V_p g$$

7.36

Solving for the settling velocity, v, we come to

$$v = \left[\frac{2(\rho_p - \rho_w) V_p g}{C_d A \rho_w} \right]^{1/2}$$

7.37

If the relationships for particle area and volume are inserted into this equation for a spherical particle, it becomes,

$$v = \left[\frac{4}{3} \left(\frac{(\rho_p - \rho_w) d_p g}{C_d \rho_w} \right) \right]^{1/2}$$

7.38

At low Reynolds numbers (for Re$_d$ < 1), which would be expected for sand particles settling in water, the drag coefficient, C_d, can be approximated by

$$C_d = \frac{24}{\text{Re}_d}$$

7.39

The Reynolds number is

$$\text{Re}_d = \frac{\rho v d}{\mu} \qquad \textbf{7.40}$$

where

μ = absolute viscosity of the fluid, in this case water [centipoise or 10^{-2} gm/cm · s]

Using these relationships, we can estimate the particle settling velocity as a function of the properties of the particle and water, and the particle diameter, or

$$v_p = \frac{(\rho_p - \rho_w)d^2 g}{18\mu} \qquad \textbf{7.41}$$

This relationship is known as Stokes' law, and the velocity is known as the Stokes velocity. It is the terminal settling velocity for a particle. Typical grit chambers are designed to retain particles with a diameter greater than 0.21 mm, or 0.0083 in. The odd dimension corresponds to a standard U.S. mesh of 65.

The vertical velocity of water in a grit chamber or settling basin is often described as the overflow rate. It is usually expressed as m³/m²·day or gal/ft²-day. It is calculated as in the following way:

$$OFR = \frac{Q}{A} \qquad \textbf{7.42}$$

where

OFR = overflow rate [m³/m² · day]

Q = flow rate [m³/day]

A = clarifier area [m²]

However, we already have an expression for the settling velocity described by the Stokes equation. Thus, the maximum overflow rate and the Stokes settling velocity are the same for a grit chamber. Therefore, we can calculate the area required for grit removal knowing the critical settling velocity v_c (or OFR).

EXAMPLE 7.13	**ESTIMATION OF STOKE'S VELOCITY AND GRIT CHAMBER AREA**

Estimate the settling velocity of sand (density = 2650 kg/m³) with a mean diameter of 0.21 mm. Assume the sand is approximately spherical. Using a safety factor of 1.4 to account for inlet and outlet losses, estimate the area required for a grit chamber to remove the sand if the flow rate is 0.10 m³/s.

SOLUTION: From the Appendix, the density of water at 20°C is 998 kg/m³. Also from the Appendix, the viscosity of water at 20°C is 1.01×10^{-3} N·s/m² (newton = kg·m/s²). The Stokes settling velocity can thus be calculated from Equation 7.41:

$$v_s = OFR = \frac{\left(2650 \ \frac{kg}{m^3} - 998 \ \frac{kg}{m^3}\right)(2.1 \times 10^{-4} \ m)^2 \left(9.8 \ \frac{m}{s^2}\right)}{18\left(1.01 \times 10^{-3} \ \frac{kg}{m \cdot sec}\right)}$$

$$= 0.039 \ m/s = 3.9 \ cm/s$$

Knowing the overflow rate, we can calculate the area required for the grit chamber from Equation 7.46:

$$A = \frac{Q}{OFR}(SF) = \frac{0.10 \ m^3/s}{0.039 \ m/s}(1.4)$$

where *SF* is the safety factor, 1.4.

$$A = 3.6 \ m^2$$

So the grit chamber must be 3.6 m³ to remove 0.21 mm grit from the wastewater.

Flocculant Settling

Flocculant settling occurs when the concentration of particles is sufficiently high to allow the particles to cluster. The agglomeration is the result of gentle mixing induced by paddles in some sedimentation basins and from differential settling velocities of particles of different mass and size. Aggregation results in larger particles that are often entrained with water, but such particles have higher settling velocities than would occur otherwise. Since the particle size and mass continually change, it is not possible to use Stokes' law to estimate the settling velocity. Flocculant settling is the predominant removal process in primary wastewater clarifiers, which will be discussed in Chapter 10, Wastewater Treatment. Flocculant settling is analyzed or estimated by using laboratory settling experiments. The data are then used to estimate the removal versus settling time in the settling basin.

Hindered Settling

Hindered settling occurs as the concentration of solids increases above that for flocculant settling. Particles settle as a structured mass with the water moving between them. Figure 7.12 depicts hindered settling. This type of sedimentation occurs in the lower regions of clarifiers used to settle primary and secondary wastewater and in some clarifiers used for settling chemical precipitation wastes.

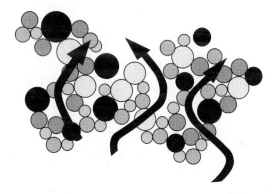

FIGURE 7.12 The movement of water through particles in hindered settling.

Compression Settling

Compression settling occurs in the bottom of many water and wastewater clarifiers where concentrations are so high that settling cannot occur without the compressive influence of the solids above. Solids at the bottom are condensed due to the weight of the mass above.

SUMMARY

The principles of mass balances, conversion, and reaction kinetics are used in a variety of applications in environmental engineering. These same concepts are also used in mechanical, chemical, and mining engineering. As you progress in your studies you will find many applications of these basic and other technical fields. They are particularly useful in analyzing the many different water and wastewater treatment plant processes, natural systems, air pollution, and air pollution control processes.

Review Questions

1. A 6-inch pipe has a flow rate of 150 gallons per minute (GPM). Calculate the velocity in the pipe in ft/min and m/min.

2. A reactor has a volume of 500 L and a flow rate of 30 L/min. What is the HRT?

3. A sedimentation basin, shown in the accompanying figure, has a diameter of 45 ft and a depth of 8 ft. The flow to the basin is 300 GPM. What is the vertical velocity of water in the basin? What is the hydraulic detention time in the basin?

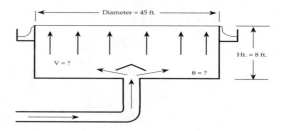

4. The purpose of a sedimentation basin is to remove solids from a water or wastewater flow and to concentrate them. For the basin shown here, determine the return solids pumping rate, Q_u, required to obtain an underflow solids concentration, C_u, of 10,000 mg/L. The flow rate into the basin, Q_i, is 0.05 m³/s. The influent solids concentration, C_i, is 4000 mg/L. The effluent solids concentration, C_e, is 15 mg/L. (*Hint:* Use two equations with two unknowns.)

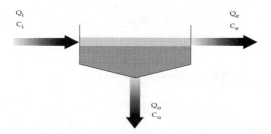

5. Solve Problem 4 assuming that the effluent solids concentration is zero. Compare the result with that for Problem 4. Is there a significant difference (>5%) between the two results?

6. Derive an expression for the concentration of P in Equation 7.3 in terms of the conversion of A, X, the initial concentrations of A and P, and the stoichiometric coefficients of A and P. (*Note:* The concentration of P increases with time.)

7. Calculate the maximum amount of ammonia that can be discharged into Spunkin Creek from the POTW at Spunkinville, given the upstream conditions shown. The concentration of ammonia nitrogen in the stream cannot exceed 15 mg/L.

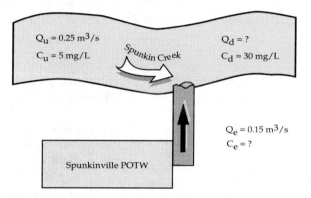

8. For the data from Problem 7, use a spreadsheet program to calculate the combination of discharge flow and concentration combinations possible while still maintaining the maximum 15 mg/L ammonia nitrogen concentration. Calculate the discharge concentrations for flows from 0.05 to 0.5 m^3/s. Plot the curve for these combinations of flow and concentration using flow on the horizontal axis (independent variable) and concentration on the vertical axis (dependent variable).

9. A CSTR aerated lagoon is to be used to treat the wastewater from a small community (population 1350). The wastewater flow is 0.15 MGD. The influent wastewater contaminant concentration is 175 mg/L. The required effluent concentration is 20 mg/L. It was found that the reaction was first order. The k value for the reaction is 0.52/day. What is the required detention time? What is the lagoon volume?

10. What is the required detention time and volume for the lagoon in Problem 9, if it was constructed to provide plug flow?

11. Compare the settling velocity of a 0.15 mm grit particle to that of a 5 mm organic particle in water at 15°C. You may assume that the specific gravity of the grit is 2.6 and of the organic matter is 1.05.

12. Using a spreadsheet program, construct a graph of overflow rate versus temperature from 0 to 30°C for three sizes of grit: 0.15, 0.20, and 0.25 mm.

13. Lagoons (large earthen or concrete ponds) are often used to treat wastewater from small municipalities (usually less than 2000 population). Lagoons are simple flow reactors. The rate of movement of material being treated does not usually approximate either plug flow or complete mix (or CSTR) flow, but is something in between. If the wastewater treatment is assumed to be first order with respect to contaminant A — that is,

$$A \xrightarrow{k} \text{Products}$$

calculate the effluent concentration of A using both plug flow and CSTR methods. The actual effluent concentration should be in that range. The lagoon treatment plant data is as follows:

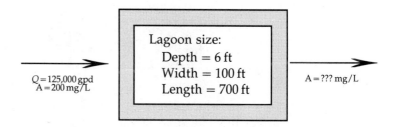

The reaction constant, k, is 0.36 mg/L · day.

References

C. D. Holland and R. G. Anthony, *Fundamentals of Chemical Reaction Engineering* (Englewood Cliffs, NJ: Prentice-Hall, 1979).

O. Levenspiel, *Chemical Reaction Engineering* (New York: Wiley, 1972).

T. J. McGhee, *Water Supply and Sewerage*, 6th ed. (New York: McGraw-Hill, 1991).

Metcalf and Eddy, Inc., *Wastewater Engineering: Treatment, Disposal, and Reuse*, 3rd ed. (New York: McGraw-Hill, 1991).

H. S. Peavy, D. R. Rowe, and G. Tchobanoglous, *Environmental Engineering* (New York: McGraw-Hill, 1985).

8  Water Quality

``...IT IS THE NATIONAL GOAL THAT
THE DISCHARGE OF POLLUTANTS
INTO NAVIGABLE WATERS BE
ELIMINATED BY 1985....''

U.S. Congress, PL 92-500, 1972

A stream in the Smoken Mountains of
North Carolina.

INTRODUCTION

In this chapter we will explore the characteristics of natural waters, water pollution sources, and how pollution affects water quality and use. The topics include water and its availability, common parameters for measuring water quality, typical qualities of rivers and lakes, sources of water pollution, characteristics of wastewater, and finally, stream pollution and its modeling. In this discussion we will be considering water as an endangered natural resource—one that can and must be protected by current technologies.

About 40% of the U.S. population of 260 million use public water supplies utilizing groundwater [1], and most people in rural areas rely exclusively on groundwater, making a total of about 53%. Other people use surfacewater [2]. Before they are consumed, surfacewater and some groundwater must be treated to remove harmful or objectionable impurities and render the water safe and acceptable for human consumption. This process is termed potable water treatment, which we will discuss in Chapter 9. **Wastewater** is the term for discarded or previously used water from a municipality or industry. Such water usually contains dissolved and/or suspended matter and must be treated prior to its discharge into a natural water. Wastewater treatment will be explained in Chapter 10. The amount of water used each day for human consumption and in industry and agriculture is staggering: 1.7 trillion liters (450 billion gallons) in the United States alone. This equates to 6400 liters (1700 gallons) per person per day [3]. To maintain a supply of water of acceptable quality for our population, the pollution of groundwater, rivers, and lakes must be controlled.

What Is Water Pollution?

Water pollution can take many forms, and how we define it—not the actual level of a contaminant but simply whether we consider a given water "polluted"—determines the extent of pollution. A possible definition for water pollution is: any condition caused by human activity that adversely affects the quality of a stream, lake, ocean, or source of groundwater. An equally acceptable definition is: the presence of any harmful chemical or other constituent present in concentrations above the naturally occurring background level. A third could be: any contaminants that adversely affect the use of the natural water for human consumption or that hurt any aquatic life or other wildlife that may rely on the water. The first will be the preferred definition for us.

Effects on Aquatic Life

An unpolluted natural stream enables a wide diversity of aquatic organisms to thrive. It contains enough dissolved oxygen to support life and is free of substantial organic and inorganic pollutants, suspended matter, and toxics. In contrast, a polluted stream inhibits the growth of many aquatic organisms, reducing natural diversity. Only a few types of organisms, usually those considered less desirable, will flourish. So there will be a large number of organisms but few species. In extreme conditions only anaerobic organisms, which need no oxygen, may be able to live in polluted water.

BENEFICIAL USES OF WATER

Water is the most widely used chemical on earth. It is necessary for all life as we know it. Other than the obvious use for drinking, water has myriad uses from irrigating farms to cooling industrial processes. Within the home, water is used for washing clothes, cooking, bathing and showering, and removing wastes, and in the American Southwest, for cooling homes by evaporation. We consume a much smaller amount of water than we withdraw. As an example, we withdraw about 200 gallons per person per day for domestic water use. However, less than 25% of this is actually consumed—used in some manner that prevents it from being returned to the water supply.

Power Plant Uses

Electric power plants use more water than does other human activity. The major use of the water is to remove waste heat from coal fired or nuclear powered boilers. However, almost all of this water is returned to the environment after use (at an elevated temperature). Daily water use (as withdrawal) is approximately 200 billion gallons of water, or 800 gallons per person per day, just for electric power.

Industrial Uses

Water is a necessary component of almost all industrial processes. It is a part of many chemical products, and it is used as a process component, for cooling, and as a solvent. Many products we purchase—soft drinks, beer, wine, and window cleaners, for example—contain mostly water. Even considering all the products that contain water, industrial consumption is only a small fraction of withdrawal. Industrial water withdrawal is about 50 billion gallons per day or 200 gallons per person per day. Consumption is less than 15% of that amount.

Agricultural Uses

And water, whether it is supplied by natural precipitation or through irrigation, is vital in agriculture. American farms and ranches withdraw approximately 150 billion gallons per day, or 600 gallons per person per day, for irrigation, and because irrigation is so inefficient, more than half of this amount is lost to evaporation.

Recreational Uses

Water is a part of many recreational activities from swimming to water skiing (and snow skiing), canoeing, ice skating, sailing, surfing, and boating. The U.S. Army Corps of Engineers and the Department of Interior, along with state and local governments, have constructed many lakes that serve the dual purpose of water supply and recreation.

MEASURING WATER QUALITY

To examine water pollution in more specific, practical terms, we must precisely define the characteristics of the water in question. Quantitative assessments of the quality of natural water, potable water, wastewater, or any other type of water are made by considering many criteria, including temperature, dissolved oxygen level, concentration of organics (measured as either a gross quantity or as a specific compound), and the concentration of various inorganic compounds. The most frequently cited parameters are defined in discussions that follow.

Dissolved Oxygen

Dissolved oxygen (DO) is the amount of molecular oxygen dissolved in water. Because oxygen is required for most aquatic life, DO is one of the most important criteria in determining the quality of a natural water. If we consider that the air we breathe is approximately 20% oxygen and the water fish "breathe" is less than $10\,mg/L$ (or 0.001%) oxygen, we can see clearly that the balance between aquatic life and dissolved oxygen is exceedingly delicate. DO also affects wastewater treatment processes that utilize microorganisms, corrosion chemistry, and other chemical reactions.

The amount of DO decreases with increasing water temperature. For example, the amount of oxygen dissolved in water at 20°C is about $9.2\,mg/L$, while at 30°C, it is only $7.6\,mg/L$. So a cool or cold water can contain much more dissolved oxygen than a warm water. As a result, aquatic life in streams and lakes is placed under more oxygen stress during summer months than during the other seasons. Figure 8.1 shows the change in oxygen solubility versus temperature. This information is also provided in the Appendix.

DO can be measured in the laboratory using chemicals ("wet chemistry") or with a special meter. In the field it is measured almost exclusively with a DO meter.

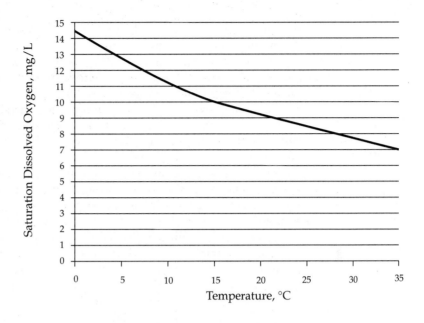

FIGURE 8.1 Solubility of dissolved oxygen in distilled water as a function of temperature.

Oxygen Demand

Oxygen demand is a term for the amount of oxygen required to oxidize a waste. Thus, it is an indirect measure of the amount of organic (or carbon-containing) impurities in a water. Oxygen demand is an important consideration because it can deplete the oxygen in a stream or lake, harming aquatic life. Oxygen demand can be estimated or measured in several ways. The most common are biochemical oxygen demand, chemical oxygen demand, and theoretical oxygen demand.

Biochemical Oxygen Demand. **Biochemical oxygen demand** (BOD) is the amount of oxygen required to oxydize any organic matter present in the water biochemically. So BOD is an indirect measure of the concentration of organic contamination in the water. The more organic matter present, the greater the amount of oxygen that microorganisms will consume in oxidizing the wastes to CO_2 and H_2O. (This process is called waste stabilization.) This measurement is frequently termed **carbonaceous biochemical oxygen demand** (CBOD) to indicate that it is the oxygen demand only from carbon-containing compounds and not from inorganics such as ammonia or ferrous iron. BOD analysis does not oxidize all of the organic matter present in the waste. Only the organics that are biochemically degradable during the standard 5-day (or other specified) time period are oxidized.

The standard conditions for BOD analysis are a temperature of 20°C, darkness (to prevent algae from producing oxygen), and an excess of nutrients for the microorganisms (so that lack of nutrients will not limit growth or stabilization). The duration of the analysis is usually 5 days, but it may be continued for 20 or 30 days. In scientific notation the time is usually subscripted as in BOD_5 for the 5-day analysis or BOD_u for an ultimate test of 20 to 30 days.

A microbial population that is acclimated to metabolize the type of wastes contained in the water sample is required for waste stabilization to occur. For many domestic wastewaters there are enough microbes present, and such wastes do not require added microbes. For other wastewaters, such as those containing industrial or chemical wastes, a microbial "seed" population is normally added to the sample to assure proper oxidation.

Many wastewaters contain ammonia. In these situations, laboratory technicians add a special inhibitor to the water to prevent the organisms from oxidizing the ammonia to nitrate. Otherwise, the oxygen depleted in this ammonia reaction would be falsely interpreted as oxygen used to stabilize organic matter. The test result would be higher. The ammonia concentration (or its corresponding oxygen demand) must be measured separately. (See the discussion of ammonia nitrogen on page 236.)

To determine the BOD of a water sample, place an appropriate dilution in a BOD bottle and find the initial DO. Then incubate the sample (usually for 5 days) under the conditions previously described. The BOD of the sample is calculated as follows:

$$BOD_t = \frac{DO_i - DO_f}{\left(\dfrac{V_s}{V_b}\right)} \qquad\qquad \text{8.1}$$

where

BOD_t = biochemical oxygen demand at t days [mg/L]

DO_i = initial dissolved oxygen in the sample bottle [mg/L]

DO_f = final dissolved oxygen in the sample bottle [mg/L]

V_b = sample bottle volume, usually 300 or 250 mL [mL]

V_s = sample volume [mL]

Note that V_b/V_s is the sample dilution.

EXAMPLE 8.1	BIOCHEMICAL OXYGEN DEMAND ANALYSIS

A wastewater sample is collected from the influent to a treatment plant. Technicians place 10 mL of the sample in a 300-mL BOD bottle. They then fill the bottle with dilution water. The initial DO of the mixture is 8.5 mg/L. After incubating the mixture at 20°C in the dark for 5 days, the technicians measure the DO at 3 mg/L. What is the BOD_5 of the sample?

SOLUTION: The dilution is 300 mL/10 mL, or 30 to 1. The BOD_5 is

$$BOD_5 = \frac{DO_i - DO_f}{\left(\frac{V_s}{V_b}\right)} = \frac{8.5\,\text{mg/L} - 3\,\text{mg/L}}{10\,\text{mL}/300\,\text{mL}}$$

$$= 165\,\text{mg/L}$$

If we look more closely at the microbial process during the BOD test, we will find that the consumption of the organic matter, which we have measured as a change in dissolved oxygen level, occurs as a function of several parameters, including the number and types of microorganisms present, the temperature, the amount of dissolved oxygen available, and others. The rate of removal of BOD (rate of oxidation of organic matter) can be approximated as a function of the BOD (organic matter) remaining, or

$$\frac{dL}{dt} = -kL \qquad \qquad \textbf{8.2}$$

where

L = BOD remaining at time t [mg/L]

k = BOD rate constant [time^{-1}]

Separating the variables and noting that at $t = 0$, $L = L_0$, we obtain

$$\int_0^{L_t} \frac{dL}{L} = -\int_0^t k\,dt \qquad \qquad \textbf{8.3}$$

Integrating yields

$$\ln\left(\frac{L}{L_0}\right) = -kt \qquad\qquad\textbf{8.4}$$

where

L_0 = initial BOD at time = 0 [mg/L]

L_t = BOD at time t [mg/L]

or, in logarithmic form:

$$L_t = L_0 e^{-kt} \qquad\qquad\textbf{8.5}$$

However, we are more often interested in the amount of oxygen consumed in some time period t, rather than the BOD remaining. The oxygen consumed is the difference between the ultimate or total BOD (L_0), and the BOD remaining (L_t), or

$$y_t = L_0 - L_t = L_0 - L_0 e^{-kt} \qquad\qquad\textbf{8.6}$$
$$y_t = L_0(1 - e^{-kt}) \qquad\qquad\textbf{8.7}$$

where

y_t = oxygen (BOD) consumed at time t [mg/L]

Example 8.2 demonstrates the use of these equations.

EXAMPLE 8.2	**BOD CONSUMED AT DIFFERENT TIMES**

A typical wastewater has a BOD_5 of approximately 220 mg/L. If the k for it is 0.23/day (base e), what is the ultimate BOD? What is the 3-day BOD?

SOLUTION: By rearranging Equation 8.7 we can obtain an equation for the ultimate BOD, L_t. To find the BOD_3 (or y_3), we simply substitute into rearranged Equation 8.7 and solve for y_t.

$$L_0 = \frac{y_t}{(1 - e^{-kt})}$$

$$L_0 = \frac{220 \text{ mg/L}}{1 - e^{(-0.23/\text{day})(5\text{days})}} = \frac{220 \text{ mg/L}}{0.68}$$

$$= 322 \text{ mg/L}$$

$$y_3 = L_0(1 - e^{-kt}) = (322 \text{ mg/L})[(1 - e^{(-0.23/\text{day})(3\text{days})})$$

$$= 161 \text{ mg/L}$$

Figure 8.2 shows the change in oxygen consumed and the oxygen demand remaining versus time. A major drawback of BOD analysis is the 5 days required to complete the test. The water discharged when the sample was taken is many miles downstream by the time the analysis is complete. Another problem is that the test is not repeatable. A sample variation of $\pm 10\%$ is typical. Even so, BOD analysis is the best procedure available at present to approximate the amount of oxygen a given waste will consume in a natural water. The EPA and state environmental protection agencies use it almost exclusively to determine the oxygen demand of various wastewaters.

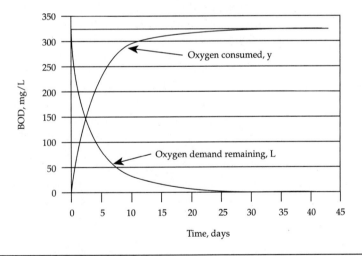

FIGURE 8.2 Biochemical oxygen demand versus time for a typical domestic wastewater.

Chemical Oxygen Demand. The equivalent amount of oxygen required to oxidize any organic matter in a water sample by means of a strong chemical oxidizing agent is called **chemical oxygen demand** (COD). The analysis for this is performed using a strong oxidizing agent, chromic acid, which is a mixture of potassium dichromate and sulfuric acid. Add a catalyst, silver sulfate, to assist in the reaction, and a complexing agent, mercuric sulfate, to prevent chloride from being oxidized. Then heat the chromic acid mixture, with the water sample, to boiling for 2 hours—a process termed chemical digestion. This oxidizes practically all of the organic carbon to CO_2 and H_2O. However, as in the BOD analysis, ammonia is not oxidized. Organic nitrogen remains as unoxidized ammonium. To determine the amount of COD present in the sample, measure the chromic acid remaining after the digestion and compare this amount to the initial amount of chromic acid. The difference between the two is the COD of the sample.

The COD analysis is relatively fast compared with the BOD analysis (3 hours versus 5 days), and it is relatively repeatable. Its disadvantages are that it does not approximate the conditions of a stream or other water body, and the COD of a sample is almost always significantly higher than the BOD. Wastewater treatment personnel frequently use COD analysis to obtain information quickly for plant operation, but the BOD analysis is used for EPA-required monitoring.

Theoretical Oxygen Demand.   **Theoretical oxygen demand** (ThOD) is the amount of oxygen required to oxidize a known compound completely to CO_2 and H_2O. It can also be used to calculate the amount of oxygen required to oxidize the ammonia present in the water or wastewater (NOD). The ThOD cannot be calculated for an unknown compound or mixture. A complete analysis of the water is required before the ThOD can be determined.

To determine the ThOD of a compound, write and balance the chemical oxidation reaction with end products of CO_2 and H_2O. Then calculate ThOD based on the mass of contaminant in the water. For carbonaceous oxygen demand, any nitrogen present in the compound becomes ammonia as an end product. The nitrogenous oxygen demand results from the oxidation of ammonia to nitrate, NO_3^-.

EXAMPLE 8.3	ThOD ANALYSIS

A chemical plant produces the amino acid glycine ($C_2H_5O_2N$). Wastewater from the facility contains approximately 30 mg/L of this acid. Calculate both the carbonaceous and nitrogenous ThOD for the wastewater.

SOLUTION: First, to calculate the carbonaceous oxygen demand, we balance the equation with CO_2, H_2O, and NH_3 as end products:

$$C_2H_5O_2N + 1\tfrac{1}{2}O_2 \rightarrow 2CO_2 + H_2O + NH_3$$

There are 30 mg/L of glycine. Each mole of glycine (75 g/mol) reacts with $1\tfrac{1}{2}$ mol of oxygen (32 g/mol). Thus, the amount of oxygen required to oxidize the carbonaceous portion is

$$ThOD = 30 \, \text{mg/L glycine} \times \frac{1.5 \, \text{mol oxygen} \times 32 \, \text{g oxygen/mol oxygen}}{1 \, \text{mol/L} \, 75 \, \text{g glycine/mol glycine}}$$

$$= 19.2 \, \text{mg/L oxygen (carbonaceous)}$$

One mole of ammonia is produced for each mol of glycine oxidized. The equation for the oxidation of the ammonia is

$$NH_3 + 2O_2 \rightarrow NO_3^- + H_2O + H^+$$

So we can see that each mole of glycine requires 2 mol of oxygen to oxidize the organic nitrogen to nitrate. The ThOD is calculated as follows:

$$ThOD = 30 \, \text{mg/L glycine}$$

$$\times \frac{2 \, \text{mol oxygen} \times 32 \, \text{g oxygen/mol oxygen}}{1 \, \text{mol glycine} \times 75 \, \text{g glycine/mol glycine}}$$

$$= 25.6 \, \text{mg/L oxygen (nitrogeneous)}$$

The amount of oxygen required to oxidize the glycine is then the sum of both the carbonaceous and nitrogeneous oxygen demands, or

$$ThOD = 44.8 \, \text{mg/L oxygen (total)}$$

Organic Content

The organic content of water is important because of the oxygen it will consume and because some organics are toxic. The different methods of measuring oxygen demand, which were previously discussed, indirectly measure the organic content of the water in question. As with oxygen demand, there are several ways to measure organic content.

Total Organic Carbon. The amount of organic carbon in a sample is called **total organic carbon** (TOC). This measurement provides an estimate of the amount of organic contamination contained in a water or wastewater sample. Several companies manufacture instruments for this. These instruments automatically oxidize the organic matter to carbon dioxide and then measure the CO_2 content. The inorganic carbon (CO_2) is either measured separately and subtracted from the reading, or it is removed prior to the oxidation step. Although expensive,* these instruments are valuable where rapid analyses are needed. TOC can be determined in 3 to 5 minutes.

Oil and Grease. The measure of oily and greasy substances in a water sample, without determination of the exact chemical compounds involved, is referred to simply as **oil and grease**. In the analysis for these substances, any organic compound that is soluble in the solvent trichlorotrifluoroethane is measured. Analysis is normally performed on wastewater samples only. Excessive amounts of oil and grease in a wastewater can have a detrimental effect on treatment processes. In some cases the oil and grease must be removed prior to processing.

Solids

Solids are matter in a wastewater. The total solids content comprises both a dissolved fraction and a suspended fraction. Each of these further divides into either volatile (organic) or fixed (inorganic) fractions. The relationships of the different solids' fractions are shown in Figure 8.3.

Total Solids. **Total solids** (TS) is the amount of organic and inorganic matter in a water. To determine TS, evaporate and dry a known volume of the sample in a preweighed crucible dish at 103°C. Weigh the crucible again after cooling in a desiccator. The mass of residue remaining divided by the initial volume is the TS, or

$$TS = \frac{m_{cf} - m_{ci}}{V}$$

8.8

where

TS = total solids [mg/L]

m_{ci} = initial crucible mass [mg]

m_{cf} = crucible mass after drying at 103°C [mg]

V = sample volume [L]

---

* A TOC analyzer costs about as much as a Mazda Miata or Toyota MR-2!

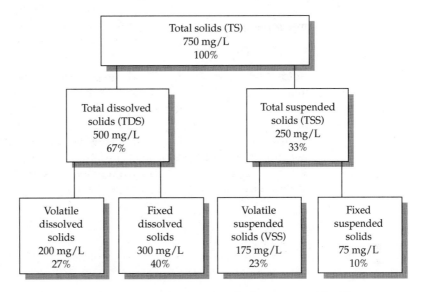

Note: These are typical values. Actual values vary considerably.

FIGURE 8.3 The components of total solids.

Total solids is an indicator of gross levels of water contaminants. As shown in Figure 8.3, it breaks into several components, including fixed and volatile or dissolved and suspended fractions.

Volatile Solids. **Volatile solids** (VS) is the amount of matter that volatilizes (or burns) when heated to 550°C. After completion of the TS test, the crucible containing the total solids mass is heated at 550°C until all volatile matter has been ignited and burned. This amount is then figured as

$$VS = \frac{m_{cf} - m_{cx}}{V} \qquad\qquad \textbf{8.9}$$

where

VS = volatile solids [mg/L]

m_{cx} = crucible mass after ignition at 550°C [mg]

Volatile solids is a useful approximation of the amount of organic matter present in water.

Fixed Solids. Solids that do not volatilize at 550°C are known as **fixed solids** (FS). This measure is used to gauge the amount of mineral matter in a sample. FS is then the difference between the TS and VS, or

$$FS = TS - VS \hspace{4cm} \textbf{8.10}$$

Total Suspended Solids. **Total suspended solids** (TSS) is the amount of matter suspended in the water. It is determined by taking a measured quantity of water and filtering it through a preweighed glass microfiber filter (effective retention, $1.5\,\mu m$). The material remaining on the filter after it dries is the TSS. The procedure is to take a glass fiber filter of known weight, filter a sample of known volume, dry the filter at 103°C and then again weigh the filter. The difference in the initial and final weights, divided by the sample volume, is the TSS:

$$TSS = \frac{m_{ff} - m_{fi}}{V} \hspace{4cm} \textbf{8.11}$$

where

TSS = total solids [mg/L]

m_{fi} = initial filter mass [mg]

m_{ff} = filter mass after drying at 103°C [mg]

Volatile Suspended Solids.   Burning the residue from the TSS determination at 550°C, cooling it and again weighing the filter will yield **volatile suspended solids** (VSS). The VSS is the difference between the initial filter weight (from the TSS determination) and the filter weight after ignition at 550°C divided by the sample volume, or

$$VSS = \frac{m_{ff} - m_{fx}}{V} \hspace{4cm} \textbf{8.12}$$

where

VSS = total solids [mg/L]

m_{fx} = filter mass after ignition at 550°C [mg]

Fixed Suspended Solids. **Fixed suspended solids** (FSS) is the matter remaining from the suspended solids analysis. In other words, it is unburnable at 550°C. FSS represents the nonfilterable inorganic residue in a sample. It is determined indirectly as the difference between the TSS and the VSS, or

$$FSS = TSS - VSS$$

From these relationships, we can deduce that TS is the sum of volatile plus fixed solids. However, TS is also the sum of the suspended plus the dissolved solids.

EXAMPLE 8.4	SOLIDS ANALYSIS

Some 20 mL of a well-mixed wastewater sample is placed in a crucible dish that weighs 49 g. After evaporation and drying at 103°C, the crucible is cooled in a desiccator and reweighed. The weight is then 49.014 g. Following this, the crucible is fired at 550°C for 1 hour. The final weight after cooling is 49.004 g. What are the TS and VS of the sample?

SOLUTION:

$$TS = \frac{m_{cf} - m_{ci}}{V} = \frac{49.014\,g - 49.000\,g}{20\,mL} \times \frac{10^3\,mg}{g} \times \frac{10^3\,mL}{L}$$

$$= 700\,mg/L$$

To obtain the volatile solids, we solve in this way:

$$VS = \frac{m_{cf} - m_{cx}}{V} = \frac{49.014\,g - 49.0004\,g}{20\,mL} \times \frac{10^3\,mg}{g} \times \frac{10^3\,mL}{L}$$

$$= 500\,mg/L$$

The fixed solids can also be obtained from this information:

$$FS = TS - VS = 700\,mg/L - 500\,mg/L$$

$$= 200\,mg/L$$

Total Dissolved Solids. The amount of matter dissolved in water is **total dissolved solids** (TDS). Components of TDS typically include dissolved salts such as calcium and magnesium chloride and calcium and magnesium sulfate. TDS also includes any dissolved organic matter present in the water. Human contact with natural waters usually results in a gradual increase in the TDS whether people are treating the water for use as a supply, actually using the water, or working with the resulting wastewater. A high TDS is undesirable for potable water because of its associated taste. The recommended maximum TDS for potable water is usually 500 mg/L.

Dissolved solids may corrode potable water distribution systems, and they can have a detrimental effect on industrial processes.

TDS is usually measured indirectly as the difference between the total solids and the total suspended solids, or

$$TDS = TS - TSS \qquad\qquad\qquad 8.14$$

Nutrients

Nutrients are important because they are required for growth of the microorganisms used in wastewater treatment processes and because, if not removed, they can lead to excess algal growth, particularly in lakes. The major nutrients in water and wastewater processes are nitrogen and phosphorus. In addition, organic nitrogen and ammonia, if not oxidized to nitrate, can consume copious amounts of oxygen. Chapters on Microbiology and Wastewater Treatment will present discussions of other nutrients.

Nitrogen. Nitrogen is an important parameter in both natural water and wastewater because it can act as a nutrient stimulating algal growth, and because oxidation of reduced nitrogen forms can consume considerable amounts of oxygen. Nitrogen occurs in water and wastewater in five forms: organic, ammonia, nitrite, nitrate, and molecular. Microorganisms can interconvert all of these forms. We will discuss the forms of nitrogen that are important to environmental engineers—ammonia, nitrate, and organic nitrogen.

Ammonia Nitrogen. Commonly measured for domestic wastewater, **ammonia nitrogen** comes from either ammonia or ammonium. Both are degradation products of proteins as well as urea, a compound contained in urine. Ammonia is commonly found in municipal wastewater, natural streams, and some industrial wastewater. In a simple acid–base reaction ammonia [NH_3] and ammonium [NH_4^+] interconvert, depending on the pH of the water. Each milligram of ammonia or ammonium nitrogen consumes approximately 4.5 mg of oxygen. Besides consuming oxygen in the water, ammonium is toxic to fish, especially fry (small or baby fish). The reaction for the oxidation of ammonium to nitrate is

$$NH_4^+ + 2O_2 \rightarrow 2H^+ + H_2O + NO_3^- \qquad\qquad 8.15$$

Nitrate Nitrogen. **Nitrate nitrogen** encourages the production of bacteria and algae in water. It is commonly measured in drinking water and occasionally in wastewater. It is especially important in drinking water because high levels of nitrate are associated with methemoglobinemia (blue baby syndrome). Nitrate can be measured by either wet chemistry or a specific ion electrode. Regardless of the method, however, accurate measurement is difficult because other substances interfere with the process and the various methods have limited concentration ranges.

Organic Nitrogen. **Organic nitrogen** is nitrogen contained as amines in organic compounds such as amino acids and proteins. During the oxidation of organic compounds the carbon and hydrogen are converted to CO_2 and H_2O, and the amines (organic nitrogen) are converted to ammonia. Subsequent oxidation of the ammonia to nitrate consumes significant oxygen (see ammonia nitrogen). Organic nitrogen is measured by oxidizing the organics to CO_2 and H_2O, leaving the organic nitrogen in the ammonia form that can be measured. The sum of the ammonia and organic nitrogen is known as the total Kjeldahl nitrogen.

Phosphorus. **Phosphorus**, which stimulates rapid growth of algae in water, can be measured by several chemical methods. This nutrient normally occurs in both natural water and wastewater as phosphates and organophosphates. Laundry detergents often are a major source of phosphates.

Metals

Metals are present in all natural waters. Household activity adds small amounts of some naturally occurring metals to the water. Sodium, potassium, and calcium are examples. However, other metals—including some toxic metals such as chromium, mercury, zinc, lead, and copper—are added to wastewater through industrial processes and household activities. Even metals essential for human growth are pollutants if present in excessive amounts.

Metals are usually detected by an atomic absorption spectrophotometer, a common instrument in environmental laboratories. The EPA does not normally regulate the concentration of metals in domestic wastewater, but they are regulated in industrial wastwater. Typical limits for metals, after treatment, are in the sub mg/L range.

Other Water Quality Parameters

Many other determinants of water quality are of interest to environmental engineers. They include turbidity, hardness, alkalinity and pH.

Turbidity. **Turbidity** is a measurement of the clarity of water. It is predominantly used for potable water monitoring, although it is occasionally used to assess wastewater treatment processes. Clouded water is caused by suspended particles scattering or absorbing the light. Thus, turbidity is an indirect measurement of the amount of suspended matter in the water. However, since solids of different sizes, shapes, and surfaces reflect light differently, turbidity and suspended solids do not correlate well. Turbidity is normally gauged with an instrument that measures the amount of light scattered at an angle of $90°$ from a source beam. Turbidity is important in potable water because microorganisms attach to suspended particles.

Hardness. The concentration of multivalent cations measured as calcium carbonate [$CaCO_3$] is **hardness**. Hard water requires increased amounts of soap for bathing or washing clothes, and forms a scale on piping, cooking vessels, boilers, and heat exchangers. The major cations responsible for hardness are usually calcium [Ca^{2+}] and magnesium [Mg^{2+}], although ferrous iron [Fe^{2+}], ferric iron [Fe^{3+}], strontium [Sr^{2+}], and manganese [Mn^{2+}] can be significant contributors in some water. Total hardness is often thought of as having two components: carbonate hardness and noncarbonate hardness. Carbonate hardness is that which is less than or equal to the carbonate alkalinity. This is easier to precipitate in hot water pipes, boilers, heat exchangers, or other surfaces in contact with hot water. Noncarbonate hardness is any hardness that exceeds the carbonate alkalinity. Thus, if the hardness is less than the carbonate alkalinity, all hardness will be in the carbonate form. There will be no noncarbonate hardness.

EXAMPLE 8.5	DETERMINATION OF HARDNESS

A water has been found to contain the following ions. Determine the total hardness, carbonate hardness, and noncarbonate hardness.

CATION	CONCENTRATION, mg/L	ANION	CONCENTRATION, mg/L
Na^+	15	HCO_3^-	75
Ca^{2+}	12	NO_3^-	10
Mg^{2+}	15	Cl^-	25
Sr^{2+}	3	SO_4^{2-}	41
K^+	15	CO_3^{2-}	0

SOLUTION: The first task is to convert each of the divalent cation concentrations from mg/L of the cation to mg/L of $CaCO_3$. (The monovalent cations do not contribute to hardness.)

$$12\,mg/L\,Ca^{2+} \times \frac{50\,mg/eq\,CaCO_3}{20\,mg/eq\,Ca^{2+}} = 30\,mg/L\,Ca^{2+} \text{ as } CaCO_3$$

$$15\,mg/eq\,Mg^{2+} \times \frac{50\,mg/L\,CaCO_3}{12.2\,mg/eq\,Mg^{2+}} = 61.5\,mg/L\,Mg^{2+} \text{ as } CaCO_3$$

$$3\,\text{mg/L}\,\text{Sr}^{2+} \times \frac{50\,\text{mg/eq}\,\text{CaCO}_3}{43.8\,\text{mg/eq}\,\text{Sr}^{2+}} = 3.4\,\text{mg/L}\,\text{Sr}^{2+}\ \text{as}\ \text{CaCO}_3$$

The total hardness is the sum of the divalent cations expressed in terms of $CaCO_3$, or

$$\text{Total Hardness} = 30\,\text{mg/L} + 61.5\,\text{mg/L} + 3.4\,\text{mg/L}$$
$$= 94.9\,\text{mg/L}\ \text{as}\ \text{CaCO}_3$$

Since the carbonate concentration is zero, all of the carbonate is in the bicarbonate form. The next task is to convert the bicarbonate concentration to mg/L as $CaCO_3$.

$$75\,\text{mg/L}\,\text{HCO}_3^- \times \frac{50\,\text{mg/eq}\,\text{CaCO}_3}{61\,\text{mg/eq}\,\text{HCO}_3^-} = 61.5\,\text{mg/L}\,\text{HCO}_3^-\ \text{as}\ \text{CaCO}_3$$

Thus, the carbonate hardness is equal to the bicarbonate concentration measured as $CaCO_3$, or 61.5 mg/L as $CaCO_3$. The noncarbonate hardness is the difference between the total hardness and the carbonate hardness, or

$$\text{Total hardness} = \text{Carbonate hardness} + \text{Noncarbonate hardness}$$
$$\text{Noncarbonate hardness} = 94.9\,\text{mg/L} - 61.5\,\text{mg/L}$$
$$= 33.4\,\text{mg/L}\ \text{as}\ \text{CaCO}_3$$

Had the total hardness been only 20 mg/L as $CaCO_3$ and the carbonate concentration remained the same, the total hardness would equal the carbonate hardness, and both would have been 20 mg/L as $CaCO_3$. (There would have been no noncarbonate hardness.)

Alkalinity. **Alkalinity** is the capacity of a water to neutralize acids. The condition is due primarily to the presence of the salts of weak acids, although strong bases can contribute to alkalinity in some cases. Most alkalinity in natural water is produced by bicarbonate since this is the dominant form of carbonate in most natural water. However, alkalinity can also result from the presence of carbonate, hydroxide, phosphate, borate, and other ions. Alkalinity is usually measured by titrating a water sample with $0.02\,N\,\text{H}_2\text{SO}_4$ to a specified final pH, usually 4.5. The alkalinity is then the amount of acid required, expressed as $CaCO_3$.

pH. The negative log of the hydrogen ion concentration is called **pH**. Because this affects biological as well as chemical reactions, it is meeasured for most waters. This factor, also important in determining the solubility of many species (especially metallic ions), is normally measured with a pH meter.

More complete descriptions of all these parameters, as well as exact methods for measuring them, are contained in two common reference books, *Standard Methods for the Examination of Water and Wastewater* [4] and *Methods for Chemical Analysis of Water and Wastes* [5]. Table 8.1 presents some of the parameters used to indicate the quality of various waters. Typical concentrations in lakes and streams are compared to the current EPA drinking water standards.

SOURCES AND QUANTITIES OF WATER POLLUTION

Since passage of the Federal Water Pollution Control Act Amendments in 1972 (PL 92-500) and the Clean Water Act in 1977 (PL 95-217), there has been a gradual move to reduce the amount of pollutants discharged into our waterways. Pollution in natural waters comes from many sources. The most commonly reported and discussed sources are discharges of municipal and industrial wastewater. These are termed point sources because they normally have a definite point of origin—a pipe from a wastewater treatment plant, for example. It has been estimated that from 1974 to 1988, our nation expended approximately $200 billion to reduce the amounts of point source pollution. The city of Milwaukee, Wisconsin, alone spent more than $2.5 billion [6].

Large amounts of pollution are also generated by agricultural activities and construction. These are termed nonpoint sources since they usually have no definite point of entry or discharge. Our nation has just begun to address the problems associated with nonpoint source pollution.

Municipal Wastewater Pollution

Publicly owned treatment works (POTWs) receive wastewater from the sewer pipes of homes, businesses, and in many cases, industries. Typical characteristics of raw municipal wastewater are shown in Table 8.2. Municipal wastewater contains much suspended and dissolved organic matter, and it usually contains trace amounts of myriad organic and inorganic toxins. After treatment to remove contaminants, the water must meet EPA requirements for discharge into natural waters: no more than 30 mg/L of BOD_5, no more than 30 mg/L of suspended matter, and a pH of 6 to 9. Current treatment practices typically reduce the organic contaminants from about 250 mg/L of BOD_5 to around 15 mg/L BOD_5, and suspended matter is reduced similarly. This level of treatment is often termed secondary. Many POTWs or industries that discharge into lakes or

TABLE 8.1   Typical Drinking Water Quality Parameters and their Associated Health Effects

PARAMETER	HEALTH EFFECTS	TYPICAL CONC. LAKE OR STREAM	DRINKING WATER STANDARD
Microbiological:			
Total Coliforms, no./100 mL	Indicator organisms, not necessarily disease causing	<100	1
Turbidity, NTU	Interferes with disinfection	1–20	1–5
Inorganic:			
Arsenic, mg/L	Nervous system and skin problems	<0.01	0.05
Barium, mg/L	Circulatory problems	<0.01	1
Cadmium, mg/L	Kidney trouble	<0.01	0.01
Chromium, mg/L	Liver/kidney problems	<0.01	0.05
Lead, mg/L	Nervous system and kidney problems (highly toxic to infants and pregnant women)	<0.01	0.05
Mercury, μg/L	Nervous system and kidney problems	<0.01	2
Nitrate, mg/L	Methemoglobinemia	<1.0	10
Selenium, μg/L	Gastrointestinal effects	<1	10
Silver, μg/L	Skin discoloration	<1	0.05
Fluoride, mg/L	Skeletal damage	<1	4
Organic:			
Endrin, μg/L	Nervous system/kidney problems	<1	0.2
Lindane, μg/L	Nervous system/kidney problems	<1	4
Total trihalo-methanes, μg/L	Cancer risk	<50	100
Benzene, μg/L	Cancer	<1	5
Other:			
pH	Corrosivity (not health)	6–8	6.5–8.5

TABLE 8.2 Typical Munipal Wastewater Characteristics, EPA Municipal Discharge Standards, and Typical Stream Characteristics

PARAMETER	TYPICAL WASTEWATER CHARACTERISTICS, mg/L EXCEPT pH	EPA DISCHARGE STANDARDS mg/L EXCEPT pH	TYPICAL CONCENTRATIONS IN LAKES OR STREAMS, mg/L EXCEPT pH
BOD_5	150–300	30	2–10
Total Suspended Solids	150–300	30	2–20
COD	400–600	N/A	5–50
DO	0	4–5	4–saturation
NH_3-N	15–40	*	<1
NO^{-3}	0	*	<1
pH	6–8	6–9	6–8

*No uniform standard, but discharge limits are required for some POTWs or industries, depending on receiving water quality.

pristine streams have much more stringent discharge requirements. In these cases the EPA or states may require a BOD_5 of 10 mg/L or less, suspended solids of 10 mg/L or less, and limits on nitrogen and/or phosphorus. This is often referred to as advanced wastewater treatment. However, the ammonia and organic materials still present in the treated wastewater, will in some instances cause a significant decrease in the stream DO. (See the case study at the end of this chapter.) There are presently more than 8500 POTWs with secondary treatment in the United States. These municipal sewage plants treat 16 billion gallons of wastewater each day. In addition, there are more than 3400 POTWs with advanced wastewater requirements, treating 15 billion gallons of wastewater each day. Unfortunately, there are also about 1900 facilities that have less than secondary treatment or no treatment (pump and discharge only). However, these account for only 5 billion gallons of wastewater daily [7]. The water received by (and discharged from) municipal wastewater treatment plants is from 500 to 600 L/capita·day (125 to 150 gal/capita-day). When you consider that there are now more than 260 million people in the United States, you can begin to understand the magnitude of the water pollution problem.

The wastewater treatment process produces several low-volume (or flow) waste streams that contain most of the organic matter removed in treatment. The resulting sludges are treated and then applied on agricultural land, buried, or incinerated. They are high in organic solids and contain low levels of many toxins. The characteristics and treatment of these sludges will be discussed in Chapter 10, Wastewater Treatment.

Industrial Wastewater Pollution

American industries use and then discharge vast quantities of water each day. Much of it is used only for cooling purposes—carrying away waste heat. However, much unwanted or dangerous material is also present in many industrial wastewaters. Some industries treat their water and then discharge it directly into streams, rivers, lakes, or marine waters. Others discharge into POTWs. The EPA requires that industries discharging directly into receiving waters treat their wastewater prior to discharge. The limits for industries are more detailed than those for POTWs because of the more varied materials contained in industrial wastewater. Industries discharging into municipal sewer systems are required to pretreat wastewater to remove toxic substances that might interfere with municipal plant operation or that would pass through plants unaffected.

Many different categories of industrial wastes are recognized. The EPA groups industries according to the federal government's Standard Industrial Classification (SIC) number. A few common industries and the pollutants they typically produce are shown in Table 8.3. The EPA requires industries to reduce these pollutants substantially prior to discharge into receiving waters, but, as with municipal treatment, some contaminants remain in the discharged water.

TABLE 8.3 Typical Industrial Discharges

INDUSTRY	TYPICAL WASTEWATER
Metal finishing, metal plating	Oil, grease, cyanides, acids, bases, metals
Automobile manufacturing	Soluble and insoluble oils, cyanides, acids, bases, metals, paints, others
Pulp and paper making	Sulfites, organics, chlorinated organics, acids, fiber solids
Dairy production	Solids, high BOD_5, organic nitrogen
Crude refining	High BOD_5, phenols, organic sulfur, organic nitrogen, naphthenic acids, inorganic acids, alkalis, inorganic salts

Nonpoint Source Pollution

Much of the pollution that makes its way into U.S. streams and rivers is from nonpoint sources. A **nonpoint source** is any origin that cannot be attributed to a particular discharge point. While a wastewater treatment plant usually discharges its water into a stream or lake from an effluent pipe located at a particular point, this is not true of polluted rainwater. Rainwater that is not absorbed into the soil by infiltration may become contaminated with fertilizers, herbicides, insecticides, and other pollutants before it runs off the land, eventually finding its way to a ditch, canal, stream, or lake. This water is considered nonpoint source pollution (or NPSP). A listing of several nonpoint sources and the amounts of nitrogen and phosphorus they typically generate is shown in Table 8.4. NPSP occurs

TABLE 8.4 Concentration of Nutrients from Nonpoint Pollution Sources at Low Flows

SOURCE	NITROGEN CONCENTRATION, mg/L	PHOSPHOROUS CONCENTRATION, mg/L
Atmospheric deposition	1–10	0.02–2
Surface Animal feedlot runoff	>30	>10
Urban stormwater runoff	1–10	0.2–2
Irrigation excess flow	10–30	2–10
Septic tank surface leakage	>30	>10
Construction site runoff	1–30	0.2–10
Combined sewer overflow	>30	>10
Subsurface Drainage from agriculture	10–30	0.2–2
Drainage from intensive agriculture	>30	0.2–2
Septic tank leakage	>10	0.2–2
In situ Production Nitrogen fixation	Unknown	N/A
Anaerobic sediments	0.2–2	1–10

SOURCE: V. N. Novotny and G. Bendoricchio, "Linking Nonpoint Pollution and Deterioration, *Water Environment and Technology* (November 1989).

not only during precipitation events but also from base (or groundwater) flows during dry weather. Thus, it is not simply an occasional occurrence.

Agricultural usage of fertilizers and pesticides has increased markedly during the past 20 years. For example, fertilizer usage in Iowa has tripled since 1955, and pesticide usage has grown sixfold [6]. Excess nitrogen and phosphorus from agricultural activities often ends up in nearby streams where it stimulates algal production. Although algae produce oxygen by photosynthesis, they also consume oxygen for respiration. During the night, on cloudy days, or in shaded areas of a stream, algal respiration can consume large amounts of oxygen. In addition, as algae die and add to the sediment in the stream, they become an additional organic load that requires oxygen for stabilization. It has been estimated that 79% of the nitrogen, 74% of the phosphorus, and 41% of the oxygen-demanding materials in our waters are attributable to NPSP [8].

Through the 1970s and 1980s the nation was much more interested in controlling point source pollution from municipalities and industries than in stemming nonpoint source pollution, but this emphasis may change in the 1990s or 2000s as we gain more knowledge about nonpoint source discharges.

STREAM POLLUTION MODELING

Many equations and computer programs available today attempt to describe the quality of water in streams, rivers, and lakes. Regulatory agencies, consultants, and industries use these models to predict the effects of different levels of wastewater treatment on receiving water quality. These constructions can also estimate the impact of new pollution sources on water quality, and some of the more complex models can approximate the effects of many different parameters through a given reach (or section) of a stream, or in a lake or estuary. One of the most fundamental and frequently used stream or river models is the Streeter Phelps equation. It is the only model we will discuss in this text.

When wastewater is discharged into a stream, the DO typically begins a slow decrease to some minimum, followed by a gradual increase in DO up to near the saturation level of dissolved oxygen. This plot of dissolved oxygen versus time or stream distance, shown in Figure 8.4, is the DO sag curve. Given the characteristics of the stream and the wastewater, the Streeter Phelps model can approximate this curve for a given stream.

To derive the DO sag curve, we must understand that there are two competing groups of reactions occurring within a stream. One set of reactions is simply the dissolving of molecular oxygen from the atmosphere into the water, or **reaeration**. Reaeration is a function of several parameters, including stream temperature, velocity, depth, and turbulence, as well as the difference between the actual dissolved oxygen level and the potential dissolved oxygen level at the saturation point.

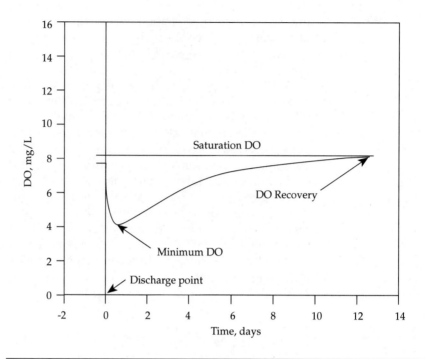

FIGURE 8.4   A typical DO sag curve.

In direct competition with reaeration is the consumption of oxygen by the various stream organisms as they oxidize materials in the stream. This is **deoxygenation**. In reality, there are other sources and consumers of dissolved oxygen, but for simplicity we will consider only these.

Before explaining the model, we need to ensure familiarity with the concept of stream or oxygen deficit. The oxygen deficit is the difference between the potential saturated oxygen level at a given water temperature and the actual dissolved oxygen level. The saturation level of dissolved oxygen depends on temperature and the amount of dissolved solids in a stream. However, for most freshwater streams the level of dissolved solids is low enough to be ignored. Thus,

$$D = \mathrm{DO_{sat}} - \mathrm{DO_{act}} \qquad\qquad 8.16$$

where

D = oxygen deficit [mg/L]

$\mathrm{DO_{sat}}$ = saturation value of dissolved oxygen [mg/L]

$\mathrm{DO_{act}}$ = actual dissolved oxygen value for the stream [mg/L]

Appendix A gives the saturation DO level versus temperature. Example 8.6 shows how to calculate the stream's oxygen deficit.

EXAMPLE 8.6	OXYGEN DEFICIT

If the stream temperature is 25°C and the DO is 6.0 mg/L, what is the oxygen deficit?

SOLUTION: First, find the saturation level of dissolved oxygen at 25°C from Appendix A. This is 8.4 mg/L. Then solve in this manner:

$$D = DO_{sat} - DO_{act} = 8.4\,mg/L - 6.0\,mg/L$$
$$= 2.4\,mg/L$$

Let us assume that the rate of oxygen entering the stream through the atmosphere is proportional to the dissolved oxygen deficit in the stream. Similarly, let's assume that the rate of oxygen consumed or leaving the stream is proportional to the amount of organic matter in the stream, expressed as BOD_u (ultimate BOD). From this we can express the change in the stream deficit with time as

$$\frac{dD}{dt} = k_1 L - k_2 D \qquad \qquad \textbf{8.17}$$

where

t = time [days]

L = ultimate stream BOD [mg/L]

k_1 = deoxygenation constant [day^{-1}]

k_2 = reaeration constant [day^{-1}]

However, since Equation 8.17 is an ordinary differential equation in three variables, we must eliminate one variable to solve it. To do this, we substitute for L using Equation 8.15, thus obtaining an ordinary linear differential equation in two variables, D and t.

$$\frac{dD}{dt} = k_1 L_0 e^{-kt} - k_2 D \qquad \qquad \textbf{8.18}$$

This can be solved by separation of variables and integration, or by using an integrating factor. The boundary condition is $t = 0$ at $D = D_0$. This yields the DO sag equation

$$D = \frac{k_1 L_0}{k_2 - k_1}(e^{-k_1 t} - e^{-k_2 t}) + D_0 e^{-k_2 t}$$ **8.19**

where

> D = stream deficit at time t [mg/L]
>
> D_0 = initial oxygen deficit (at $t = 0$) [mg/L]

It is important to note two points about this relationship. First, the equation is valid only if the effluent discharged into the stream mixes completely with the stream in a short time (compared to the time of interest). Second, this equation yields the oxygen deficit, not the dissolved oxygen. To obtain the dissolved oxygen, DO, we must use Equation 8.16.

The most stress is placed on the aquatic life in a stream when the DO is at a minimum, or the deficit, D, is a maximum. This occurs when $dD/dt = 0$. We can obtain the time at which the deficit is a maximum by taking the derivative of the DO sag equation with respect to t and setting it equal to zero, then solving for t. This yields

$$t_{crit} = \frac{1}{k_2 - k_1} \ln\left[\frac{k_2}{k_1}\left(1 - \frac{D_0(k_2 - k_1)}{k_1 L_0}\right)\right]$$ **8.20**

where

> t_{crit} = time at which maximum deficit (minimum DO) occurs [days]

The deoxygenation rate for the Streeter Phelps equation can be determined by a series of experiments similar to the BOD analysis. The reaeration rate can be determined empirically from a knowledge of the stream velocity, slope, and depth. Or, given sufficient data, the coefficients can be treated as fitting parameters and found by trial and error or by regression techniques. In this process the stream is sampled just up from the discharge point, the discharge is sampled, and the stream is sampled at multiple points downstream. The DO and BOD_u at various points along the stream are then used to "calibrate" the DO sag equation. The preferred time to take the samples is during the late summer after a continued dry period. The low flows usually occurring then reduce the amount of water the stream has for dilution. High temperatures increase the microbial activity, which consumes the oxygen and decreases the solubility of oxygen in the water. Such surveys are sometimes done at night to avoid the effect of algal oxygen production. This method of sampling better approximates the stream at its worst possible conditions—that is, low flow, no algal oxygen production, and high water temperatures. With the stream k rates determined, we can model the stream for different conditions of flow, discharge strength, and dissolved oxygen levels.

CASE STUDY

The West Fork of the White River, Indiana

The West Fork of the White River receives treated effluent from Southport Wastewater Treatment Plant, one in a network of facilities serving the city of Indianapolis. During the low-flow conditions of summer months, the effluent comprises approximately 75% of the stream flow downstream from the treatment plant, and another wastewater plant discharges into the river upstream from this facility. Until early 1983 the river measured far below the required minimum of 4 mg/L of dissolved oxygen in summer. Then utilities officials brought advanced wastewater treatment processes on line at both plants. This significantly decreased the BOD_u loading to the stream, resulting in marked improvement of stream quality. The discharge conditions for both years are shown in Figure 8.5. An additional schematic for the plant appears in Figure 8.6.

Estimate the minimum dissolved oxygen expected in the stream for both the summer of 1982 and again after advanced treatment was implemented in the summer of 1983. Also, compare the actual dissolved oxygen data [9] to the Streeter Phelps model predictions.

The plant flow is approximately 575,000 m^3/day and the low stream flow is approximately 195,000 m^3/day just up from the plant's discharge point. The stream velocity below the treatment plant is approximately 18.1 km/day. The levels of dissolved oxygen for the summer of 1982 and 1983 are shown in Table 8.5. Previous studies have determined that the stream deoxygenation coefficient is 0.4/day (base e) and that the reaeration rate is 2.0/day (base e).

1982—Before Plant Improvements. Using a mass balance, determine initial conditions in the stream immediately after mixing (just downstream of the discharge point). Next, find the minimum dissolved oxygen in the stream due to this discharge and the time at which this occurs. Finally, determine how far downstream from the discharge point this minimum DO occurs.

Note: *The subscript (o) will be used to denote initial conditions after mixing, the subscript (u) to denote upstream conditions, and the subscript (w) to denote the waste discharge.*

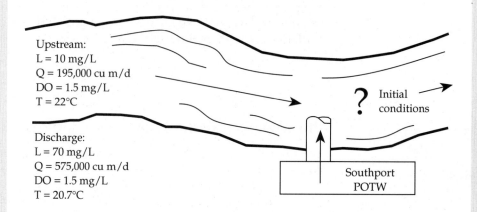

1982 Conditions, Before Improvements

Upstream:
L = 10 mg/L
Q = 195,000 cu m/d
DO = 1.5 mg/L
T = 22°C

? Initial conditions

Discharge:
L = 70 mg/L
Q = 575,000 cu m/d
DO = 1.5 mg/L
T = 20.7°C

Southport
POTW

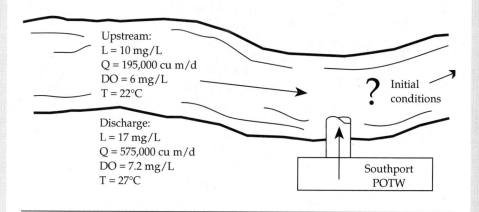

1983 Conditions, After Improvements

Upstream:
L = 10 mg/L
Q = 195,000 cu m/d
DO = 6 mg/L
T = 22°C

? Initial conditions

Discharge:
L = 17 mg/L
Q = 575,000 cu m/d
DO = 7.2 mg/L
T = 27°C

Southport
POTW

FIGURE 8.5   Stream conditions before and after improvements to the South-port POTW.

We must first find the stream temperature, T_o (by using an energy balance), the flow rate, Q_o, the dissolved oxygen, DO_o, and ultimate biochemical oxygen demand, L_o (using mass balances).

$$Q_o = Q_u + Q_w = 195,000 \, \text{m}^3/\text{day} + 575,000 \, \text{m}^3/\text{day}$$

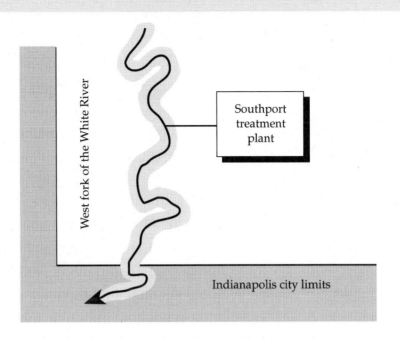

FIGURE 8.6   Southport Treatment Plant, Indianapolis, Indiana.

$$Q_o = 770{,}000 \, \text{m}^3/\text{day}$$

$$T_o Q_o = T_u Q_u + T_w Q_w$$

$$T_o = \frac{(295 \, \text{K})(195{,}000 \, \text{m}^3/\text{day}) + (294 \, \text{K})(575{,}000 \, \text{m}^3/\text{day})}{770{,}000 \, \text{m}^3/\text{day}}$$

$$= 294 \, \text{K} = 21°\text{C}$$

$$L_o Q_o = L_u Q_u + L_w Q_w$$

$$L_0 = \frac{(10 \, \text{mg/L})(195{,}000 \, \text{m}^3/\text{day}) + (70 \, \text{mg/L})(575{,}000 \, \text{m}^3/\text{day})}{770{,}000 \, \text{m}^3/\text{day}}$$

$$= 54.8 \, \text{mg/L}$$

$$\text{DO}_o Q_o = \text{DO}_u Q_u + \text{DO}_w Q_w$$

$$\text{DO}_o = \frac{(1.5 \, \text{mg/L})(195{,}000 \, \text{m}^3/\text{day}) + (2.5 \, \text{mg/L})(575{,}000 \, \text{m}^3/\text{day})}{770{,}000 \, \text{m}^3/\text{day}}$$

$$= 2.2 \, \text{mg/L}$$

TABLE 8.5   Dissolved Oxygen in the West Fork of the White River, Indiana

DISTANCE DOWNSTREAM, km	TIME, DAYS	1982 DISSOLVED OXYGEN, mg/L	1983 DISSOLVED OXYGEN, mg/L
Before outfall	−0	1.5	5.0
0	0	2.5	6.9
2.5	0.14	1.5	6.8
5.2	0.29	0.7	6.3
8.0	0.44	0.5	6.2
10.9	0.60	1.1	6.3
15.2	0.84	1.2	6.4
16.8	0.93	0.9	6.4
19.5	1.08	1.2	6.2
21.4	1.18	1.3	5.9
25.7	1.42	1.2	6.0
27.7	1.53	1.2	6.1
30.0	1.66	1.6	6.2
34.2	1.89	2.1	6.1
35.5	1.96	2.4	6.2

With this in mind, we must now determine the saturation dissolved oxygen level using Appendix B. For a temperature of 21°C, the DO_{sat} is 8.9 mg/L. Knowing this, we can now calculate the initial stream deficit using Equation 8.40:

$$D_o = DO_{sat} - DO_{act} = 8.9\,mg/L - 2.2\,mg/L$$
$$= 6.7\,mg/L$$

We now have the information necessary to calculate the time , t_{crit}, at which the minimum DO (and maximum deficit) occurs.

$$t_{\text{crit}} = \frac{1}{k_2 - k_1} \ln \left[\frac{k_2}{k_1} \left(1 - \frac{D_o(k_2 - k_1)}{k_1 L_o} \right) \right]$$

$$= \frac{1}{2.0/\text{day} - 0.4/\text{day}}$$

$$\times \ln \left[\frac{2.0/\text{day}}{0.4/\text{day}} \left(1 - \frac{(6.7\,\text{mg/L})(2.0/\text{day} - 0.4/\text{day})}{(0.4/\text{day})(54.8\,\text{mg/L})} \right) \right]$$

$$= 0.58\,\text{day}$$

$$X_{\text{crit}} = (0.58\,\text{day})(18.1\,\text{km/day})$$

$$= 10.4\,\text{km}$$

The maximum deficit thus occurs 0.58 day after discharge, or 10.4 km downstream. We can now use Equation 8.19 to determine the maximum deficit in the stream:

$$D = \frac{k_1 L_o}{k_2 - k_1}(e^{-k_1 t} - e^{-k_2 t}) + D_o e^{-k_2 t}$$

$$D_{\text{crit}} = \frac{(0.4/\text{day})(54.8\,\text{mg/L})}{(2.0/\text{day} - 0.4/\text{day})}(e^{-(0.4/\text{day})(0.58\text{day})} - e^{-(2.0/\text{day})(0.58\text{day})})$$

$$+ (6.7\,\text{mg/L})e^{-(2.0/\text{day})(0.58\text{day})}$$

$$= 8.7\,\text{mg/L}$$

$$\text{DO}_{\text{crit}} = \text{DO}_{\text{sat}} - D_{\text{crit}} = 9.0\,\text{mg/L} - 8.7\,\text{mg/L}$$

$$= 0.3\,\text{mg/L}$$

Thus, the minimum dissolved oxygen level is only 0.3 mg/L, far below that necessary for fish to survive.

1983—After Plant Improvements. Calculating the t_{crit} and the minimum DO for the summer of 1983 is left to the student. In Problem 12 you are asked to calculate the t_{crit} and the DO_{min}. In Problem 13 you are asked to plot the curve and compare it to the actual data in the case study. The actual data and the DO sag curves for both summers are presented in Figure 8.7. During the test period in 1982, it can be seen that the West Fork of the White River did not attain the required 4 mg/L of dissolved oxygen in the first 45 km 2.5 days after discharge. The model and data for the summer of 1983 are similarly presented, and from them it can be seen that the stream DO was much improved in 1983—and that the dissolved oxygen was above the required 4 mg/L. Score one for environmental engineers!

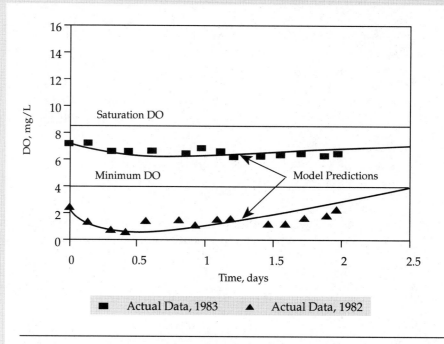

FIGURE 8.7   DO Sag Curve for Case Study of the West Fork of the White River. (Actual stream data are from Reference 9.)

Review Questions

1. Determine the total, carbonate, and noncarbonate hardness for the water described in the following table.

CATION	CONC., mg/L	ANION	CONC., mg/L
Nfra⁺	20	HCO_3^-	135
Ca^{2+}	15	NO_3^-	3
Mg^{2+}	10	Cl^-	6
Sr^{2+}	2	SO_4^{2-}	10
K^+	10	CO_3^{2-}	0

2. Define the following terms (do not write equations; describe the terms in your own words).

   a. Water pollution

   b. Point source pollution

   c. Nonpoint source pollution

   d. Reaeration

   e. Deoxygenation

   f. Alkalinity

   g. Hardness

   h. Total solids

   i. Volatile solids

   j. Fixed solids

   k. Total suspended solids

   l. Volatile suspended solids

   m. Fixed suspended solids

   n. Biochemical oxygen demand

   o. Chemical oxygen demand

   p. Turbidity

   q. Nutrient

   r. Dissolved oxygen sag curve

   s. Oxygen deficit

3. Why is the COD of a wastewater always higher than the BOD_5?

4. If the total suspended solids (TSS) of a water is 220 mg/L and the fixed suspended solids (FSS) is 85 mg/L, what is the volatile suspended solids (VSS)?

5. Fill in the missing values in Figure 8.8 (next page)

6. Using Equation 8.15, calculate how much oxygen would be required to oxidize 15 mg/L of ammonium to nitrate.

7. Choose a local industry and describe what types of wastes would be present in its wastewater. You may either talk with someone at the plant or use a reference book to determine the wastes.

8. Fill in the missing values in Figure 8.9 (next page).

9. Fill in the missing values in Figure 8.10 (page 257). Be sure to note that fixed and volatile solids, not dissolved and suspended, are shown.

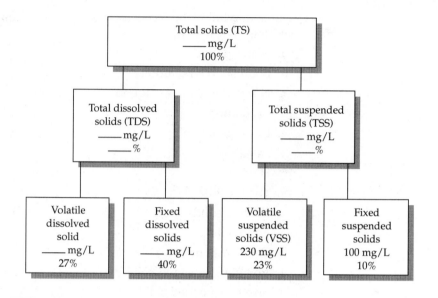

FIGURE 8.8 Total solids components.

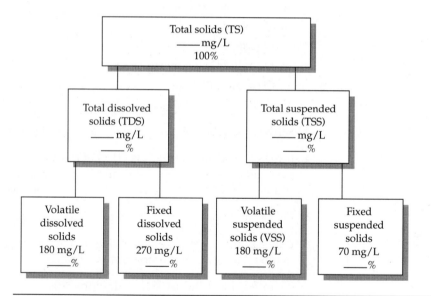

FIGURE 8.9 Total solids components.

10. The equation for the oxidation of ammonium to nitrate is shown in
    Equation 8.15. Write a balanced equation for the oxidation of ammonia
    to nitrate. How much oxygen is required to oxidize 1 mg of ammonia
    (expressed as nitrogen)?

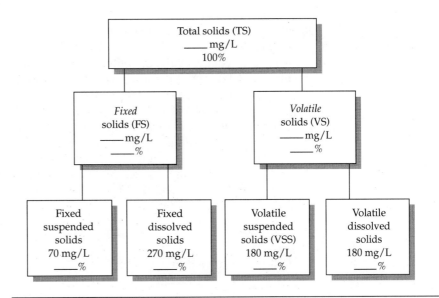

FIGURE 8.10 Total solids components

11. Under some conditions (i.e., high ultimate oxygen demand, low reaeration rate, and high deoxygenation rate) the Streeter Phelps equation will predict a DO of less than zero at t_{crit}. What is the physical significance of a DO less than zero?

12. Calculate the t_{crit} and minimum DO for 1983 (after improvements at the wastewater treatment plant) for the case study on the West Fork of the White River. (You may use the same flow rates, reaeration and deoxygenation rates, and other information.)

13. Using the data from Problem 12, write a spreadsheet program using Borland's Quattro®, Lotus 1-2-3®, or a similar program, which will draw the DO sag curve for any wastewater discharge or streamflow conditions.

14. In the White River study, the BOD_u of the wastewater is typical for effluent from a treatment plant in the 1990s. Assume the plant had only partial treatment in the 1960s, resulting in a BOD_u of 240 mg/L. Find the minimum DO for the stream (and its location) if the plant discharged only 2000 m^3/day during this period. You may use the same stream flow rate, reaeration and deoxygenation rates, temperatures, and other data. Only the BOD_u and flow rate from the wastewater treatment plant changes.

15. Using the numbers from Problem 14, write a spreadsheet program that will draw the DO sag curve for any wastewater discharge or streamflow conditions.

16. Streams are usually stressed the most during late summer and early fall. Water flows are typically at the yearly minimum, and the temperatures are usually high. In contrast, during the spring, water flows are high and temperatures are low. In this problem you will compare a stream during the late fall and early spring. The accompanying figure shows the details of the stream and discharger during both periods. Thedeoxygenation rate is 4.0/day and the reaeration rate is 7.5/day for both cases. Calculate the time of minimum DO (t_{crit}), the maximum deficit, and the minimum DO in both situations. You may do these calculations by hand or using a spreadsheet program. If you use a spreadsheet, also graph the DO sag curve and the saturation DO level.

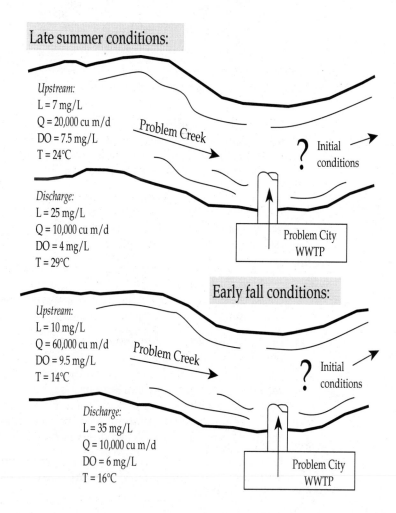

Late summer conditions:

Upstream:
L = 7 mg/L
Q = 20,000 cu m/d
DO = 7.5 mg/L
T = 24°C

Problem Creek

? Initial conditions

Discharge:
L = 25 mg/L
Q = 10,000 cu m/d
DO = 4 mg/L
T = 29°C

Problem City
WWTP

Early fall conditions:

Upstream:
L = 10 mg/L
Q = 60,000 cu m/d
DO = 9.5 mg/L
T = 14°C

Problem Creek

? Initial conditions

Discharge:
L = 35 mg/L
Q = 10,000 cu m/d
DO = 6 mg/L
T = 16°C

Problem City
WWTP

17. List several of your activities that increase the total dissolved solids of the water you use.

18. From the information given in the following table, determine the TS, TDS, and TSS of the water.

COMPOUND	CONCENTRATION mg/L	DISSOLVES?	VOLATILIZES OR BURNS AT 550°C?
Sodium Chloride	45	Yes	No
Calcium sulfate	30	Yes	No
Clay	100	No	No
Copper chloride	10	Yes	No
Acetic acid	20	Yes	Yes
Coffee grounds	25	No	Yes

19. Use the data from Protein 18 to determine the TS, TSS, VSS, and FSS.

20. Use the data from Problem 18 to create a figure similar to that in Problem 5.

21. From the information given in the following table, determine the TS, TDS, and TSS of the water.

COMPOUND	CONCENTRATION, mg/L	DISSOLVES?	VOLATILIZES OR BURNS AT 550 C?
Sodium chloride	60	Yes	No
Calcium sulfate	25	Yes	No
Sodium nitrate	10	Yes	No
Clay	10	No	No
Copper chloride	5	Yes	No
Acetic acid	15	Yes	Yes
Coffee grounds	20	No	Yes

22. Use the data from Problem 21 to determine the TS, TSS, VSS, and FSS.

23. Use the data from Problem 21 to create a figure similar to that in Problem 5.

References

1. *Progress in Ground-Water Protection and Restoration,* EPA 440/6-90-001 (Washington, DC: U.S. Environmental Protection Agency, February 1990).

2. *National Water Summary 1986–Hydrologic Events and Ground-Water Summary,* U.S. Geological Survey Water-Supply Paper 2325 (Washington: U.S. Government Printing Office, 1988).

3. *Ground-Water Protection Strategy,* Office of Ground-Water Protection (Washington: EPA, August 1984).

4. *Standard Methods for the Examination of Water and Wastewater,* 17th ed. (Alexandria, VA: Water Pollution Control Federation, 1989).

5. *Methods for Chemical Analysis of Water and Wastes,* EPA 600/4-79-020 as revised (Washington: EPA, March 1979).

6. V. N. Novotny and G. Bendoricchio, "Linking Nonpoint Pollution and Deterioration," *Water Environment and Technology,* vol. 1, no. 3 (November 1989), p. 400.

7 *1988 Needs Survey Report to Congress, Assessment of Needed Publicly Owned Wastewater Treatment Facilities in the United States,* EPA 430/09-89-001 (Washington: EPA, February 1989).

8. A. L. Alm, "Regulatory Focus: Nonpoint Sources of Water Pollution," *Environmental Science and Technology,* vol. 24, no. 7 (1990), p. 967.

9. M. S. Kennedy and J. M. Bell, "The Effects of Advanced Wastewater Treatment on River Water Quality," *Journal of the Water Pollution Control Federation,* vol. 58, no. 12 (1986), pp. 1138–44.

Additional References

W. W. Eckenfelder, *Industrial Water Pollution Control,* 2nd ed. (New York: McGraw-Hill, 1989). A summary of wastewater treatement methods applicable to industrial wastes.

N. L. Nemerow, *Industrial Water Pollution: Origins, Characteristics, and Treatment* (Reading, MA: Addison-Wesley, 1978). An excellent review of many industries and their waste characteristics.

N. L. Nemerow, *Scientific Stream Pollution Analysis* (New York: McGraw-Hill, 1974). A good presentation of stream pollution, its effects, and modeling.

C. N. Sawyer and P. L. McCarty, *Chemistry for Environmental Engineering*, 3rd ed. (New York: McGraw-Hill, 1978). Although several years old, this book provides good descriptions of the theory and practice of routine analytical procedures used in environmental engineering.

R. V. Thomann and J. A. Mueller, *Principles of Surface Water Quality Modeling and Control*, (New York: Harper & Row, 1987). Describes surfacewater quality modeling and the effects of pollution control.

9  Water Treatment

"Water contributes
much to health"

Hippocrates (460 to 354 B.C.)

The new carbon adsorption facility of the
Cincinnati Water Works. Photograph
courtesy of the Cincinnati Water Works.

INTRODUCTION

A safe supply of drinking water may be the most important difference between a developed nation and a developing nation. Waterborne diseases that have not been a problem in the United States or Western Europe for decades continue to kill millions of people, mostly children and the aged, in less-developed countries throughout the world.

History tells us that some four millennia ago humans had some level of water purification, if only for aesthetic reasons. There is evidence, for example, that in India around 2000 B.C., water was to be heated, boiled, or filtered to remove impurities. And a drawing from an Egyptian tomb dating back to 1450 B.C. depicts people siphoning liquid, presumed to be water or wine, from clarifying or sedimentation canisters [1]. So it appears that the need to purify drinking water has been recognized for a long time.

This chapter includes a discussion of current federal regulations affecting drinking water quality, the characteristics of raw water from both surface and ground supplies, the processes employed in water treatment, typical treatment steps for both surfacewater and groundwater, and disposal of water treatment sludges.

REGULATIONS

The quality of water for human consumption has been of interest since links between water and health were first discovered. Thousands, perhaps millions, of people died from waterborne disease before definitive links between disease and water were established. In fact, London, England passed a law in 1852 requiring that all water for human consumption be filtered—10 years prior to Louis Pasteur's development of the germ theory [2].

Until the past 20 or so years, the primary focus in water quality management has been on microbiological parameters, since many diseases are known to be transmitted by water. In the United States the concentration has shifted to trace contaminants. However, much of the less-developed world has yet to attain even microbiologically pure water.

Early Controls

The first water quality standards in the United States were adopted by the U.S. Public Health Service in 1914. The American Water Works Association, a trade and professional organization, comprising of public water utilities, water treatment professionals, and educators, played a significant role in establishing these criteria. The standards applied only to interstate water, and they included bacteriological considerations only. A plate count* was

---

* A plate count is the number of bacterial colonies growing on a standard agar (or other growth medium) petri dish inoculated with a prescribed volume of sample and incubated for a

not to exceed 100 bacterial colonies per mL, and no more than one in five 10-mL portions of supply water was to test positive for *B. coli*.

The procedures to be followed were those listed in *Standard Methods of Water Analysis*, published by the American Public Health Association in 1912. This was an early version of *Standard Methods for the Examination of Water and Wastewaters* [3], which is currently in use. Because of persistent disagreement on possible specifics [2], no requirements for chemical quality were involved in these standards.

The next changes in regulations occurred in 1925, when higher standards for bacteriological quality were enacted and chemical quality regulations were adopted for the first time. The rules identified maximum permissible levels of lead, zinc, and copper, and included recommended limits on several other inorganic substances. The quality of source water was also to be considered. And, for the first time, there were recommended standards for aesthetic parameters such as color, taste, and odor.

Further changes instituted in 1942 and 1946 resulted in increasingly stricter bacteriological and chemical criteria. In 1942 the first requirements for aesthetic quality were established, and other chemicals were added to the regulations. Legislation passed in 1962 added a number of pesticides to the list of regulated chemicals, as well as an alternative method of evaluating the bacteriological quality of a water—the membrane filter test, a procedure that is quicker and easier to use than the older approaches employing broth and test tubes.

All this represented considerable progress at the federal level, yet water quality standards still applied to interstate carriers only. (Few water supplies are distributed across state lines!) Fortunately, most states were adopting and enforcing the standards on their own.

Current Regulations

It was recognized early that water quality involves both aesthetic and health concerns. Thus, primary drinking water standards relate directly to health, and secondary standards relate more to the appearance and non-consumptive uses of water. Within the primary standards, **maximum contaminant levels**, or MCLs, have been established for a wide variety of organics and inorganics. **Primary standards** are mandatory. **Secondary standards** are suggested upper limits for non–health related parameters.

In 1974 the U.S. Congress passed the Safe Drinking Water Act (SDWA). This gave enforcement of drinking water quality to the U.S. Environmental Protection Agency. It also greatly expanded coverage of the federal regulations. Any water supply that served more than 25 people daily for 60 or more days was required to comply with and meet the SDWA standards. This included any water supply with more than 15 service connections as well. Table 9.1 summarizes regulatory development to this present time.

TABLE 9.1 Chronological Listing of Drinking Water Standards in the United States

ACT	AGENCY	YEAR	DESCRIPTION
U.S. Treasury Standards	PHS*	1914	Bacterial plate count and *B. coli* limit of MPN $\leqslant$ 2; no required testing frequency
	PHS	1925	*B. Coli* limit of MPN $\leqslant$ 1; maximum limits established for copper, lead, and zinc; recommended limits on some other inorganics; no required testing frequency
	PHS	1942	Entire coliform group limited to MPN $\leqslant$ 1; maximum limits established on additional inorganics; required aesthetic parameters established; required frequency of testing based on population served
	PHS	1962	Required limits on color and turbidity, maximum limits added for several pesticides; the membrane filter technique for determination of coliform bacteria was added as an acceptable alternate method
Safe Drinking Water Act	EPA	1974	Specific primary and secondary standards for drinking water, applicable to all water supplies serving more than 25 people or having 15 or more service connections

*U.S. Public Health Service.

Acting at the direction of Congress, the EPA promulgated the National Interim Primary Drinking Water Standards in 1975. These are a more encompassing set of standards than were previously in effect. And, as more information about hazardous constituents has become available, the NIPDWR has been modified. In 1979 the EPA added an MCL for total trihalomethanes. Also in 1979 the agency wrote optional secondary standards. In 1980 it established standards to prevent corrosion that might allow lead or other metals to be dissolved from distribution or consumer piping into water intended for human consumption. In 1991, additional corrosion standards relating to lead and copper were added. Thus, the NIPDWR is an ever-changing document, reflecting current knowledge about hazardous substances contained in drinking water.

SOURCE CHARACTERISTICS

The two major sources of public drinking water are surfacewater and groundwater. Treatment of these is usually quite different because the quality and characteristics of the waters are not the same. **Surfacewater** sources include streams and rivers, natural lakes, and constructed lakes. These are exposed to plant and animal life and to human influences from land. In addition, they include fish, aquatic plants, and microorganisms. Raw (untreated) surface water thus contains a wide variety of microorganisms and natural organics.

In contrast, **groundwater** is usually free of significant levels of organics (unless contaminated by human activities). Groundwater also has very low levels of miocrobial contamination compared to surfacewater. However, it can and often does contain significant levels of dissolved inorganics, particularly carbonates, iron, and manganese. Dissolved gases such as methane and hydrogen sulfide may also be a problem.

Many surface- and groundwaters do not meet the norm. Some surface water is rather hard, requiring softening. This condition comes about in two ways. Natural springs can produce hard water, or soft water may pass through limestone before being collected for use. Groundwater may also be contaminated with organics or microorganisms from human activity.

TREATMENT PROCESSES

The purpose of water treatment is to produce a safe, aesthetically pleasing water. This requires that the water be free of harmful chemicals and microbes, as well as have an acceptable taste and odor. The treatment processes necessary to attain this quality differ depending on the contaminants present. We will discuss common processes and their applications.

Gas Transfer

Gas transfer, or **gas stripping** is used to remove dissolved gases and add oxygen that will oxidize iron and manganese. Dissolved gases such as carbon dioxide, hydrogen sulfide, and methane are removed because the partial pressure of these gases is very low in air (See Henry's law, Chapter 4.) Sufficient transfer will reduce the concentration of these gases to near the equilibrium values for air. The added oxygen will also react with soluble reduced metals such as ferrous iron [Fe^{2+}] or manganous manganese [Mn^{2+}], oxidizing the metals to their less soluble forms. However, for high concentrations of these metals, oxidants such as chlorine or ozone may be required. The oxidation process is slow.

Theory of Gas Transfer. Aeration is a common term for gas transfer, although in reality aeration refers only to the addition of air to the water, not gas removal. Recall from Chapter 4 that the amount of a gas that is soluble in a liquid is proportional to the partial pressure the gas exerts on the liquid, or

$$C_A = K_H P_A \qquad\qquad 9.1$$

where

C_A = concentration of species A at equilibrium [mol/L or mg/L]

K_H = Henry's law constant for species A [mol/L·atm or mg/L·atm]

P_A = partial pressure that gas A exerts on the liquid [atm]

For a typical water system the change in concentration of the gas with time can be expressed as

$$\frac{dC_A}{dt} = -K_L a(C_s - C) \qquad\qquad 9.2$$

where

$K_L a$ = gas transfer coefficient* [time^{-1}]

C = concentration at time t [mol/L or mg/L]

C_s = saturation concentration from Henry's law.

This equation can be separated and integrated from $C = C_0$ at $t = 0$ to $C = C_t$ at $t = t$, yielding

$$\ln\left(\frac{C_s - C_t}{C_s - C_o}\right) = -K_L at \qquad\qquad 9.3$$

Or, in exponential form, the equation is

$$\frac{C_s - C_t}{C_s - C_0} = e^{-K_L at} \qquad\qquad 9.4$$

An **aeration** system can be evaluated to determine $K_L a$ by starting the system with a low dissolved oxygen concentration (below saturation) and then monitoring the concentration over time. If the term $-\ln[(C_s - C_t)/(C_s - C_0)]$ is plotted versus time, t, the slope of the line is $K_L a$. A more

---

* $K_L a$ is actually the gas transfer coefficient K_L times the specific surface area, a, where a is the bubble surface area divided by the bubble volume. It is quite difficult to determine the two parameters separately. Since they are normally used together a separate determination is not necessary.

effective gas transfer system will have a higher $K_L a$ value; a less effective system, a lower value. The following example demonstrates the determination of $K_L a$.

EXAMPLE 9.1	DETERMINATION OF $K_L a$ FOR AN AERATION SYSTEM

An aeration system has been monitored to determine $K_L a$. The dissolved oxygen versus time is shown in the following table. The saturation dissolved oxygen concentration is 9.0 mg/L. Use the data to determine $K_L a$.

t, min	0	4	8	12	16	20	24	28
C_t, mg/L	1.2	2.4	3.7	4.3	5	5.6	6	6.6

SOLUTION: To determine $K_L a$, we must plot $-\ln[(C_s - C_t)/(C_s - C_0)]$ versus t. The slope of the line of best fit is the value of $K_L a$. The best fit can be determined visually or by regression using a calculator or spreadsheet. The results of the calculations appear in the following table, along with the regression line data

t min	C_t mg/L	$\dfrac{-\ln(C_s - C_t)}{(C_s - C_0)}$	Regression
0	1.2	0	0
4	2.4	0.163	0.159
8	3.7	0.375	0.319
12	4.3	0.490	0.478
16	5	0.644	0.638
20	5.6	0.799	0.797
24	6	0.916	0.956
28	6.6	1.124	1.116

Also shown is a spreadsheet plot whose slope, which is equal to $K_L a$, is 0.040/min or 2.4/hr.

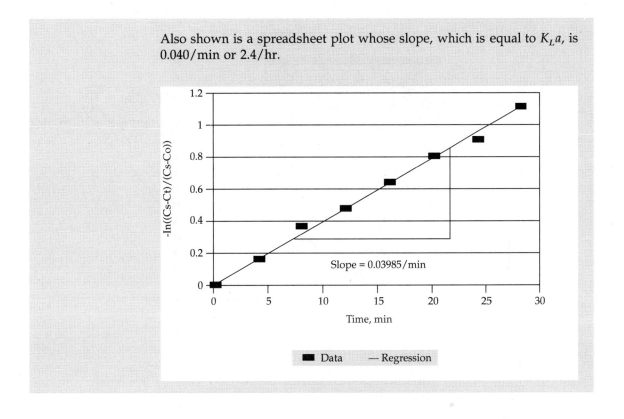

Types of Gas Transfer. Several devices can produce gas transfer. They include tray aerators, diffused aerators, cascade aerators, and spray aerators. Tray aeration, shown in Figure 9.1, is most common. It consists of several layer of trays. The water distributed over the top tray trickles or drips down to the bottom, one tray or layer at a time. Diffused aeration, also depicted in Figure 9.1, is another common gas transfer method. In this approach compressed air is forced through a porous material and into the liquid. Typical diffusers are 2 to 6 inches in diameter. To provide sufficient aeration, many diffusers are usually placed near the bottom of an aeration tank, attached to the pipe supplying the compressed air.

Oxidation. Oxygen added to water through aeration oxidizes reduced species. Where there are significant levels of reduced iron and manganese, most treatment facilities use a stronger (and faster) oxidizing agent such as chlorine. Oxidation converts $[Fe^{2+}]$ to $[Fe^{3+}]$. Ferrous iron $[Fe^{2+}]$ is relatively soluble in water. It also produces a red to reddish-brown discoloration as it is oxidized. It is not, however, a detriment to health. The oxidized form, ferric iron $[Fe^{3+}]$ will react with the hydroxide in the water

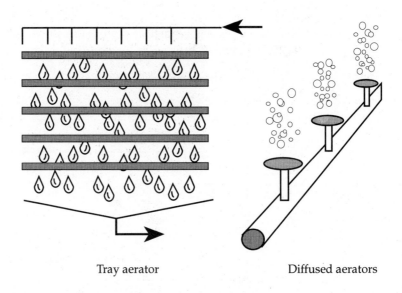

Tray aerator   Diffused aerators

FIGURE 9.1 Gas transfer devices.

to form a reddish-brown ferric hydroxide [Fe(OH)$_3$] precipitate. The reactions involved are

$$4Fe^{2+} + O_2 + 4H^+ \rightarrow 4Fe^{3+} + 2H_2O \qquad \textbf{9.5}$$

$$Fe^{3+} + 3OH^- \rightarrow Fe(OH)_3 \downarrow \qquad \textbf{9.6}$$

Reduced manganese causes a grey-to-black discoloration as it is oxidized, but this, too, is not a health hazard. The reduced manganese [Mn^{2+}] can be oxidized to [Mn^{4+}] and precipitated as manganese dioxide. The precipitates are then removed by sedimentation or filtration.

Coagulation

Large particles can easily be removed by settling. Their mass allows them to settle to the bottom of a tank and then be removed as sludge. However, colloidal particles in water have insufficient mass to overcome the fluid forces on their surfaces. Most naturally occurring particles in water are negatively charged. Since like charges repel, these small particles, or **colloids**, will remain suspended almost indefinitely if their surface charge is not reduced, allowing them to agglomerate. The addition of chemical **coagulants** induces agglomeration. The chemicals reduce colloidal surface charge and form precipitates that enhance the clustering process and sedimentation. There are two major forces acting on colloids. At greater distances (relatively speaking), electrostatic repulsion dominates. (Simply, negative colloids repel other negatively charged colloids.) However, at very short distances, intermolecular, or van der Waals, attraction occurs. This force is large compared to electrostatic forces, but it acts over a very short range. Figure 9.2 shows the two forces and their sum. It can be seen that at greater distances, the repulsive forces dominate, but at shorter distances the attractive forces dominate. Coagulants can be used to reduce the electrostatic repulsive forces, thereby increasing the distance at which attraction occurs.

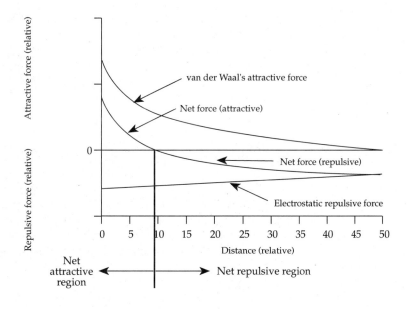

FIGURE 9.2 Attractive and repulsive forces acting on waterborne colloids.

The addition of coagulants containing divalent or trivalent cations can both reduce the negative surface charge and form a precipitate in which to trap additional particles. The reduction in the electrostatic repulsion is shown in Figure 9.3. At the top in this drawing, we see a large electrostatic repulsion due to the high negative surface charge. At the bottom of the drawing, the electrostatic repulsion reduced by the addition of counter-charged ions [Al^{3+}] is depicted. Such charge reduction can be accomplished by several coagulants, including ferric chloride [$FeCl_3$], aluminum sulfate or alum [$Al_2(SO_4)_3$], ferric sulfate, [$Fe_2(SO_4)_3$], or lime [$Ca(OH)_2$]. All four of these will reduce the surface charge on negative colloids. In

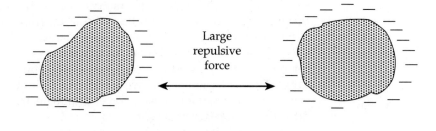

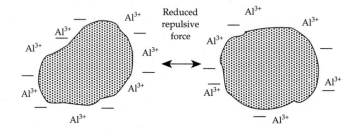

FIGURE 9.3 Reduction of colloidal electrostatic repulsion by addition of trivalent aluminum ions.

addition, each will form a precipitate. The first three precipitate as hydroxides, the latter as a carbonate. (Calcium carbonate precipitation requires that carbonate be present in or added to the water.)

The precipitation reaction for ferric chloride is

$$Fe_2(SO_4)_3 + 6OH^- \rightarrow 2Fe(OH)_3 + \downarrow 3SO_4^{2-} \qquad \textbf{9.7}$$

The precipitation reactions for alum and ferric sulfate are similar. For lime the precipitation reaction is

$$Ca(OH)_2 + CO_3^{2-} \rightarrow CaCO_3 + \downarrow 2OH^- \qquad \textbf{9.8}$$

After addition, a portion of the metallic cations will be attracted to the surface of the negative colloids. Others will precipitate as hydroxides or carbonates, sweeping colloidal particles out of the water. Gentle mixing or flocculation, a process that is discussed next, then causes the destabilized (reduced charge) colloids to cluster.

Another method of enhancing agglomeration is to add organic polymers. These compounds consist of a long carbon chain with active groups such as amine, nitrogen, or sulfate groups along the chain. The chain is long enough to allow active groups to bond to multiple colloids. In Figure 9.4,

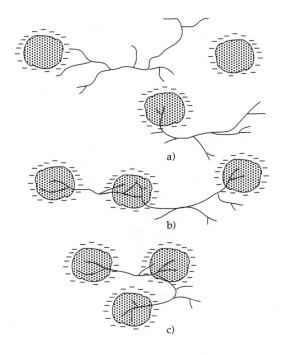

FIGURE 9.4 Agglomeration by organic polymers.

three colloids are initially separated (top), then partially joined by organic polymers (middle). Finally, all three are joined by the two polymer chains (bottom). In actual systems, many more colloids are connected—and are often swept from the water by the rapidly forming precipitates.

Organic polymers, however, are expensive compared to typical inorganic coagulants such as alum or lime. Polymers are used only in very low concentrations to assist the agglomeration process.

Flocculation

Flocculation is gentle mixing to speed the agglomeration of colloidal materials. Coagulation and flocculation are integral parts of the sedimentation process, which is discussed next. Flocculators are usually adjacent to or inside settling tanks. Coagulants are added to the water to enhance agglomeration. The water then enters a small tank or section of a tank in which paddles are turning slowly. Their movement causes the small particles to collide and stick together. (Fast or vigorous mixing would separate combined particles.)

The gentle mixing combined with the reduction in repulsive surface charges allow the particles to join rapidly. This has the effect of increasing particulate diameters and aiding in sedimentation. Recall from Chapter 7, Treatment Processes, that [the settling velocity of a particle is proportional to the square of the diameter.] Doubling the particle diameter increases its settling velocity by a factor of 4. The water then flows into the sedimentation basin where the solids settle to the bottom and are removed.

Sedimentation

The purpose of sedimentation is to remove preexisting solids, as well as the precipitates formed in coagulation and flocculation. Sedimentation immediately follows these processes. In fact, many clarifiers include both coagulation and flocculation equipment. As detailed in Chapter 7, there are four types of sedimentation. In most water treatment clarifiers, the predominant type is flocculant. However, in some systems with solids recycling, the concentration of solids can reach high enough levels to create hindered settling in the lower regions of clarifiers.

A schematic of a typical circular sedimentation tank appears in Figure 9.5. Note that the chemical addition for coagulation occurs near the center. Mixing is provided by a motorized impeller also located at the center of the tank. The sludge solids formed from the particles present in the raw water, as well as the precipitates generated in coagulation and flocculation sink to the bottom of the tank and are withdrawn there. A circular clarifier is pictured in Figure 9.6.

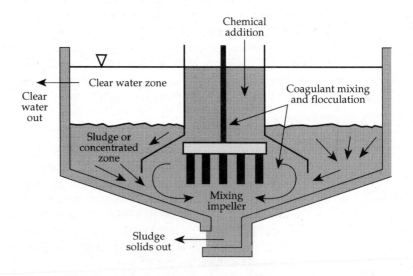

FIGURE 9.5 Cross section of a typical circular clarifier.

FIGURE 9.6 Circular water treatment clarifier showing the centerwell and a portion of the rake and skimmer. Photograph courtesy of Envirex, Inc., Waukesna, Wisconsin.

Figure 9.7 shows a typical rectangular sedimentation basin (or clarifier) with chemical addition and flocculation. Note that the chemical addition and rapid mixing (coagulation) occurs first. This is immediately followed by gentle mixing (flocculation), and then sedimentation.

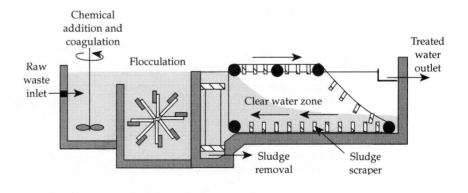

FIGURE 9.7 Cross section of a typical rectangular clarifier.

Water Softening

Water **hardness** is caused by multivalent cations, usually calcium and magnesium. Other ions that may cause hardness are strontium, manganese, and iron. The anions associated with these cations are usually carbonate and sulfate, with chloride and nitrate occurring to a lesser extent in natural waters. However, hardness is independent of the associated anions. The divalent cations in water react with soaps, reducing their effectiveness. The term hard or hardness originated because waters containing such cations were "hard" to use for laundry. Chemists originally used a standard soap solution to determine water hardness. With modern instrumentation it is relatively easy to measure the individual cation concentrations for the determination of hardness, a quantity usually expressed in mg/L calcium carbonate. This notation probably originated because calcium and carbonate are often the dominant hardness cation and associated anion.

A second problem associated with hard water is the precipitation of calcium carbonate onto the walls of water heaters. Although many substances are more soluble at elevated temperatures, calcium carbonate becomes less soluble as the temperature rises. Thus, a water heater actually precipitates calcium carbonate. Where excessive hardness occurs and the cations are not removed, water heaters lose efficiency and wear out sooner by several years.

There are no absolute standards for hardness, since it is not a health consideration. The typical calcium and magnesium cations and associated anions are not known to be harmful to humans. However, concentrations above 150 to 200 mg/L $CaCO_3$ are usually detectable by taste. Actually, hardness is a desirable property if the concentration is not too high. When hardness is below 50 mg/L $CaCO_3$, it is difficult to remove the soap after a bath or shower. Most public water supplies remove hardness in excess of about 150 mg/L $CaCO_3$. Some will add hardness if it is below 25 to 50 mg/L $CaCO_3$. Table 9.2 lists hardness ranges with typical descriptive terms.

TABLE 9.2   Hardness Concentration Ranges

DESCRIPTION OF HARDNESS	HARDNESS, AS mg/L $CaCO_3$
Soft	<50
Medium	50–149
Hard	150–300
Extremely hard	>300

Hardness can be removed by chemical treatment to precipitate the divalent cations, or by ion exchange using sodium ions to replace the divalent ions. (See Chapter 7, Treatment Processes.)

Precipitation of Hardness. Hardness can be removed by increasing the pH of water, thus causing precipitation of the divalent ions as carbonates and/or hydroxides. Bicarbonate is the dominant carbonate form at normal pH ranges, but carbonate becomes dominant as the pH approaches and exceeds 10. The most often used chemicals are lime $[Ca(OH)_2]$ and soda ash $[Na_2CO_3]$. It may seem odd to use a calcium compound to remove calcium, but lime is quite effective. As the pH is raised by the lime, the calcium added as lime as well as the preexisting calcium are precipitated as calcium carbonate. An alternative chemical is sodium hydroxide. It is usually more costly, but it is easy to handle and produces less carbonate sludge.

The precipitation reaction for calcium carbonate is

$$Ca(HCO_3)_2 + Ca(OH)_2 \rightarrow 2CaCO_3\downarrow + 2H_2O \qquad \textbf{9.9}$$

The magnesium is precipitated as magnesium hydroxide. The first step is the formation of magnesium carbonate, which is soluble:

$$Mg(HCO_3)_2 + Ca(OH)_2 \rightarrow CaCO_3\downarrow + MgCO_3 + 2H_2O \qquad \textbf{9.10}$$

This is followed by the precipitation of the magnesium with the formation of magnesium hydroxide, which is insoluble:

$$MgCO_3 + Ca(OH)_2 \rightarrow Mg(OH)_2\downarrow + CaCO_3\downarrow \qquad \textbf{9.11}$$

Note that 1 mol of calcium hardness requires 1 mol of lime for removal. But 1 mol of magnesium hardness requires 2 mol of lime.

Ion Exchange for Hardness Removal. Ion exchange, presented in Chapter 4, uses a synthetic resin to trade a monovalent cation, typically sodium or hydronium $[H^+]$, for the divalent cations. Sodium is not toxic or hazardous to humans except for its tendency to elevate blood pressure. A typical human diet contains several thousand mg of sodium per day. A restricted diet for high blood pressure typically restricts total sodium intake to 2000 to 3000 mg Na^+ per day. If 100 mg/L $CaCO_3$ of hardness is removed from a water supply by sodium ion exchange, the added sodium from the process would be 46 mg/L Na^+. Thus, for most water supplies and most people, the added sodium resulting from softening should not be of concern.

The use of $[H^+]$ for ion exchange will result in a decrease in pH during the exchange process. This is not important where there is sufficient alkalinity to neutralize the acidity.

Most ion exchange processes use sodium as the exchange ion and salt (sodium chloride) as the regenerant (material to recharge the exchange resin). Salt is cheaper and safer to use than hydrochloric acid, which is the regenerant for $[H^+]$.

Filtration

Many particles in water are too small to remove by sedimentation alone. Filtration removes microorganisms and suspended matter from water not receiving sedimentation treatment, or it eliminates precipitated particles and flocs remaining after sedimentation. Filtration was actually developed prior to the discovery of the germ theory by Louis Pasteur in France. The first sand filter beds were constructed in the early 1800s in Great Britain.

Particle removal is accomplished only when the particles make physical contact with the surface of the filter medium. This may be the result of several mechanisms, as shown in Figure 9.8. Larger particles may be removed by straining. That is, the particle is larger than the pore, so it is trapped. Particles may also be removed by sedimentation as they progress through the filter. Others may be intercepted by and adhere to the surface of the medium due to inertia. Filtration efficiency is greatly increased by destabilization or coagulation of the particles prior to filtration. This reduction in the particle charge increases particle agglomeration and reduces the forces necessary to trap particles within the filter.

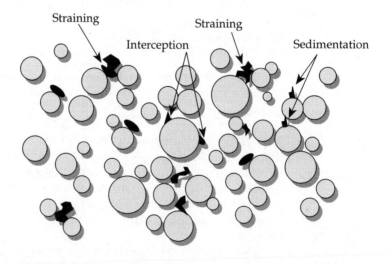

FIGURE 9.8 Filtration removal mechanisms.

Slow Sand Filtration. The early filtration units developed in Great Britain used a process in which the hydraulic loading rate is relatively low. Typical slow sand filtration velocities are only about 0.4 m/hr. At these low rates, the filtered contaminants do not penetrate to an appreciable depth within the filtration medium. The filter builds up a layer of filtered contaminants on the surface, which becomes the active filtering medium. This active filtration layer is termed a *schmutzdecke*. When the filter is first started after cleaning, the filtered water must be wasted until the filtration efficiency increases as the schmutzdecke is formed. Most of Europe has retained the lower loading rates (or slow sand filtration), whereas the United States developed and primarily uses rapid sand filtration.

Slow sand filters are cleaned by taking them off line and draining them. The organic or contaminant layer is then scraped off. The filter can then be restarted. After water quality reaches an acceptable level, the filter can then be put back on line.

Rapid Sand Filtration. In rapid sand filtration much higher application velocities are used. Filtration occurs through the depth of the filter. A comparison of rapid and slow sand filtration is shown in Table 9.3. In the United States, filter application rates are often expressed as volumetric flow rate per area, or gal/min-ft², which is actually a velocity with atypical units.

TABLE 9.3 Filtration Rates

FILTRATION TYPE	APPLICATION RATE	
	m/hr	gal/ft²-day
Slow sand	0.04 to 0.4	340 to 3400
Rapid sand	0.4 to 3.1	3400 to 26,000

Figure 9.9 shows a typical rapid sand filter in normal operation. The water above the filter provides the hydraulic head for the process. The filter medium is above a larger gravel, rock, or other media for support. Below the rock is usually an underdrain support of some type. This is simply a porous structure to support the filter medium. The water flows through the filter and support media, exiting from a pipe below.

Most modern filters employ two separate filter media in layers. The lower layer is composed of a dense, fine media, often sand. The upper layer is composed of a less dense, coarse media, often anthracite coal. The coarse

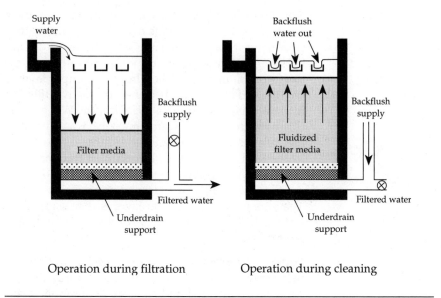

Supply water

Backflush water out

Backflush supply

Filter media

Backflush supply

Fluidized filter media

Filtered water

Filtered water

Underdrain support

Underdrain support

Operation during filtration   Operation during cleaning

FIGURE 9.9 Cross-section of a rapid sand filter.

upper layer removes larger particles before they reach the fine layer, allowing the filter to operate for a longer period before clogging.

As the filter begins to clog from accumulated solids, less water will pass through it. At some point cleaning is required. Usual filter operation before cleaning is from a few hours to 2 days. Cleaning is accomplished by reversing the flow of water to the filter, or backwashing, as shown at the right in Figure 9.9. The backwash velocity is sufficient to **fluidize** the bed—that is, to suspend the bed with the reverse flow. After backwashing, the filter is again placed in operation. In larger plants, the backwashing operation is automatic. In smaller plants, the operation is often controlled by operating personnel.

Taste and Odor Control

Control of taste and odor is an important part of water treatment. Problems in these areas can originate from dissolved gases in the water or from certain species of algae. It is seldom that a plant has not been faced with such problems at some time in its history. The best method of control, by far, is source selection. However, no matter how good the source, most treatment plants are faced with either taste or odor problems, or both, at some point. In some instances, offensive gases can be air-stripped, which eliminates or at least controls the problem. However, oxidation of the offending compound(s) is often required. Any of the oxidizers used in disinfection will also oxidize taste- and odor-causing materials.

Carbon Adsorption

Carbon adsorption, first presented in Chapter 4, can remove **trace** organic **contamination**. Taste and odor problems can be minimized or eliminated by the use of activated carbon in many instances. Adsorption can be accomplished prior to disinfection to prevent the formation of chlorinated organics, or after chlorination to remove much of the remaining organics, including the chlorinated organics formed at chlorination. Carbon adsorption can be a separate process or it can be combined with filtration. When it is combined with filtration, activated carbon becomes one of the granular media in the filter.

Disinfection

Disinfection is the destruction or killing of pathogenic microorganisms—whereas **sterilization** is the destruction or killing of all microorganisms. The objective of **potable water** disinfection is to kill or inactivate the harmful organisms present in the water and to leave a residual of disinfectant that will kill or inactivate organisms in the distribution system. The major disinfectant used in the United States is gaseous chlorine. Other chlorine disinfectants are also used, including chlorine dioxide, calcium hypochlorite, and sodium hypochlorite. Much of Europe uses ozone for disinfection.

Disinfectants can destroy microorganisms by destruction or damage to their cellular structure, by disabling their energy production, or by crippling cellular synthesis. Disinfection can be accomplished by chemical or nonchemical means. Most chemical disinfectants are oxidizing agents. Nonchemical disinfection includes the use of ultraviolet light, gamma radiation, and heat.

An ideal disinfectant must have several properties. It must kill microorganisms and should leave a residual to reduce the possibility of contamination from the distribution system. It should be inexpensive and not be harmful to humans or create harmful byproducts. It should be safe to handle and use, and it should not harm the environment or create an undesirable taste or odor.

The EPA is considering a drastic reduction in or possibly an outright ban of chlorine and chlorinated organics commonly used today. Many chlorinated organics are either known or suspected carcinogens. This policy may alter accepted disinfection practices in the United States. Ozone would be a likely replacement.

Chick's Law. In the early 1900s Harriet Chick postulated estimating the destruction of microorganisms by disin of time [4]. Dr. Chick postulated that the death of the first order with respect to time, for a given disinfectant or

$$\frac{dN}{dt} = -kN \qquad\qquad \textbf{9.12}$$

This can be separated and integrated (with $N = N_0$ at $t = 0$) to yield

$$\ln\!\left(\frac{N}{N_0}\right) = -kt \qquad\qquad \textbf{9.13}$$

where

N_0 = initial concentration of microorganisms [no./mL]

N = concentration of microorganisms at time t [no./mL]

t = time of disinfection [hr]

k = an empirical constant descriptive of the particular microorganisms and disinfectant in use [hr^{-1}]

The equation can be plotted as a straight line with $-\ln\!\left(\dfrac{N}{N_0}\right)$ versus t. The intercept is zero, and the slope is k. Thus, if laboratory experiments are used to determine N/N_0 for different times, k can then be determined.

In exponential form the equation is

$$N = N_0 e^{-kt} \qquad\qquad \textbf{9.14}$$

So we can expect an exponential decrease in the number of microorganisms in contact with a disinfectant as time progresses. The following example demonstrates Chick's law.

EXAMPLE 9.2	USE OF CHICK'S LAW TO ESTIMATE DISINFECTION PROPERTIES AND TIME

The following are actual data for the poliomyelitis virus exposed to an experimental disinfectant. Determine the k value for Chick's law. Estimate the time required to obtain a reduction of 1/10,000 the original number of viruses. (Source: R. Floyd et. al., "Inactivation of Single Poliovirus Particles in Water by Hypobromite Ion, Molecular Bromine, Dibromine, and Tribromine," *Environmental Science and Technology*, vol. 12, no. 9 (September 1978), pp. 1031–35.

TIME, sec	N/N_0
4	1/13
8	1/158
12	1/2000

SOLUTION: Equation 9.13 plots as a straight line. The y axis is $-\ln(N/N_0)$ and the horizontal axis is t. The intercept is zero. This can be done using a calculator and then plotting the coordinates by hand, or it can be worked out on a spreadsheet program. The data for the plot are as follows.

TIME, s	N/N_0	$-\ln(N/N_0)$
4	0.0769	2.56
8	0.00633	5.06
12	0.00050	7.60

These data are plotted in the accompanying figure, along with a least

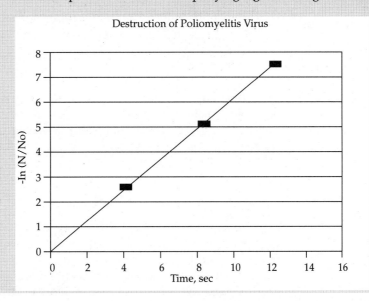

Destruction of Poliomyelitis Virus

squares regression line. The slope of the line is the disinfection constant $k = 0.634/s$. The time required for a reduction of $1/10,000$ is then

$$t = \frac{-\ln\left(\dfrac{N}{N_0}\right)}{k} = \frac{-\ln\left(\dfrac{1}{10,000}\right)}{0.634/s}$$

$$= 15\,s$$

For the experimental disinfectant to reduce the concentration of the poliomyelitis virus to $1/10,000$ of its original level, it must be operational for 15 seconds.

Concentration - Time Product.   It has been observed that the destruction of microorganisms is a function of the concentration of disinfectant and the time of contact, or

$$K = C^n t \qquad\qquad\qquad\qquad\qquad \textbf{9.15}$$

where

> K = constant for a particular disinfectant and microorganisms
>   [mg/L·min or mg/L·s]
>
> C = disinfectant concentration [mg/L]
>
> t = contact time [min or s]
>
> n = empirical constant

This relationship holds true only for a given set of conditions—for example, constant pH, temperature, and a specific microorganism.

Many studies have found that n is in the range of 0.8 to 1.2 for most microorganisms. In engineering practice it is usually assumed that n is unity. So the equation becomes

$$K = Ct \qquad\qquad\qquad\qquad\qquad \textbf{9.16}$$

The constant K varies with the type of disinfectant, species of microorganism, concentration, pH, and temperature. However, for a given set of conditions, the concentration time product, or K, is constant. Thus, if the concentration is doubled, the time required to disinfect is halved. To determine the Ct product for a different disinfection level or temperature, additional laboratory work must be performed at the desired destruction level.

The Chick–Watson Relationship. If Chick's law is combined with the concentration product concept, the relationship of time, concentration, and degree of lethality can be established [5]. H. E. Watson observed that the concentration time product was a function of the degree of disinfection as well as other factors including the microorganism present. This led him to pose the following formula expressing this relationship:

$$\frac{-\ln\left(\dfrac{N}{N_0}\right)}{\Lambda} = Ct \qquad\qquad \textbf{9.17}$$

or

$$-\ln\left(\frac{N}{N_0}\right) = \Lambda C t \qquad\qquad \textbf{9.18}$$

Where Λ is the coefficient of specific lethality [$L \cdot min/mg$].

EXAMPLE 9.3	CONCENTRATION-TIME PRODUCT, THE CHICK – WATSON RELATIONSHIP

Chlorine gas is a common disinfectant. Its most effective form is HOCl. The specific lethality of HOCl is $5\,L \cdot min/mg$ for the Polio virus. Determine the time necessary to achieve a. a 99% inactivation and b. a 99.99% inactivation of the virus.

SOLUTION: Solve the Chick–Watson equation for the time:

$$t = \frac{-\ln\left(\dfrac{N}{N_0}\right)}{\Lambda C}$$

a. For a 99% decrease in activity,

$$\text{Removal} = 99\% = 0.99$$

$$\text{Active} = \frac{N}{N_0} = 1 - \text{Removal} = 0.01$$

$$t = \frac{-\ln(0.01)}{\left(5\dfrac{L}{mg \cdot min}\right)\left(1\dfrac{mg}{L}\right)}$$

$$= 0.92 \, min$$

b. For a 99.99% decrease in activity,

$$t = \frac{-\ln(1 - 0.9999)}{\left(5\dfrac{L}{mg \cdot min}\right)\left(1\dfrac{mg}{L}\right)}$$

$$= 1.8 \, min$$

Thus, the disinfection time approximately doubles for a 100-fold decrease in microbial activity.

Chlorine. Cl_2 gas has been used since the end of the last century as a disinfectant in the United States. It is effective and inexpensive. A problem with it, however, is the formation of trihalomethanes (THMs)—chemicals formed when organic matter in the water reacts with the chlorine. Trichloromethane, a THM, is a carcinogen [6]. The current limit for THMs is 100 ppb. Other disinfectants produce lower amounts of THMs or do not produce THMs.

When chlorine is injected into water it forms

$$Cl_2 + H_2O \rightarrow HCl + HOCl \qquad\qquad \textbf{9.19}$$

The hypochlorous acid ionizes to hypochlorite:

$$HOCl \rightleftharpoons H^+ + OCl^- \qquad\qquad \textbf{9.20}$$

Although both hypochlorous acid and hypochlorite are disinfectants, hypochlorous acid is much more powerful. The equilibrium reaction is

$$K_a = 10^{-7.5} = \frac{[H^+][OCl^-]}{[HOCl]} \qquad\qquad \textbf{9.21}$$

From the relationship

$$pH = pK_a + \log\frac{[A^-]}{[HA]} \qquad\qquad \textbf{9.22}$$

we can see that HOCl and OCl^- are equally distributed when the pH equals the pK_a. Thus, at pH = 7.5 there are equal amounts of both HOCl and OCl^-. At lower pH values there is progressively more HOCl, and disinfection is more effective. A graph of the distribution of HOCl appears in Figure 9.10.

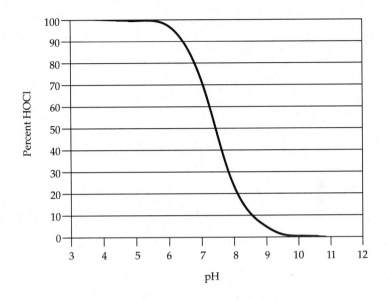

FIGURE 9.10   Distribution of HOCl versus pH.

| EXAMPLE 9.4 | EFFECTIVENESS OF HOCI VERSUS OCI⁻ FOR DISINFECTION |

Determine the time required to obtain a 99.99% inactivation of the Polio virus using both HOCl and OCl^- at concentrations of 0.5 mg/L. The specific lethality of HOCl is $5 \, L \cdot min/mg$. The specific lethality of OCl^- is $0.5 \, L \cdot min/mg$.

SOLUTION: **a.** For the HOCl,

$$t = \frac{-\ln\left(\dfrac{N}{N_0}\right)}{\Lambda C}$$

For a 99.99% decrease in activity,

$$\frac{N}{N_0} = 1 - \text{Removal} = 0.0001$$

$$t = \frac{-\ln(0.0001)}{\left(5\dfrac{L}{mg \cdot min}\right)\left(0.5\dfrac{mg}{L}\right)}$$

$$= 3.7\,min$$

b. For OCl^-,

$$t = \frac{-\ln(0.0001)}{\left(0.5\dfrac{L}{mg \cdot min}\right)\left(0.5\dfrac{mg}{L}\right)}$$

$$= 36.8\,min$$

Thus, the hypochlorite ion is only one-tenth as effective as hypochlorous acid.

Calcium Hypochlorite. Calcium hypochlorite is also used as a commercial disinfectant. It is more commonly used to disinfect swimming pools, but it is used in some smaller water treatment plants and a few larger plants as well. Calcium hypochlorite is more expensive than chlorine, but otherwise very similar. It produces the same two active chemicals as chlorine. The calcium hypochlorite ionizes, forming hypochlorite ions:

$$Ca(OCl)_2 \rightarrow Ca^{2+} + 2OCl^- \qquad\qquad \textbf{9.23}$$

The hypochlorite combines with H^+, forming HOCl:

$$H^+ + OCl^- \rightleftharpoons HOCl \qquad\qquad \textbf{9.24}$$

The distribution of HOCl and OCl^- has the same pH dependence as with Cl_2. Calcium hypochlorite is safer and requires less equipment than chlorine. The disinfection (and propensity to produce THMs) is similar to gaseous chlorine since the active disinfectants are HOCl and OCl^- in both agents.

Chlorine Dioxide. Chlorine dioxide [ClO_2] has become more common in the last 10 years because it produces lower levels of THMs. At typical pH values encountered in water treatment systems, the chlorine dioxide forms chlorite [ClO_2] in solution. Its major disadvantage is that it must be produced on site. Chlorine dioxide is an explosive gas at elevated temperatures or when exposed to light or organics [7]. Another problem with it is that it always contains some chlorine gas as an impurity. Thus, the production of chlorine dioxide results in a product that was to be avoided. However, new processes are being developed that promise to provide much higher levels of chlorine dioxide, and only minor contamination with chlorine gas [7]. We will almost certainly see increased usage of chlorine dioxide in coming years.

Chloramines. Chlorine combined with ammonia produces chloramines. Ammonia can be present in the water supply or added to enhance the formation of chloramines. Like chlorine dioxide, chloramines have been used more extensively in recent years because they reduce the formation of THMs. As noted, when chlorine is added to water, it produces hypochlorous acid [HOCl]. The HOCl reacts with ammonia [NH_3] to produce monochloramine [NH_2Cl]:

$$HOCl + NH_3 \rightarrow NH_2Cl + H_2O \qquad\qquad 9.25$$

The monochloramine can react with ammonia to produce dichloramine:

$$NH_2Cl + HOCl \rightarrow NHCl_2 + H_2O \qquad\qquad 9.26$$

Similarly, the dichloramine can react with additional ammonia to produce trichloramine:

$$NHCl_2 + HOCl \rightarrow NCl_3 + H_2O \qquad\qquad 9.27$$

The pH of the water determines which chloramine is formed. At pH values above 7, monochloramine is favored. At lower pH values either di- or trichloramine is preferentially formed. Since chloramines are not as effective as chlorine as a disinfectant, higher levels must be added to the water. Even so, chloramines have the advantage of producing much lower levels of THMs.

Ozone. Ozone, [O_3], is produced by electric discharge or ultraviolet radiation. It is continuously generated in the upper atmosphere and exists at low concentrations (about 0.05 ppm) in the lower atmosphere. It is also produced naturally by lightning discharges. The fresh odor after a thunderstorm is partly the result of ozone. In water or wastewater treatment, ozone is produced on site by high-voltage electric discharge through air. It cannot be produced off site because it rapidly degrades back to O_2. Commonly used in Europe, ozone has historically seen little use in the United States.

The gas does not produce trihalomethanes, but it has drawbacks: the need to produce it on site, its high cost, and the absence of a residual to react with any contamination elsewhere in the piping system.

PROCESS SUMMARY

We have examined processes used to treat most potable waters. Selection of appropriate methods is based on the characteristics of the source water. Where significant numbers of particles exist in it, sedimentation followed by filtration is required. Where the water is too hard, softening is required. In cases where dissolved gases or reduced metals are present, gas transfer is needed. The following sections discuss treatment processes for typical surfacewater and groundwater.

SURFACEWATER TREATMENT

Surfacewater contains dissolved organics and inorganics as well as suspended and colloidal particles composed of organics, inorganics, and microorganisms. Since filtration was first introduced in the early 1800s, the focus of water treatment has been to remove or inactivate pathogenic organisms. With the increased concern over chronic or long-term health hazards, the focus has changed in the last 40 years. Where the major concern in the 1950s was supplying a water free of microbial contamination, it is now to supply a water that presents a low risk of cancer or other chronic health hazards many years in the future. This change has been facilitated, at least in part, by increased technology, which makes possible the detection of **trace contaminants** in not only the ppb (parts per billion) range, but also the ppt (parts per trillion) range. A schematic of a typical surfacewater treatment system is shown in Figure 9.11.

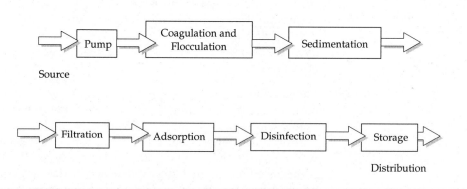

FIGURE 9.11   Schematic of a surfacewater treatment plant.

GROUNDWATER TREATMENT

Groundwater has historically been considered free of microbiological contamination. Many municipal water supplies using a groundwater source did not even attempt to disinfect their waters through much of this century. With the knowledge that we have contaminated many of our groundwater supplies with hazardous wastes and the goal of an even safer water supply, the federal government has begun to establish more stringent treatment requirements for groundwater. Also, in coastal areas some groundwater supplies have been contaminated with saltwater intrusion. This occurs where groundwater is pumped at rates greater than natural recharge of the aquifer can occur. The difference between the volume of water pumped and the volume of recharge water is replaced by saltwater in these areas. This contaminates the aquifer. Saltwater intrusion has occurred in Florida, New York, and other coastal states. A schematic of a typical groundwater treatment plant is shown in Figure 9.12. At the top the water quality is such that only disinfection is required. Below the line the water requires aeration, sand filtration, and disinfection.

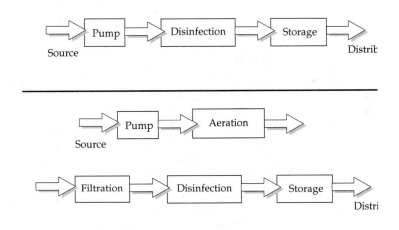

FIGURE 9.12 Schematic of a groundwater treatment plant.

RESIDUALS MANAGEMENT

Until passage of the Clean Water Act (PL 92-500) in 1972, water treatment plant wastes were usually run back to the original water supply (downstream if a river, away from the intake if a lake). In cases where the source was groundwater, the wastes were often discharged to the nearest stream. However, the Clean Water Act classified water treatment residuals as industrial wastes. Although no concerted effort has been made to force existing plants to comply with the regulations, as new facilities are built or existing facilities are expanded, they are usually required to comply with the regulations. This typically means removing most of the water from the sludges and then either sending the wastes to a municipal wastewater treatment plant or landfilling them.

Source of Residuals

Residuals result from several processes in a water treatment plant. As shown in Figure 9.13, the primary sources are sedimentation basin sludges, including softening sludges, filter backwash waters, brines from ion exchange softening, and spent carbon adsorption media. Table 9.4 provides approximate volumes and concentrations of typical water treatment residuals [2, 8]. Sludges from sedimentation contain most of the solids but result

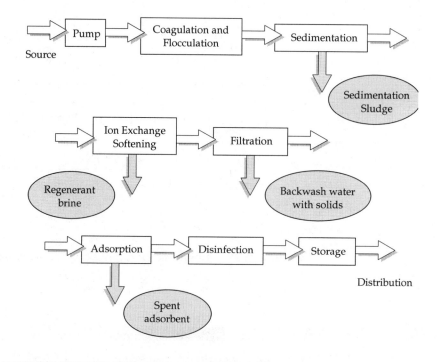

FIGURE 9.13   Sources of water treatment residuals.

TABLE 9.4   Volumes and Concentrations
of Water Treatment Plant Residuals

SOURCE	PERCENT OF FLOW	TOTAL SOLIDS PERCENT
Sedimentation without softening	0.1 to 5	0.2 to 2
Sedimentation with softening	0.3 to 6	2 to 33
Filter backwash	2 to 5	Low
Softening brine	3 to 10	1.5 to 3.5

in a volume of 0.1 to 0.3% of the plant flow. In contrast, filter backwash is usually 3 to 5% of the plant flow, but contains little solids compared to sedimentation sludge. Ion exchange brines are produced during regeneration. They typically have high levels of sodium and chloride, along with some calcium and magnesium. Activated carbon must be replaced every few months to few years, depending on the concentration and characteristics of the organics being removed.

Characteristics of Residuals

Sedimentation sludge has different properties, depending on the type of coagulant used and whether softening is required. In general, waste streams containing solids can be dewatered or concentrated with mechanical filtration devices or centrifuges. Mechanical filtration usually results in a semisolid cake that can easily be transported and landfilled. Earthen sludge lagoons are also used at some plants to allow the solids to settle to the bottom, producing clear water that can be drained either to a nearby stream or into the wastewater sewer. The concentrated solids can then be landfilled.

Some larger plants practice reuse of softening chemicals. The carbonate sludges, for example, can be heated, driving off the CO_2 and producing quick lime [CaO]. The reaction is

$$CaCO_3 \xrightarrow{\Delta} CaO + CO_2 \uparrow \qquad\qquad 9.28$$

Equipment costs usually prevent smaller facilities from using this process.

Ion exchange brines can be injected into the ground in some instances. Although not toxic, brines must be inserted in strata that will not come into contact an aquifer with the potential for human use. In other situations brines may be discharged into a wastewater plant. However, such facilities do not have processes to remove brines. This practice simply dilutes the brine before discharge to the receiving water.

Spent activated carbon can be regenerated and reused. However, only very large facilities can justify the equipment necessary for this. More often, the company supplying the activated carbon contracts to accept the used material. The company then regenerates the carbon and sells it again.

CASE STUDY

Water Treatment Plant, Carbondale, Illinois

The city of Carbondale, Illinois, used a water treatment plant constructed in the 1940s until 1994. During the early 1990s it became apparent that more stringent water quality regulations would make it impossible for the city to supply water of acceptable quality using its old plant. The city, in conjunction with a consulting engineering firm, decided to construct a new facility located near their water source, a small lake south of Carbondale.

The population of Carbondale is approximately 25,000. Southern Illinois University at Carbondale has an enrollment of approximately 24,000 students during the academic year. However, some of these students take classes off campus, others commute from surrounding communities, or are residents or Carbondale (and thus counted in the city population). The combined city and University total is approximately 40,000 people.

The new facility incorporates clarification using a combination of caustic soda [NaOH] and lime [Ca(OH)$_2$] as coagulants. There are three 51.5-ft diameter clarifiers. The overflow rate for the clarifiers is 1280 gal/ft^2-day.

Following clarification there are six high-rate filters. The filters use a dual media of sand and anthracite. Each filter has a diameter of 18 ft, providing an area of 254 ft^2 each. The total filter area is then 1527 ft^2 with a design loading of 5200 gal/ft^2-day.

A schematic of the new facility is shown in Figure 9.14. the new water treatment plant has a design capacity of 8 million gallons per day or 200 gal/person-day. The total cost of the plant was $10,000,000.

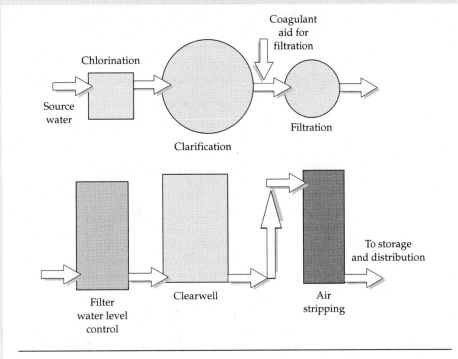

FIGURE 9.14 A schematic of the treatment steps for the new Carbondale, Illinois, water treatment plant.

Review Questions

1. Define the following:
   a. potable water
   b. water softening
   c. gas stripping
   d. coagulation
   e. flocculation
   f. sedimentation
   g. disinfection
   h. colloid
   i. hardness
   j. trace contaminants

2. Two aeration systems have been tested for their gas transfer efficiency. Dissolved oxygen values during the test are shown in the following

table. The initial conditions for the tests were different. For System A, the initial DO concentration was 2.0 mg/L, and the saturation DO was 9.2 mg/L. For System B, the initial DO concentration was 0.7 mg/L, and the saturation DO was only 8.5 mg/L. Determine the value of K_La for each system. Which system has better gas transfer ability?

	DO, mg/L	
t, min	SYSTEM A	SYSTEM B
0	2.0	0.7
2	2.4	1.4
4	2.9	2.1
6	3.3	2.7
8	3.7	3.2
10	4.1	3.7
12	4.4	4.2
14	4.7	4.6
16	5.0	4.9
18	5.3	5.3
20	5.6	5.6

3. A water contains 150 mg/L of hardness as $CaCO_3$. If ion exchange is used to remove all the hardness, with sodium as the exchange ion, what is the sodium concentration expressed as $mg/L\,Na^+$?

4. The following are actual data for an experimental disinfectant used to destroy the poliomyelitis virus [9]. Determine the k value in Chick's equation. Then estimate the time required to reduce the activity to 1/100,000 of the original concentration.

TIME, s	N/N_0
0.5	0.125
1	0.0158
1.5	0.0010
2	0.00030

5. Compare the settling velocity of a bacterium with a specific gravity of 1.02 and a diameter of 1.5 μm with that of a chemical floc with a specific gravity of 1.05 and a diameter of 1 cm. Use a water temperature of 20°C.

6. A water treatment plant has been evaluating a new disinfectant. The following data are for *E. coli* with an initial density of 100,000/mL. Use Chick's law to estimate the time required to reduce the number of *E. coli* to 50/mL.

TIME, min	*E. coli*/mL
0	100,000
2	20,000
4	4,300
6	790
8	150

7. An equation was developed in the text for the determination of the aeration gas transfer parameter $K_L a$. In this case, the initial dissolved gas concentration (oxygen concentration) is lower than the final gas concentration. Derive a similar expression for gas stripping—that is, where the initial gas concentration (carbon dioxide, methane, say) is higher than the final concentration. Be sure to define terms. *Hint:* First, write the initial conditions. The equation will be of a form similar to the one for aeration. The variables can be separated, and the equation can be integrated. Integral calculus is required, not differential equations.

References

1. M. N. Baker *The Quest for Pure Water* (New York: American Water Works Association, 1949).

2. *Water Quality and Treatment*, 3rd ed. (Denver, CO: American Water Works Association, 1971). (This contains an excellent description of the early history of water quality regulations in the United States.)

3. *Standard Methods for the Examination of Water and Wastewater*, published jointly by the American Water Works Association, American Public Health Association, and Water Pollution Control Federation, 1989.

4. H. Chick, "An Investigation of the Laws of Disinfection," *Journal of Hygiene,* vol. 8 (January 1908), p. 92.

5. H. E. Watson, "A Note on the Variation of the Rate of Disinfection with Change in the Concentration of the Disinfectant," *Journal of Hygiene,* vol. 8 (1908), p. 536.

6. S. Budavari, ed., *The Merck Index,* 11th ed. (Merck & Co., 1989).

7. J. M. Montgomery Consulting Engineers, Inc., *Water Treatment Principles and Design* (New York: Wiley, 1985).

8. American Water Works Research Foundation Report, "Disposal of Wastes from Water Treatment Plants—Part 3," *Journal of the American Water Works Association,* vol. 61, no. 11 (November 1969), pp. 619–38.

9. R. D. Floyd, G. Sharp, and J. D. Johnson, "Inactivation of Single Poliovirus Particles in Water by Hypobromite Ion, Molecular Bromine, Dibromamine, and Tribromamine," *Environmental Science & Technology,* vol. 12, no. 9 (September 1978), pp. 1031–35.

10 🏭 Wastewater Tr

``THESE POLICIES SIMPLY MEAN THAT
STREAMS AND RIVERS ARE
NO LONGER TO BE CONSIDERED PART
OF THE WASTE
TREATMENT PROCESS.''

**SENATOR EDMUND MUSKIE,
REFERRING TO THE
CLEAN WATER ACT**

The Bissell-Point Treatment Plant in St. Louis
Missouri is designed to treat 150 million
gallons of wastewater per day.
Photograph ourtesy of the St. Louis
Metropolitan Sewer District.

▰▰▰▰▰▰▰▰▰▰▰▰▰▰▰▰▰

Surfacewater quality in the United States has improved more in the last 20 years than any other area of the environment. Beginning with the passage of the Clean Water Act (officially the Federal Water Pollution Control Act Amendments of 1972), our nation embarked on a miraculous period of water quality improvement. Prior to the 1970s water quality in the United States was abysmal. Many of our best known streams, rivers, and lakes were nothing more than dumping grounds for municipal and industrial pollutants. Today most, although not all, waters are in far better shape. This shows that when a nation sets such goals and works diligently toward them, they can be achieved. That is not to say we need no further improvement but the water quality picture is clear evidence that we can substantially reduce human pollution of the environment. Nonpoint source pollution, as discussed in Chapter 8, Water Quality, is still in need of control. Some municipalities and industries have still not met the standards. But overall, point source pollution is being constrained, and the gains are both commendable and conspicuous.

In this chapter we will discuss wastewater characteristics, federal water quality regulations relating to wastewater treatment, municipal and industrial wastewater treatment, sludge treatment and disposal, stormwater treatment and disposal, and advanced wastewater treatment.

WASTEWATER CHARACTERISTICS AND QUANTITIES ▰▰▰▰▰▰▰

The wastewater received at a typical municipal treatment plant comes from many different sources, including homes, apartments, businesses, and industries. And there is street and parking lot runoff. The concentration of the contaminants and the influent flow rate varies substantially from day to day as well as during the day. In addition, domestic wastewater varies in concentration and flow rate from city to city. These variations reflect such factors as the affluence of the area, the use of garbage disposals, the amount of infiltration and inflow of stormwater and groundwater, and the amount and type of industries in the area. Table 10.1 lists typical values for some of the common characteristics of wastewater. An estimate of the flow rate can be obtained by multiplying the city population by 600 L/capita·day (150 gal/capita-day). That's right, 600 liters for each person, each day. And that excludes the flow from industries.

REGULATIONS ▰▰▰▰▰▰▰▰▰▰▰▰▰▰▰▰▰▰▰▰

Federal water quality regulations were first instituted at the end of the last century with the passage of the Refuse Act of 1899. This was intended to prevent persons or industries from placing objects in the water that could impede river or harbor traffic. It was not really intended to control water

TABLE 10.1 Typical Wastewater Characteristics

PARAMETER	CONCENTRATION, mg/L
Biochemical oxygen demand (BOD_5)	250
Total suspended solids	250
Chemical oxygen demand	500
Nitrogen, total	40
Ammonia	30
Organic	10
Nitrate	0
Phosphorus, total	10
Ortho	6
Organic	4
Total organic carbon	150
Chloride (above water supply level)	50

quality. Today our nation has a well developed set of regulations to promote water quality and prevent the discharge of excessive pollutants. Table 10.2 presents important federal water quality laws. This section describes the history of water quality and water pollution control legislation in the United States.

After passing the Refuse Act, Congress did little else of environmental consequence until 1948, when it adopted the Water Pollution Control Act. Although this was the first federal legislation to address water pollution and water quality in the nation, it did little to correct the rapidly worsening problems. The act recognized states' rights to control their water quality in a manner which was best for the states. No federal water quality standards or wastewater discharge standards were legislated. In fact, it did not even contain federal goals.

The WPCA did provide a 5-year program of federal grants to municipalities for construction of wastewater treatment facilities, but total funding for it was only $22.5 million. In addition, municipalities could receive $1 million per year to plan and administer these improvement programs. The same yearly amount was available for the study of industrial pollution and its control. However, the most far-reaching part of the act was probably the

TABLE 10.2   Federal Water Pollution Control Legislation

TITLE	NUMBER	FUNDING, millions/ year
Refuse Act of 1899		
Water Pollution Control Act of 1948	PL 80-845	$24
Federal Water Pollution Control Act of 1956	PL 84-660	$24
Amendments to the FWPCA of 1961	PL 87-88	$50
Water Quality Act of 1965	PL 89-234	
Clean Water Restoration Act of 1966	PL 89-753	
Federal Water Pollution Control Act Amendments of 1972 (Clean Water Act)	PL 92-500	$7000
Clean Water Act of 1977	Pl 95-217	$5000
Municipal Wastewater Treatment Construction Grant Amendments of 1981	PL 97-117	
The Water Quality Act of 1987	PL 100-4	$2400

$4 million allocated for the development of a national water pollution control research facility in Cincinnati, Ohio. This is now the Robert A. Taft Engineering Center, one of the EPA's five major research facilities.

During the 1950s and 1960s federal funding and involvement continued to increase, but Congress was careful not to infringe upon states' rights. The attitude among federal lawmakers was that it was the states' prerogative to decide if they wished to require dischargers to treat their wastewater, and it was up to the states to say what the level of treatment would be. Some federal legislation was passed in 1956, 1961, and 1966. With each successive act, there were increases in the level of funding and the percentage of construction costs the federal government would pay. Yet, although wastewater treatment was encouraged, there was no all-encompassing federal plan. And there was essentially no federal intervention to require pollution control. That "hands off" attitude ended with the Clean Water Act in 1972.

The Clean Water Act

In 1972 the Nixon administration opposed measures to create a comprehensive federal water pollution law. However, most members of the U.S.

House and Senate felt that passage of such a bill was in the national interest. With the presidential election coming in November, the administration needed to establish some record of environmental support. In an effort to prevent Congress from passing such a broad bill and at the same time show support for environmental problem solving, the administration declared that it would use provisions in the Refuse Act of 1899 to authorize the Corps of Engineers to enforce water quality. Since the Refuse Act of 1899 contains no water quality provisions, it was unclear how this effort could be effective. Congress did not agree with the president's approach. Led by Senator Edmund Muskie of Maine, Congress passed a bill that set ambitious goals, including the elimination of all pollutant discharges into our nation's waters within 10 years. President Nixon vetoed the bill. However, both houses of Congress overrode the veto, and on October 18, 1972, the Federal Water Pollution Control Act Amendments of 1972 (PL 92-500) became law.

PL 92-500, often called the Clean Water Act (CWA), established national water quality goals. Its objective was to "restore and maintain the chemical, physical, and biological integrity of the Nation's waters." The CWA set two ambitious water quality goals: elimination of the discharge of all pollutants to our nation's waters by 1985, and an interim goal of swimmable and fishable waters by 1983. Although these goals have not been attained today, and probably will never be, they expressed the national consensus that it was time to reverse the trend of decreasing water quality. The CWA also directed the EPA to establish criteria for water quality and discharge limits and a procedure for the permitting of all dischargers into navigable waters. The result was the National Pollutant Discharge Elimination System (or NPDES), the provisions of which were to be enforced by the states. Unlike previous laws dealing with water quality, however, the standards were not options or suggestions. They were mandates. The requirements for conventional pollutant discharges, the secondary treatment standards, are shown in Table 10.3. Developed by the EPA at the direction of Congress, these were applied to municipal dischargers. All municipal wastewater treatment plants were to meet the standards by July 1977.

Penalties.   Failure to comply with CWA standards could carry a heavy price for both municipalities and industries. Fines for willful noncompliance ranged from $2500 to $25,000 per day, and imprisonment was possible for willful failure to meet the requirements or for falsifying reports submitted to the EPA. Thus, for the first time, polluting a water was considered a crime.

Industrial Dischargers.   Industrial facilities either discharge their wastewater into a **POTW** or directly into a receiving water. Industries pouring wastes directly to receiving waters were required to have their own **NPDES permits**. The CWA set two standards for conventional pollutants from such

TABLE 10.3 Initial Requirements for Secondary Treatment Required by PL 92-500, the Clean Water Act

PARAMETER	REQUIREMENT
BOD_5	30 mg/L
TSS	30 mg/L
pH	6–9
Microbial	200 coliform colonies/100 mL

direct dischargers. The first, to be attained by all industries by July 1, 1977, was "the best practical control technology currently available," or BPT. This level of treatment was to be increased to "the best available technology economically achievable," or BAT, by July 1, 1983. The act also contained the beginning of toxic discharge enforcement, although actual standards and enforcement would be delayed due to the emphasis on conventional pollutants.

Industries sending wastes to municipal treatment facilities were required to pretreat their discharges. Materials that could not be removed by the POTW (and thus would pass through the system) were to be eliminated prior to discharge. Additionally, industries could not get rid of materials that would interfere with the operation of the POTW. These criteria are often referred to as "pretreatment standards."

Municipal Dischargers. PL 92-500 also established a massive construction grants program, providing several billion dollars of federal money each year for the construction of municipal plants. This program has continued through the 1980s until today, although during the Reagan administration it was modified into a revolving loan program rather than outright grants. Even with this massive infusion of federal monies, municipal accomplishments lagged behind the progress made in industry. By July 1977, 85% of industrial dischargers were in compliance with the BPT requirements of the CWA, but only about a third of major cities had attained secondary treatment [1].

The Clean Water Act of 1977

In 1977 the CWA was amended. The changes, officially titled the Clean Water Act of 1977, recognized that the nation was not prepared to commit $7 billion per year to municipal wastewater facilities in the construction grants program. The 1977 bill reduced the yearly figure to $5 billion. Left intact, however, were the national goals of "swimmable and fishable" waters by 1983 and achieving "zero discharge" by 1985.

There were also refinements to the management of toxic discharges. The EPA was instructed to develop criteria for conventional, nonconventional (nitrogen and phosphorus), and toxic pollutants. As the agency promulgated (officially issued) standards for toxic pollutants, affected industries would have from 1 to 3 years to comply with the more stringent requirements.

Another major change in the CWA was in the construction grants program. It was modified to allow states to use 2% of their funds for administering the grants program. Also, the EPA could issue compliance extensions for cities not able to obtain construction grant funding. Some 75% of municipalities were in compliance with secondary treatment standards by 1980. The federal share of funding for plants approved for construction after 1985 was reduced from 75% to 55%. Congress also required engineering firms designing and constructing facilities to help municipalities solve start-up problems and train workers for 1 year after plants went on line.

Interim Measures

Coupled with other federal requirements, the construction grants program was so cumbersome and paperwork-oriented that the time involved from application to completion of a POTW had increased from 2 to 3 years to 7 to 9 years [1]. With support and cooperation from the Reagan administration, Congress passed the Construction Grant Amendments of 1981, PL 97-117. Its intent was to simplify the construction grants application and approval process and restrict the applicable areas for construction grant funding. Also, several processes used by smaller POTWs—including lagoons, oxidation ponds, and trickling filters—were defined as secondary treatment as long as discharges did not decrease water quality. This provision was intended to prevent high-cost, complex treatment systems from being required in smaller communities [1].

The Water Quality Act of 1987

In the fall of 1986, Congress was considering several important environmental issues, including revisions in the Clean Water Act and reauthorization of the Superfund, a toxic waste cleanup law that had expired in 1984. In October, lawmakers passed both measures. Just before the election, however, President Reagan vetoed the CWA amendments, stating that they were too costly. He signed the Superfund Reauthorization into law, though. When Congress reconvened in early January, the CWA bill was reintroduced without going to committee (an unusual measure for Congress) and was passed as the Water Quality Act of 1987. President Reagan again vetoed it. On February 4, 1987, Congress overrode the veto and the amended CWA became PL 100-4.

The law converted the construction grants to a revolving state loan program over a period of 4 years. The funding authorized by PL 100-4 is shown in Figure 10.1 [1]. The purpose of revolving loans was to end the large construction grant program began with PL 92-500, the Clean Water Act of 1972. In the revolving loan program, the states authorize loans to municipalities for construction of wastewater treatment plants, as well as other water quality measures (such as nonpoint source control or estuary pollution control). Cities may no longer obtain outright grants. As the fund is reimbursed by some cities, other cities may apply for loans. Thus, it is a revolving program.

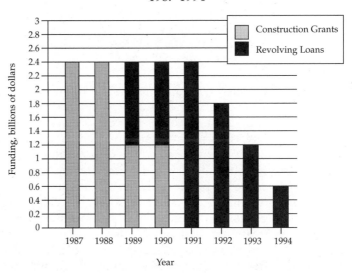

FIGURE 10.1 Grants and loans under the Water Quality Act of 1987, PL 100-4.

Other key provisions of the Water Quality Act of 1987 include the permitting of stormsewer outfalls and the imposition of limits of toxics in sludge. These requirements will be phased in over several years, with larger cities affected first.

The Water Quality Act of 1987 gave extensions to industries having serious difficulties meeting BAT and BPT requirements, but in an anti-backsliding provision, EPA is not to rewrite existing industrial standards to make them less stringent.

Summary of Water Quality Regulations

Implementation of the Clean Water Act and its subsequent amendments has made major improvements in water quality. To accomplish this feat, the federal government has expended some 7.5 billion dollars. State and local governments have probably spent an additional 15 to 25 billion dollars, and industries have most certainly expended an amount greater than all municipalities combined.

WASTEWATER TREATMENT

The objective of wastewater treatment is to reduce the amount of pollutants in the water to such a level that the water can be returned to the environment without causing stress on aquatic life and be of sufficient quality for subsequent (downstream) users. Direct reuse of wastewater is limited, at present, to some industrial applications. However, in view of the significant water shortages predicted for the future, direct reuse of domestic wastewater may become reality. Even now we practice a form of indirect reuse. Cities along the Mississippi River withdraw the water, treat it, consume it, and then treat the wastewater and put it back in the river. This occurs all along the river's path. By the time Mississippi water reaches New Orleans, it has been used multiple times. So treatment of wastewater is important not only from an environmental quality standpoint, but also for human health.

Conventional wastewater treatment is a multistep process: (1) removal of materials that will interfere with pumping and later treatment steps, (2) removal of the solid materials that will settle by gravity under quiescent conditions, (3) conversion of the remaining soluble and colloidal material into microbial solids, (4) removal of most of the remaining pollutant materials in a second sedimentation, and (5) treatment and disposal of the residual solids and sludges generated in the other steps. Figure 10.2 depicts the basic stages in the wastewater treatment process. Figure 10.3 shows a typical wastewater treatment facility.

Design Methods

There are several ways to determine the scope of various treatment processes. Commonly used methods include the use of loading criteria, empirical formulas, and derived equations. Many of today's wastewater treatment methods have been in use for the past 10 to 75 years. A considerable amount of experience has been gained during those years. Most domestic wastewater falls within a predictable concentration range.

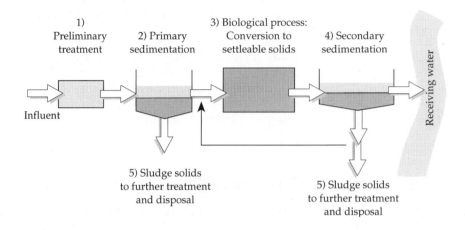

FIGURE 10.2 Steps in the wastewater treatment process.

FIGURE 10.3 The San Jose wastewater treatment plant in California. Photograph courtesy of Envirex, Inc., Waukesha, Wisconsin.

The daily flow, on a per person basis, is also predictable. Loading criteria can be typical hydraulic detention times, or unit sizes divided by flow or population. See Example 10.1. The design criteria developed by the Great

Lakes–Upper Mississippi River Board of State Sanitary Engineers illustrate the use of loading factors [2]. Often referred to as the "Ten States Standards," these criteria can be used to size most, if not all, operations and processes in a conventional wastewater treatment process.

EXAMPLE 10.1	THE USE OF LOADING CRITERIA

One wastewater treatment process, activated sludge, which will be discussed later, requires a detention time of 4 hr, or approximately 20 gal/capita. If a city has a population of 10,000 and an average flow of 1.2 MGD, what approximate tank volume is required?

SOLUTION: The tank size can be estimated based either on the flow, using the typical detention time, or on population using the size per capita.

a. The typical detention time, θ is 4 hours. Thus, the tank volume is

$$V = Q\theta$$

$$= \left(1.2 \times 10^6 \frac{gal}{day}\right)\left(4\,hr \times \frac{1\ day}{24\ hr}\right)$$

$$= 200{,}000\ gal$$

b. Based on population requirements, the volume is

$$V = \left(20 \frac{gal}{capita}\right)(10{,}000\ capita)$$

$$= 200{,}000\ Gal$$

Other design methods use empirical formulas developed from practical experience and years of data. Still others employ formulas derived from the principles of reactor design and mass balances.

Preliminary operations such as bar screening and grit removal are usually scaled to the hourly peak flow rate, whereas most major treatment operations and processes—primary and secondary sedimentation and secondary biological treatment—are sized based on average flows. To prevent flooding and overflows, all pumping and piping must be sized with peak flows taken into account.

The average flow and peak flow are related by the peaking factor

$$Q_{peak} = PF \cdot Q_{avg}$$

where

Q_{avg} = average volumetric flow [volume/time]

Q_{peak} = peak volumetric flow [volume/time]

PF = peaking factor

The peaking factor, typically 2 to 5, can be established statistically from previous flow data for a particular sewer system or estimated from the population. Small communities have larger peaking factors. Large cities have smaller peaking factors.

Preliminary Operations

Preliminary operations are designed to remove materials that will interfere with subsequent treatment processes. These include large foreign objects such as sticks, logs, shoes, and occasionally even dead animals, which enter sewer systems in a variety of ways. Grit is also removed in preliminary operations because it will cause undue wear on piping and pumping systems, as well as accumulate in some processes.

Bar Racks. Large objects like tree limbs tennis shoes, and animals get into wastewater influent because sewer access covers are left off or storm sewer entrances are too large. An 18-ft diameter main feeder line to a wastewater treatment plant in St. Louis was once blocked with 72 shopping carts! Such obstructions (hopefully most are not as large as shopping carts) are prevented from entering pumping and treatment systems by placing metal bars, spaced from 1 cm to a few cm apart, across the water flow. Foreign objects can create serious problems for equipment at a treatment plant. Possible problems are clogging pumps or smaller piping within the plant.

Bar racks can be cleaned either mechanically or manually. Only very small facilities have manually cleaned racks. The bar spacing ranges from 0.5 cm to 3 or 4 cm. Some facilities have one bar rack with a wide spacing followed by a second rack with a smaller spacing. Figure 10.4 depicts a mechanically cleaned bar rack. Bar racks are available as purchased items from a variety of companies. The engineer must determine the size of the unit(s) based on an area requirement. This is estimated by limiting the approach velocity (at peak) flow to the range of 0.6 to 1 m/s for mechanically cleaned bar racks and 0.3 to 0.7 m/s for manually cleaned bar racks.

FIGURE 10.4 A mechanically cleaned bar rack and a schematic of the principal interior components. Photograph courtesy of Envirex, Inc., Waukesha, Wisconsin.

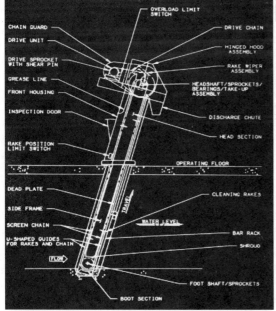

EXAMPLE 10.2	SIZING BAR RACKS BASED ON APPROACH VELOCITY

A municipality with a population of 50,000 people has an average design flow of 6.5 MGD. The peaking factor for the plant is 2.8. Estimate the area required for two mechanically cleaned bar racks.

SOLUTION: Our first task is to estimate the peak hourly flow:

$$Q_{peak} = PF \cdot Q_{avg} = (2.8)(6.5\,\text{MGD})$$

$$= 18.2\,\text{MGD}$$

If we wish to limit the approach velocity to 0.8 m/s, we use the relationship between the flow rate, area and velocity, or

$$Q = AV$$

Since we know the required approach velocity and the peak flow, we can solve for the required area:

$$A = \frac{Q}{V} = \frac{\left(1.82 \times 10^7 \dfrac{\text{gal}}{\text{day}}\right)\left(\dfrac{3.78\,\text{L}}{\text{gal}} \times \dfrac{\text{m}^3}{10^3\,\text{L}}\right)}{(0.8\,\text{m/s})\left(86{,}400\dfrac{\text{s}}{\text{day}}\right)}$$

$$= 1.00\,\text{m}^2$$

Note: *Since one bar rack may be required to be off line for maintenance or repairs, each unit must be able to handle the peak flow. Therefore, each bar rack must have an area of 1 m².*

Grit Removal. Grit, which is composed primarily of sand, cinders, and gravel, enters the wastewater collection system due to cracks in pipes, improper or poorly fitting pipe joints, poorly fitting or missing manhole covers, and stormwater runoff from streets and parking lots. Grit is removed primarily because it is harmful to later treatment processes. It causes excessive wear in pipes and pumps, and it accumulates in downstream tanks where flow velocities are insufficient to keep it in suspension. As grit accumulates, it reduces the effective tank volumes and thus the treatment effectiveness. The objective of grit removal operations is to remove as much material as possible while minimizing the removal of organic matter that should be removed and treated in later processes. The high specific gravity of grit (compared to most organic matter) makes its removal by settling fairly easy. Grit chambers are relatively small in comparison to those for subsequent processes, and their retention times are on the order of 1 to 5 minutes, compared with several hours for many downstream processes.

Grit particles settle individually without agglomerating. It is possible to estimate the settling velocity of discrete particles using the basic principles of physics discussed in Chapter 7, Analysis of Treatment Processes.

Primary Sedimentation

Sedimentation is the gravity settling, and thus removal, of materials more dense than a suspending fluid. Suspended solids in wastewater include organic matter, grit, clay, sand, and bacteria. Sedimentation is accomplished in large circular or rectangular tanks (**clarifiers**), as shown in Figure 10.5. Such processes can typically remove about one-third of the BOD_5 and two-thirds of the suspended solids. Sedimentation tanks are designed so that the water velocity is reduced enough that much of the suspended matter will settle to the bottom of the tanks where they are collected in sludge hoppers and removed (see Figure 10.6). The sludge then receives further treatment prior to disposal.

FIGURE 10.5 Circular sedimentation tank. Photograph courtesy of Envirex, Inc., Waukesha, Wisconsin.

In a grit chamber, the settling is discrete. That is, the particles settle individually. In primary and secondary sedimentation basins at wastewater treatment plants, the settling is more complex. The detention time is long enough and the concentration of the particles is high enough that

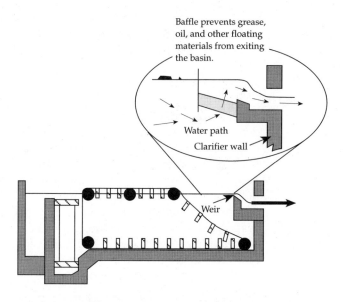

Baffle prevents grease,
oil, and other floating
materials from exiting
the basin.

Water path

Clarifier wall

Weir

FIGURE 10.6   Schematic of a typical rectangular sedimentation basin.

differential settling causes particles to agglomerate; thus, particle mass, number, and settling velocity change continuously during the process. Flocculant settling is the most common approach in primary clarifiers. We will leave analysis of this to other texts dealing specifically with wastewater treatment. (See References or Additional References at the end of the chapter.)

Typical design parameters for primary clarifiers are presented in Table 10.4. Of major importance is the surface overflow rate (vertical velocity), the weir loading (flow rate of water per foot of weir), and detention time in the basin. Such parameters are often used to estimate the size of sedimentation basins.

The surface overflow rate is the volumetric flow rate into the basin divided by the basin surface area. For volumetric flow rate in gal/day and area in ft^2, the surface overflow rate is expressed as gal/ft^2-day. In metric it is m^3/m$^2 \cdot$day. Using proper conversions, these units become velocity—length over time. The surface overflow rate is simply the vertical velocity of the liquid in the basin.

A weir is a baffle over which the water flows as it leaves the basin, as shown in the inset in Figure 10.6. Most clarifiers used in wastewater have a trap to prevent floating materials from exiting the unit. As the flow rate increases, the water velocity near the weir increases. In extreme cases this can reduce clarifier efficiency. Thus, there is a maximum amount of water

TABLE 10.4 Typical Design Data for Primary Sedimentation Tanks

PARAMETER	DESIGN RANGE AT AVERAGE FLOW	TYPICAL DESIGN VALUES
Surface overflow rate, m³/m²·day, (Gal/ft²-day)	35–45 (800–1200)	40 (1000)
Average detention time, hr	1.5–2.5	2.0
Weir loading, m³/m·day (gal/ft-day)	125–500 (10,000–40,000)	275 (20,000)

SOURCE: Ten State Standards and *Wastewater Engineering.*

(flow rate) that can flow over a given length of weir. The weir loading is the volumetric flow rate divided by the length of weir. Typical values for weir loading appear in Table 10.4. The following example illustrates the use of design loading criteria to size a primary clarifier.

EXAMPLE 10.3 **SIZING A PRIMARY CLARIFIER FOR A WASTEWATER TREATMENT FACILITY**

Use the typical design values in Table 10.4 to estimate the size (both diameter and depth) for two clarifiers used to treat wastewater at a design flow of 16 MGD. Each clarifier is to treat half the flow.

SOLUTION: Each clarifier should receive half the flow, or 8 MGD. Since the surface overflow rate is the velocity of the water rising in the clarifier, we can compute the area from the flow rate:

$$Q = AV$$

where

Q = flow rate to the clarifier [gal/day or m³/day]

A = required clarifier area [ft² or m²]

V = overflow rate [gal/ft²-day or m³/m²·day]

Solving for the area yields

$$A = \frac{Q}{V} = \frac{8 \times 10^5 \, \text{gal/day}}{1000 \, \text{gal/ft}^2\text{-day}}$$

$$= 8000 \, \text{ft}^2$$

From the area we can calculate the clarifier diameter, which is the term normally used to signify the size of a particular unit. The diameter is then

$$d = \sqrt{\frac{4A}{\pi}} = \sqrt{\frac{4(8000 \, \text{ft}^2)}{\pi}}$$

$$= 101 \, \text{ft}$$

Clarifiers are generally available in diameters that are multiples of 5 ft in the United States or multiples of 2 m outside the United States. So the clarifier needed would be the next size larger, or 105 ft in diameter in the United States. To determine the detention time, we must first calculate the clarifier volume:

$$V = Q\theta$$

where θ is the detention time [hr]

$$V = \left(8 \times 10^6 \, \frac{\text{gal}}{\text{day}}\right)\left(\frac{\text{ft}^3}{7.48 \, \text{gal}} \times \frac{\text{day}}{24 \, \text{hr}} \times 2.0 \, \text{hr}\right)$$

$$= 89{,}100 \, \text{ft}^3$$

Clarifiers should be as shallow as possible, but not less than 2 m [7 ft] deep. So the approximate depth, h, is obtained from

$$V = Ah$$

or

$$h = \frac{V}{A} = \frac{89{,}100 \, \text{ft}^3}{\pi(52.5 \, \text{ft})^2}$$

$$= 10 \, \text{ft}$$

The final dimensions are a diameter of 105 ft and a depth of 10 ft.

Secondary (or Biological) Treatment

Following **primary treatment**, biological treatment is used to convert most of the remaining soluble and colloidal organics into settleable microbial solids. In essence, the microorganisms are fed the remaining wastewater organics. The microbes grow and reproduce. The resulting microbial mass is then removed by a second sedimentation process, creating a second sludge. **Secondary treatment** includes both the biological process and its associated sedimentation operation, as shown in Figure 10.7.

Biological treatment can be accomplished by both attached growth and suspended growth microorganisms. In **attached growth reactors**, the microorganisms are provided a surface on which to grow. They remain attached to the surface and the wastewater flows over the surface. Collectively, the attached microorganisms are called a **biofilm**. Excess microorganisms are "sloughed off" the surface and are separated from the water in the final clarifier. In **suspended growth reactors**, the microorganisms remain in the wastewater rather than attached to a fixed surface. They move with the wastewater through the treatment process, are separated by sedimentation, and then are recycled to treat additional wastewater.

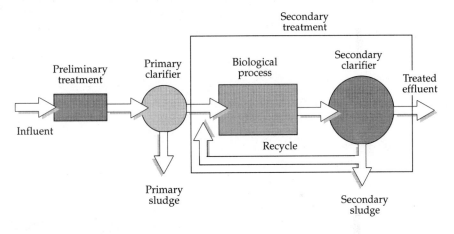

FIGURE 10.7 Secondary treatment, including the biological process and secondary sedimentation.

Trickling Filters. **Trickling filters** (or **packed towers**) are the most common attached growth process. They were developed and first used at the end of the last century. One of the first was constructed by Joseph Corbett in Salford, England, in about 1893 [3]. Trickling filters were a widely used form of wastewater treatment during the first half of this century, and many are still in use. In a trickling filter, the wastewater is sprayed or poured over rock or a synthetic medium on which a microbial population grows. The name is a bit misleading, for no filtration takes place. The wastewater simply flows (or trickles) down through the medium and collects in a trough at the bottom. It is then recycled to increase the contact with the biofilm and to provide a constant shear force to prevent an excessive buildup of biomass (see Figure 10.8). Earlier versions of the trickling filter were circular tanks with rock (5 to 10 cm in size) used as the medium for bacterial attachment. Tank depth was usually limited to 2 or 3 m to obtain sufficient oxygen penetration. More recent versions of the

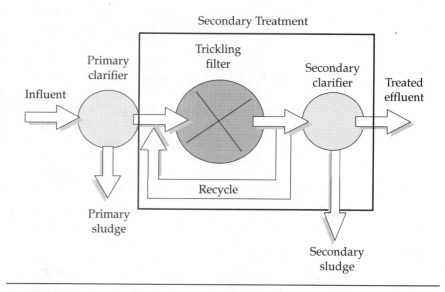

FIGURE 10.8 Schematic of a trickling filter wastewater treatment facility.

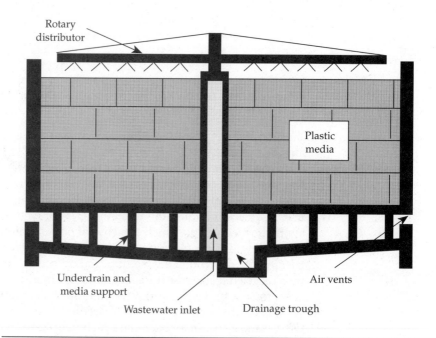

FIGURE 10.9 Cross-sectional view of a plastic-medium trickling filter.

FIGURE 10.10 Rotary distributor and rock media of a conventional trickling filter. Photograph courtesy of Envirex, Inc., Waukesha, Wisconsin.

trickling filter (often called packed towers) use synthetic media that maximize the surface area-to-volume ratio, allowing greater biomass concentrations (and thus improved treatment). In addition, packed towers can be much deeper because they use forced ventilation to increase oxygen penetration. Other depictions of trickling filters appear in Figure 10.9 and Figure 10.10.

There are several ways of estimating the required size of a trickling filter. Some methods use loading factors—either organic load or flow rate. Others employ formulas based on either empirical data or theoretical considerations relating to biofilm thickness and oxygen and substrate penetration.

Schultz [4] developed an empirical design expression for trickling filters that was later modified by Germain [5] for use with synthetic media filters or packed towers. The equation for the removal of soluble BOD is

$$\frac{L_i}{L_0} = e^{(-kh(A/Q)^{0.5})}$$

10.2

where

L_0 = effluent BOD$_5$ from the trickling filter [mg/L]

L_i = influent BOD$_5$ into the trickling filter [mg/L]

k = a biodegradability constant that must be determined in laboratory studies (varies with filter depth)

$$[\text{ft}^{-1/2}\text{-day}^{-1/2}, \text{m}^{-1/2} \cdot \text{day}^{-1/2}]$$

h = filter depth [m]

Q = volumetric flow rate to the filter [m^3/day]

A = filter area [m^2]

If this equation is solved for the required area, it becomes

$$A = \frac{Q\left[\ln\dfrac{L_0}{L_i}\right]^2}{k^2 h^2}$$

10.3

EXAMPLE 10.4	**USE OF THE SCHULTZ – GERMAIN EQUATION TO DETERMINE THE REQUIRED SIZE OF A PACKED TOWER**

Six packed towers are to be used in upgrading the Bissell Point Wastewater Treatment Plant for the St. Louis Metropolitan Sewer District. The design flow is 150 MGD (25 MGD each). The influent wastewater concentration is 300 mg/L, and the required effluent concentration is 50 mg/L. (A second biological process will further reduce the wastewater concentration before discharge to the Mississippi River.) Designers plan to make the towers 32 ft high. Extensive laboratory studies have found the k value to be 0.865 ft$^{-1/2}$-day$^{-1/2}$. Find the area required for the packed towers and estimate the diameter.

SOLUTION: We have all of the information we need to calculate the required area. All that is necessary is to substitute into Equation 10.3 and convert units into a consistent system:

$$A = \frac{\left(25.0 \times 10^6 \dfrac{\text{gal}}{\text{day}} \times \dfrac{\text{ft}^3}{7.48\,\text{gal}}\right)\left[\ln\left(\dfrac{300\,\text{mg/L}}{50\,\text{mg/L}}\right)\right]^2}{(0.865/\text{ft}^{1/2}\text{-day}^{1/2})^2(32\,\text{ft})^2}$$

$$= 14{,}000\,\text{ft}^2$$

An area of 14,000 ft^2 results in a diameter of 134 ft for each filter. Thus, six filters are required, each 32 ft high with a diameter of 134 ft. (Six packed towers were built at the St. Louis MSD Bissell Point Plant during the period 1991 to 1993 with these dimensions. They use forced draft ventilation and have an odor control system for the ventilation air prior to exit.)

RBCs.   **Rotating biological contactors** (or RBCs) are a more recent attached growth process. Developed in Europe during the 1960s, they use a rotating drum of synthetic media to maximize surface area. The microorganisms attach to the media surface. The drum rotates, partially submerged, in the wastewater as it flows through the treatment tank(s). Exposure of the biofilm to the atmosphere at each revolution improves oxygen transfer to the wastewater. Even so, oxygen transfer limits first-stage treatment, and means that aerators must be placed in some first stage RBC tanks. Figure 10.11 pictures an operational RBC, and Figure 10.12 is a cross-sectional view of an RBC. The design provides better contact between microorganisms and wastewater than in trickling filters. The shear force from each submerged cycle causes a relatively even or steady loss of biomass and prevents an excessive buildup of biofilm.

FIGURE 10.11   Rotating biological contactors enclosed in a building to prevent freezing. Photograph courtesy of Envirex, Inc., Waukesha, Wisconsin.

RBC design allows for modular construction. Typical modular units in use today have a drum diameter of 10 to 12 ft. The drums, usually submerged about 40%, have a rotational speed of approximately 1.5 RPM. Plants are constructed of multiple identical units. Treatment typically involves two or more "stages" of units, as depicted in Figure 10.13. The advantages of RBCs include reduced energy costs since flows are not

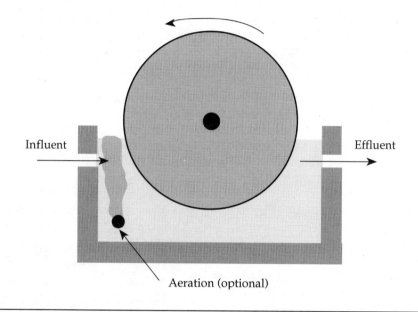

FIGURE 10.12 Cross-sectional view of an RBC unit.

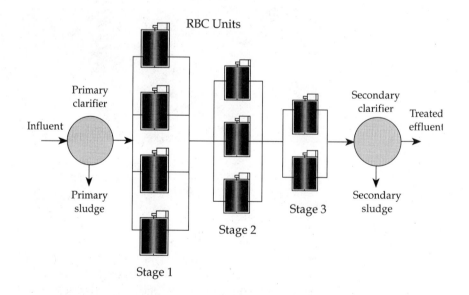

FIGURE 10.13 Schematic of an RBC plant.

recycled, simplicity of operation, and reduced detention times. Some earlier RBCs, used in this country several years ago, experienced mechanical problems with shaft fatigue and bearing failure due to excessive weight from biomass buildup. These problems have been corrected with more recent designs.

The design of RBCs relies on extensive pilot scale studies to determine the removal rates and oxygen requirements. This is true of units used for municipal wastewater as well as industrial wastewater. Some manufacturers have developed loading curves based on the organic load (BOD or COD) per unit surface area. These can be used in some instances, although it is always better to confirm the design where possible by pilot studies.

Activated Sludge. Activated sludge, a suspended growth process, is the most widely used form of wastewater treatment in the United States. It uses an active population of microorganisms suspended in the water (as opposed to attached to a surface). Air is added to the suspension both to mix the liquid and provide oxygen to the microorganisms. Microbial growth performs two functions. First, the microbes oxidize a fraction of the waste materials to carbon dioxide, nitrate, and water. Second, they assimilate most of the remaining colloidal and soluble organics. The microbial population is separated from the water in the final sedimentation process, leaving a water that is low in soluble and suspended organics. The basic form of an activated sludge plant is shown in Figure 10.14.

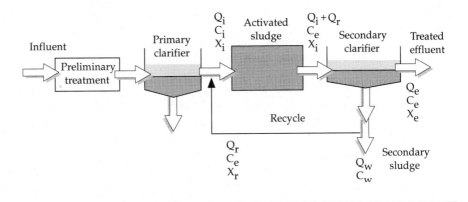

FIGURE 10.14 Schematic of the activated sludge wastewater treatment process.

Several activated sludge process modifications have evolved over the years. The two basic variations are the conventional process or plug flow reactor (PFR), and the high-rate, or complete mix, stirred tank reactor (CSTR). Other modifications are variations of these two. In the conventional activated sludge process (Figure 10.15), the wastewater entering the tank is

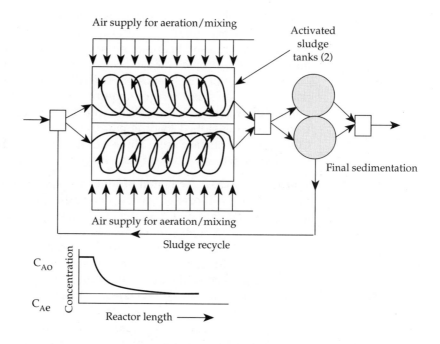

FIGURE 10.15 Conventional activated sludge plant layout.

aerated by submerged diffused aerators (similar in concept to those in common home aquariums) along one side or the bottom of the tank. This placement maximizes radial mixing while minimizing axial mixing. In the high-rate activated sludge process, the tank is mixed by either mechanical surface aerators or submerged diffused aerators in such a way that the concentration is considered uniform throughout. Thus, the influent wastewater is almost immediately mixed and diluted to the effluent concentration. Figure 10.16 is a layout for a high-rate activated sludge secondary plant with two aeration tanks and two final clarifiers. In theory, the conventional process should provide better removal of contamination where the flow is relatively uniform and no toxic compounds are introduced. With its complex mixing, the high-rate process should also provide better dispersal of toxics or **shock loads**. However, in practice, these differences have proven to be much less than expected. The processes are comparable.

As wastewater moves through a plug flow reactor, the organic contaminants are reduced. Since the amount of oxygen required is proportional to the organic contaminant concentration, the amount of oxygen required in the process is initially high but gradually decreases as the water progresses through the tank. Tapered aeration activated sludge is a modi-

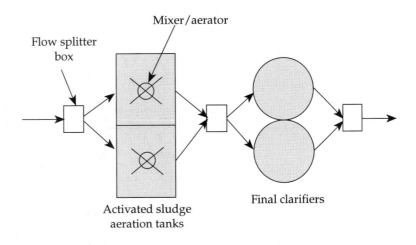

FIGURE 10.16   High-rate activated sludge secondary treatment plant.

fication intended to provide additional aeration capacity to the first part of the tank and less in later stages. In other respects the process is identical to the conventional activated sludge process. A schematic appears at the top in Figure 10.17.

Step feed activated sludge has several feed points along the plug flow reactor, in order to disperse the organic load more evenly throughout the reactor. A schematic of this modification is shown at the bottom in Figure 10.17.

Extended aeration activated sludge uses the same flow pattern as conventional activated sludge, but the objective is to reduce the sludge wasting. The only differences are the increased hydraulic and solids detention times. The process is applicable only to low flow rates, normally under 1 or 2 MGD. The process has been marketed in "packaged treatment plants" with the entire tank system prefabricated of steel and delivered by truck.

Analysis of the activated sludge process is quite complicated. It includes determinations of the rate of generation of solids in the reactor (sludge production) and wasting, the flow rate and concentration of return sludge, the amount of organics oxidized and converted to cellular mass, and other parameters. Advanced courses in wastewater treatment or biological processes usually feature thorough analyses of activated sludge reactors. We will present a brief introduction to the process.

Analysis involves the fundamental principles of mass balance and reactor design presented in Chapter 7. Although many parameters must be determined for an activated sludge plant design, the mean cell residence

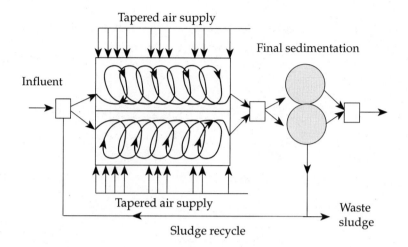

Tapered aeration activated sludge.

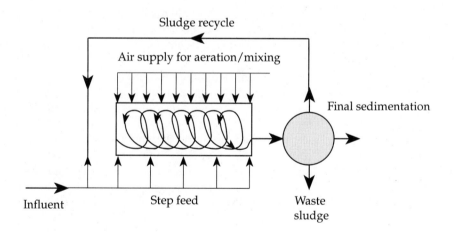

Step feed activated sludge.

FIGURE 10.17 Common activated sludge process modifications.

time and the required hydraulic detention time (determines the reactor volume) are vital. The rate at which the microbial population grows determines the sludge wasting rate and, in part, the reactor detention time.

The mean cell residence time (MCRT, or θ_c) is the average (or mean) time that the microbial cells remain in the system. This is equivalent to the mass of cells in the system divided by the mass of cells leaving the system each day. Since activated sludge systems use sludge or cell recycle, the MCRT is not the same as the hydraulic retention time, θ. The MCRT can be determined by dividing the average mass of cells in the system (the VSS or **MLVSS**) by the mass of cells wasted or lost from the system each day, or

$$\theta_c = \frac{\text{Mass of cells in the reactor}}{\text{Mass of cells wasted per day}} \qquad \textbf{10.4}$$

The mass of cells in the system is the concentration of cells, X, times the reactor volume, V. The mass of cells wasted per day is the concentration of cells wasted, X_w, times the wasting flow rate, Q_w. Cells are also lost in the effluent. This is the concentration of the cells in the effluent, X_e, times the plant effluent flow, Q_e. Thus, the MCRT is

$$\theta_c = \frac{XV}{X_wQ_w + X_eQ_e} \qquad \textbf{10.5}$$

However, for a properly operating system the concentration of cells in the effluent is small. Thus, the mass of cells lost in the effluent is small in relation to the amount wasted. So the equation reduces to

$$\theta_c = \frac{XV}{X_wQ_w} \qquad \textbf{10.6}$$

To determine the hydraulic retention time, θ, we can use a mass balance on the microbial mass in the recycle CSTR. In word form the mass balance equation is

$$\begin{bmatrix} \text{Cell} \\ \text{accumulation} \\ \text{rate} \end{bmatrix}_i = \begin{bmatrix} \text{Cell} \\ \text{input} \\ \text{rate} \end{bmatrix}_i - \begin{bmatrix} \text{Cell} \\ \text{output} \\ \text{rate} \end{bmatrix}_i + \begin{bmatrix} \text{Cell} \\ \text{production} \\ \text{rate} \end{bmatrix}_i \qquad \textbf{10.7}$$

In symbol form it is

$$V\frac{dX}{dt} = \sum_{i=1}^{n} (X_iQ_i)_{\text{in}} - \sum_{j=1}^{n} (X_j\,Q_j)_{\text{out}} + r_{ng}V \qquad \textbf{10.8}$$

If the mass balance is taken around the entire secondary system, the equation becomes

$$V\frac{dX}{dt} = X_oQ - [X_wQ_w + X_eQ] + r_{ng}V \qquad \textbf{10.9}$$

where

X_o = inflow volatile solids concentration [mg/L VSSL]

X_e = effluent volatile solids concentration [mg/L VSS]

X_w = waste sludge volatile solids concentration [mg/L VSS]

Q = volumetric flow rate [L/day]

Q_w = waste sludge flow rate [L/day]

V = reactor volume [L]

r_{ng} = net rate of microbial growth [day^{-1}]

Further, if the system is at steady state, the time-dependent term is zero. And in comparison to the concentration of microorganisms in the system, the inflow concentration of microorganisms is very small. It can be assumed to be zero. Thus, equation 10.9 reduces to

$$X_w Q_w = r_{ng} V \qquad \qquad \textbf{10.10}$$

We found in Chapter 6, Microbial Growth, that the net rate of bacterial growth, r_{ng}, is

$$r_{ng} = -Yr_{su} - k_d X \qquad \qquad \textbf{10.11}$$

If we substitute for r_{ng} in Equation 10.10 we obtain

$$X_w Q_w = (-Yr_{su} - k_d X)V \qquad \qquad \textbf{10.12}$$

If we divide Equation 10.12 by XV, we get

$$\frac{Q_w X_w}{XV} = -\frac{Yr_{su}}{X} - k_d \qquad \qquad \textbf{10.13}$$

The term on the left is $1/\theta_c$. Thus, Equation 10.13 becomes

$$\frac{1}{\theta_c} = -\frac{Yr_{su}}{X} - k_d \qquad \qquad \textbf{10.14}$$

Also in Chapter 6, we found that the rate of substrate utilization, r_{su}, is

$$r_{su} = -\frac{C_{A0} - C_A}{\theta} \qquad \qquad \textbf{10.15}$$

When this is substituted for r_{su} in Equation 10.14, the expression becomes

$$\frac{1}{\theta_c} = \frac{Y(C_{A0} - C_A)}{\theta X} - k_d \qquad \textbf{10.16}$$

Rearranging and solving for θ, we obtain

$$\theta = \frac{\theta_c Y(C_{A0} - C_A)}{X(1 + k_d\theta_c)} \qquad \textbf{10.17}$$

Equation 10.17 can be used to calculate the required hydraulic retention time for a particular wastewater. Knowing the flow rate, we can calculate the volume. Typical detention times for activated sludge systems are from 4 to 6 hours.

EXAMPLE 10.5	ESTIMATION OF ACTIVATED SLUDGE REACTOR VOLUME

Estimate the sludge wasting rate and the required reactor volume of an activated sludge plant. The flow to the plant is 25 MGD. The influent soluble BOD_5, C_{A0}, is 200 mg/L, and the effluent soluble BOD_5, C_A, is 8 mg/L. It is estimated that the plant will have a mean cell residence time of 10 days. The maximum yield coefficient, Y, is 0.6 mg cells/mg substrate, and the endogenous decay coefficient, k_d, is 0.09/day. The reactor is to operate at an MLVSS of 2500 mg/L VSS, and the return sludge concentration (same as the waste sludge concentration) is 9000 mg/L VSS.

SOLUTION: The hydraulic detention time can be computed directly from Equation 10.17:

$$\theta = \frac{\left(10\,\text{days} \times 0.6\frac{\text{mg}}{\text{mg}}\right)(200\,\text{mg/L} - 8\,\text{mg/L})}{(2500\,\text{mg/L})[(1 + (0.09/\text{day})(10\,\text{days})]} = 0.242\,\text{day}$$

$$= 5.8\,\text{hr}$$

The reactor volume can be computed from

$$V = Q\theta = \left(25 \times 10^6\frac{\text{gal}}{\text{day}}\right)\left(0.242\,\text{day} \times \frac{3.78\,\text{L}}{\text{gal}}\right)$$

$$= 22.9 \times 10^6\,\text{L}$$

The sludge wasting rate can now be computed using Equation 10.5. Solving it for Q_w, we get

$$Q_w = \frac{XV}{X_w \theta_c} = \frac{(2500 \, \text{mg/L})(22.9 \times 10^6 \, \text{L})}{(9000 \, \text{mg/L})(10 \, \text{days})}$$

$$= 635,000 \, \text{L/day}$$

So the sludge wasting represents only 2.7% of the plant flow.

Disinfection

Disinfection of wastewater is sometimes required prior to discharge. The process for wastewater is similar to that used in potable water treatment. (See Chapter 9, Water Treatment.) The benefits of wastewater disinfection are the destruction of potentially hazardous pathogenic organisms that are potentially hazardous to humans should they come in contact with the water after it is discharged. (If the water is later used as a water supply source, it would be treated and disinfected prior to distribution.)

Although several disinfection processes are available for wastewater, the most common by far is chlorination. This includes the use of chlorine gas $[Cl_2]$, calcium hypochlorite $[Ca(OCL)_2]$, and sodium hypochlorite $[NaOCl]$. There are several negative aspects to wastewater chlorination, however. Chlorinated water is toxic to aquatic life, particularly small fish. And the process produces chlorinated organics that can be detrimental to downstream users.

Ozonation is an alternative. However, its cost is significantly higher than for chlorination. Also, ozone degrades rapidly after production and must be produced on site. It does not cause chlorinated organics, but extensive research into the compounds created by ozonation have not been completed to date. Ozonation may create other toxic compounds.

Another choice is chlorine dioxide. As with ozone, chlorine dioxide must be generated on site, and it, too, is costly compared to chlorination. It produces fewer chlorinated organics, however.

SLUDGE TREATMENT AND DISPOSAL

Though certainly not glamorous, sludge treatment and disposal is one of the most challenging areas of wastewater treatment for both practicing engineers and operating personnel. Professor Aarne Vesilind of Duke University, who has studied sludge disposal methods and problems extensively, identifies the procedure as the most expensive component of wastewater treatment, representing approximately 30 to 40% of the capital costs, 50% of the operating costs, and 90% of the headaches at a typical treatment plant [6].

Sludge Characteristics

Wastewater **sludge** is a mixture of organic and inorganic solids in water. It is often offensive in character and odor, although fresh waste activated sludge has the odor of moist earth. The color of sludge varies from brown to black. The amount of solids contained in it varies with origin and the amount of processing it has received. Typical concentrations for some sludges are shown in Table 10.5.

TABLE 10.5 Typical Sludge Concentrations

SOURCE	TYPICAL CONCENTRATION, %
Primary sludge, without thickening	2–7
Waste activated sludge	0.5–1.5
Waste trickling filter sludge	1–5
Digested sludge	4–10
Dewatered sludge	12–50

Treated sludge is often used commercially as a soil conditioner. This is advantageous in that it eliminates the need for other ultimate disposal methods such as incineration or landfilling. A negative, however, is that waste sludges contain low levels of toxic compounds, including heavy metals. Table 10.6 lists toxic materials commonly found in municipal sludge and their representative concentrations [7, 8]. It can be seen that typical concentrations are rather low. However, where repeated applications occur, these components tend to build up over time. The increase can eventually reach hazardous levels.

Sludge Treatment

Sludge treatment and disposal divides into several operations: concentrating (thickening), stabilizing, dewatering, and drying the solids and ultimately disposing of them. As indicated in Figure 10.18, most small wastewater plants use either land application or landfilling to dispose of sludge because land is more readily available and land prices are lower in rural areas, which have most of the smaller treatment facilities. Another

TABLE 10.6   Toxic Constituents in Municipal Sludges

CONSTITUENT	RANGE mg/dry kg	TYPICAL mg/dry kg
Chromium	10–99,000	500
Copper	84–17,000	800
Nickel	2–5300	80
Zinc	101–49,000	1700
Cadmium	1–3410	10
PCBs	1.5–9.3	3.8
Lindane		0.8
Chlordane	0.6–19	4.8
Hexachlorobenzene		0.6

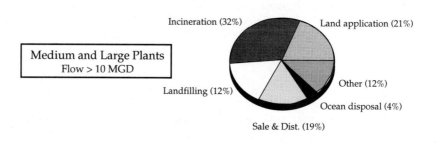

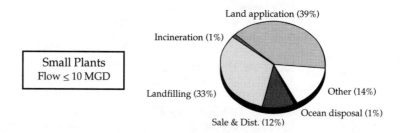

FIGURE 10.18   Ultimate disposal methods for municipal wastewater sludges.

reason is the high cost and mechanical complexity of incineration. In contrast, larger plants often incinerate because it minimizes the use of land and eliminates problems associated with land disposal methods.

A typical example of sludge processing and disposal at a large plant is shown in Figure 10.19, top. In contrast, the process at smaller plants might entail anaerobic digestion followed by drying on sand dewatering beds, and then landfilling, as shown at the bottom in the figure. The following is a brief discussion of some of the more common sludge treatment and disposal operations and processes.

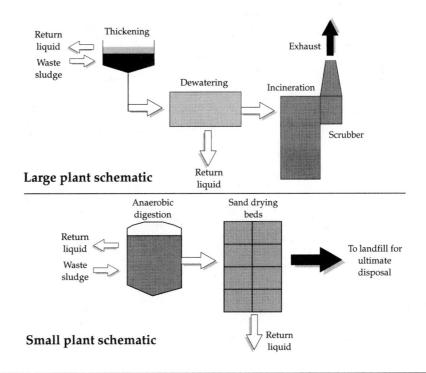

FIGURE 10.19   Sludge treatment and disposal schematics for both a large and a small wastewater treatment plant.

Thickening. Sludge thickening reduces the water in primary and secondary sludges. Due to capital and operating costs, thickening is seldom used at smaller wastewater treatment plants. However, it is almost always used at large facilities.

Thickening can be accomplished in several ways. Common methods include gravity thickening—using an additional clarifier to remove more water. The thickened sludge then proceeds to further processing. The overflow is returned to the head of the plant (before or after primary treatment). Gravity thickening is used more for primary sludge than for secondary sludge.

Dissolved air flotation (DAF) is often used to concentrate secondary sludges. In this process sludge is pressurized and injected with air. After remaining under pressure for 1 to 2 minutes, it is released into a settling tank. When the pressure is released, the extra air comes out of solution (in accordance with Henry's law), attaching to the sludge particles as microscopic bubbles. The sludge is thus floated to the surface in a concentrated form. The underflow, or subnatant, returns to the head of the plant. DAF can obtain concentrations of 3 to 6% (versus 0.5 to 1.5% before thickening).

Sludge Stabilization.   Stabilization alters the characteristics of sludge so it can be returned to the environment with a minimum of environmental and health risks. The most common stabilization methods are **anaerobic** and **aerobic** digestion. Lime stabilization and heat treatment are used to a lesser extent.

Anaerobic Digestion.   Anaerobic biological treatment of sludge will decrease the volatile organics by 40 to 50% and reduce the numbers of pathogenic organisms in sludge. It is accomplished by holding the sludge in closed tanks for periods of 10 to 90 days. Older versions of anaerobic digestion used unmixed, unheated tanks. This results in very long detention times (30 to 90 days). However, more recent processes involve complete mixing and heating to temperatures of 35 to 40°C, reducing detention time to 10 to 20 days (and thus bringing about a significant reduction in digester volume). A schematic of a typical complete mix, or high-rate, anaerobic digester appears in Figure 10.20.

The advantage of anaerobic digestion include the production of usable energy (in the form of methane gas), low solids production, very low energy input (if the methane gas is used to heat the digester). Disadvantages include high capital costs, susceptibility to upsets from shock loads or toxics, and complex operation requiring skilled operators.

Aerobic Digestion.   In aerobic treatment the sludge is aerated for an extended time, typically 12 to 20 days. During this time the amount of biological material is reduced to about half its original amount. The tanks used for aerobic digestion are usually constructed identical or very similar to those used for the activated sludge process. The benefits of aerobic digestion are its ease of operation and process stability. The drawbacks include high energy input for the aeration and mixing.

Lime Stabilization.   Adding lime to raise the pH of the sludge high enough to reduce biological activity is another approach. With biological activity held to a minimum for a long enough period, the sludge can be dewatered and landfilled. The pH must be elevated to above 12 [9]. However, the

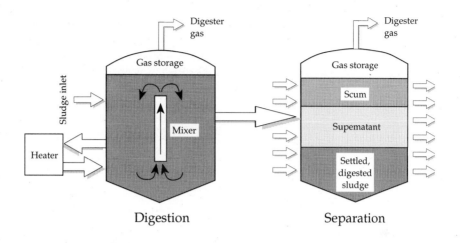

FIGURE 10.20 Complete mix, or high-rate, anaerobic digester.

process does not sterilize the sludge. So, as the pH comes down due to reaction with materials in the sludge, organisms will become active again, producing noxious odors. Lime stabilization also does not reduce the quantity of sludge, as does biological stabilization. The process is simple and requires little equipment. Its disadvantages are the relatively short time it can prevent biological activity and its lack of solids reduction.

Heat Treatment. Heating sludge under pressure to temperatures in the range of 200 to 300°C for a few minutes can effectively sterilize it and convert it to a form that is easily dewatered. The major disadvantages of heat treatment are its high energy requirement and the production of a high-strength return liquid from the dewatering process. Heat treatment is used only in a few large POTWs.

Dewatering and Drying. Where sludge is not incinerated or land applied, it must be dewatered and/or dried. This can be achieved by applying the sludge to sand drying beds or by using mechanical dewatering equipment.

Sand drying beds consist of a layer of sand with an underdrain system. The sludge is pumped onto the bed. Much of the water then drains through the sand and is returned to the head of the plant. The sun and wind dry the material further. This method is frequently used at smaller POTWs. It is less successful in areas with long, cold winters, or with high annual precipitation since the drying beds are usually not covered or enclosed. The beds also require that there be enough storage capacity in the digesters, sludge storage tanks, or lagoons to hold the sludge during wet or extremely cold periods.

Mechanical dewatering equipment is often used at medium-sized plants, and almost always used at larger POTWs. In this operation, the sludge is applied to a metal, cloth, or synthetic rubber surface. The liquid is then pressed out of the sludge and returned to the head of the plant. This equipment is automated but an experienced, well-trained operator is required. Mechanical dewatering is often preceded by sludge thickening when incineration is used. When digestion is used, dewatering is the last step before disposal. (See Figure 10.19, top, for a typical process that includes mechanical dewatering.)

Ultimate Disposal

Ultimate disposal is the return of the material to the environment. This can be in the form of landfilling, land application, incineration, or other methods, as shown in Figure 10.18. Following either aerobic or anaerobic digestion, the sludge is usually land applied in wet form (not dewatered), dewatered and landfilled, or used as a sludge conditioner. As noted earlier, care must be taken in applying sludge to land, so that excessive concentrations of heavy metals or other toxic materials do not accumulate in the soil.

Incineration is often used by larger municipalities. Its advantages include maximum volume reduction, detoxification, and energy recovery. On the down side, both capital and operating costs are high, and there are environmental effects (air discharges, scrubber sludge generation), operational problems, and the continuing need for trained operating personnel.

ADVANCED WASTEWATER TREATMENT

Advanced wastewater treatment is the removal of any dissolved or suspended contaminants beyond secondary treatment. This is often the removal of the nutrients nitrogen and/or phosphorus, but it can also include the removal of toxic metals, toxic organics, or the attainment of very low levels of BOD or suspended solids. Advanced wastewater is now used for 26% of the U.S. population. It is expected to be required for 42% of the population when all documented needs are met [10]. Methods of advanced treatment vary with the materials or contaminants being removed. We will examine the removal of nitrogen and phosphorus here.

Nitrogen Removal

As with phosphorus, nitrogen is removed from wastewater to reduce algal growth in reservoirs and pristine streams. However, unlike phosphorus, nitrogen in the ammonia (or ammonium) from also represents a considerable oxygen demand on a receiving water when not removed or converted to nitrate. Thus, nitrogen treatment can take two forms depending on the objective. Conversion of the ammonium into nitrate will remove the oxygen demand of the ammonia. Removal of the nitrogen using nitrification–dentrification, air stripping, or adsorption/ion exchange will remove the oxygen demand of the ammonium and eliminate the nutrient effect as well. We will consider these two possibilities for nitrogen treatment in turn.

Nitrification. Ammonium in wastewater can be biologically oxidized to nitrate. The wastewater so treated will not require oxygen from the stream to oxidize the ammonia. This process, termed **nitrification**, results from the activity of several groups of microorganisms naturally present in surface-waters and which can be grown in wastewater treatment plants. One group, *Nitrosomonas*, converts ammonium to nitrite:

$$NH_4^+ + 1\tfrac{1}{2}O_2 \rightarrow NO_2^- + H_2O + 2H^+ \qquad\qquad \textbf{10.18}$$

The second step, which converts the nitrite to nitrate, is performed by *Nitrobacter* and other organisms. This reaction is

$$NO_2^- + \tfrac{1}{2}O_2 \rightarrow NO_3^- \qquad\qquad \textbf{10.19}$$

The overall reaction is

$$NH_4^+ + 2O_2 \rightarrow NO_3^- + H_2O + 2H^+ \qquad\qquad \textbf{10.20}$$

Nitrification often occurs to some extent in wastewater treatment plants during the summer months when water temperatures are sufficiently high for the nitrifying population to grow rapidly enough to sustain itself. During winter months, particularly in the northern half of the United States and in Canada, nitrification does not occur at most plants not specifically intended for the purpose.

Nitrification requires relatively light loading on the biological treatment system, but processes designed for nitrification are otherwise quite similar to conventional biological treatment systems [11]. Most biological treatment processes, including both fixed film and suspended growth, can be used for nitrification. It is usually accomplished with a postsecondary treatment system (separate stage), but some systems have been designed as an integral part of the secondary treatment system. A typical separate stage biological nitrification process (as well as denitrification) is indicated in Figure 10.21.

Denitrification. **Denitrification** is the anoxic conversion of nitrate to nitrogen gas. Thus, it must be preceded by nitrification. It occurs naturally under some conditions and can be accomplished by design in wastewater treatment systems. Since the oxygen demand has already been met, the only purpose of denitrification is the removal of the nutrient properties of the nitrate. The overall reaction for denitrification is

$$NO_3^- + organics \rightarrow N_2 \uparrow + CO_2 + H_2O \qquad\qquad \textbf{10.21}$$

Although secondary treated wastewaters contain residual organics, the general practice in engineered denitrification is to supply methanol as an organic source. The process requires an additional reactor and sedimentation basin since it occurs only at very low oxygen levels (or no oxygen)—situations inconsistent with nitrification. The process is shown schematically in Figure 10.21.

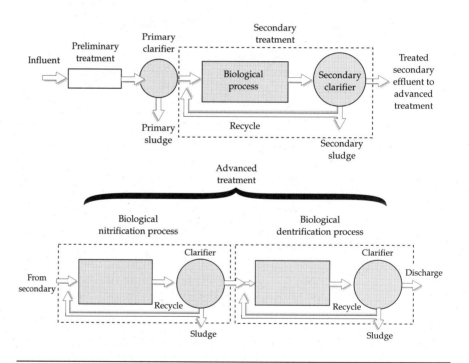

FIGURE 10.21 Biological nitrification.

Air Stripping. Ammonia can also be removed by air stripping to the atmosphere [11]. Ammonium cannot be removed directly because it is a charged ionic species. The form of ammonia (either ammonia or ammonium) is pH-dependent. Most ammonia exists in wastewater in the ammonium form because the pH is usually less than 9.3. Removing ammonium requires converting it to the uncharged ammonia by adding a base, usually lime or sodium hydroxide. In practice the pH is usually raised above 10.5. The high-pH water is then sprayed into the air with a device similar to a cooling tower. Since the solubility of ammonia in water is very low (Henry's law again), the ammonia comes out of solution and into the air. The equilibrium reaction for ammonium conversion to ammonia is

$$NH_4^+ + OH^- \rightarrow NH_3\uparrow + H_2O \qquad\qquad \textbf{10.22}$$

The equilibrium reaction for the solubility of ammonia in water, using Henry's law, is

$$NH_3(g) \rightleftarrows NH_3(aq) \qquad\qquad \textbf{10.23}$$

After the stripping operation, the pH of the water is adjusted back down by adding an acid such as sulfuric acid. This process has two serious negative aspects: It is difficult to operate in cold weather because the air stripping towers tend to ice, and it uses considerable amounts of chemicals. However, it has been used in a few applications.

Phosphorus Removal

Phosphorus is taken out of some waters to prevent excess growth of algae, particularly where the wastewater is discharged into a reservoir or pristine stream. Phosphorus exists in wastewater as orthophosphate ($H_2PO_4^-$, HPO_4^{2-}, PO_4^{3-}), polyphosphates, and organic phosphates. The phosphate and polyphosphates can be removed from wastewater by chemical precipitation, accomplished by the addition of multivalent metallic ions (Ca^{2+}, Al^{3+}, or Fe^{3+}) that form insoluble phosphates. Precipitation can be done at primary or secondary sedimentation stages. The addition of the coagulant at primary results in an increase in the efficiency of the primary sedimentation operation, but care must be taken not to remove so much phosphorus that the microorganisms in the secondary biological process are limited in growth by phosphorus. However, this has not generally been a problem.

Some coagulants used for phosphorus removal are ferric chloride, aluminum sulfate (alum), and calcium hydroxide (lime) [12]. Although other reactions occur with the phosphorus and the metals, the following reactions are generally accepted as being predominant in phosphorus precipitation.

For aluminum ions:

$$Al_2(SO_4)_3 + 2HPO_4^{2-} \rightarrow 2AlPO_4\downarrow + 3SO_4^{2-} + 2H^+ \qquad\qquad \textbf{10.24}$$

For ferric ions:

$$FeCl_3 + HPO_4^{2-} \rightarrow FePO_4\downarrow + 3Cl^- + H^+ \qquad\qquad \textbf{10.25}$$

With lime precipitation, the lime reacts preferentially with the alkalinity in the water, producing calcium carbonate. After the alkalinity is removed, the phosphorus is precipitated as calcium hydroxylapatite. The phosphorus precipitation reaction is as follows:

$$3HPO_4^{2-} + 5Ca^{2+} + 4OH^- \rightarrow Ca_5OH_4(PO_4)_3 + H_2O \qquad \textbf{10.26}$$

EXAMPLE 10.6	PHOSPHORUS REMOVAL USING FERRIC CHLORIDE

Estimate the amount of ferric chloride, in kg/yr, required to remove 8 mg/L of phosphorus (as P) from the wastewater of a 30-MGD plant. Assume that a 50% excess of ferric chloride is required.

SOLUTION: The first step is to determine the mass of phosphorus to be removed each year.

$$m_P = CV = \left(8\frac{mg\,P}{L} \times \frac{Kg}{10^6\,mg}\right) \times$$

$$\left(30 \times 10^6\,\frac{gal}{day} \times \frac{365\,days}{yr} \times \frac{3.78\,L}{gal} \times 1\,yr\right)$$

$$= 3.31 \times 10^3\,Kg\,P$$

The amount of ferric chloride required per unit mass of phosphorus can be obtained from Equation 10.25:

$$FeCl_3 + HPO_4^{2-} \rightarrow FePO_4\downarrow + 3Cl^- + H^+$$

162.3   30.1

(Note that since the mass of phosphate is given as P, the molecular weight of phosphorus, not phosphate, is used in the calculation above.) The mass of ferric chloride is then

$$M_{FeCl_3} = (3.31 \times 10^3\,Kg\,P)\left(\frac{162.3\,Kg\,FeCl_3}{30.1\,Kg\,P}\right)$$

$$= 1.79 \times 10^6\,Kg/yr$$

If a 50% excess is required, the total ferric chloride used per year would be 2.68×10^6 kg.

The Bissell Point Wastewater Treatment Plant, St. Louis, Missouri

The Bissell Point Wastewater Treatment Plant in St Louis, Missouri, is designed to treat approximately 150 million gallons of wastewater per day. Under a legal consent decree, the St. Louis Municipal Sanitary District agreed to upgrade the facility from primary to secondary treatment.* The engineering firm contracted for the design, Sverdrup Corporation, selected a combination biological treatment process which includes packed towers followed by activated sludge aeration. The combined biological treatment is then followed by secondary clarifiers. Waste sludge is dewatered and then incinerated. A flow schematic of the facility is shown in Figure 10.22.

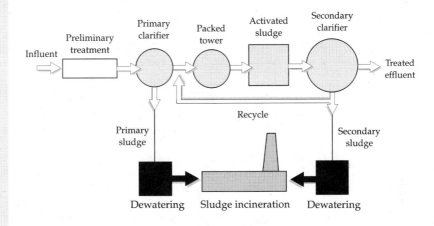

FIGURE 10.22 Schematic of the Bissell Point Wastewater Treatment Facility in St. Louis, Missouri.

Sverdrup and St. Louis MSD determined that six packed towers would be required. Each packed tower is 32 ft in height with a diameter of 134 ft. The packed towers are followed by six activated sludge aeration tanks with a volume of 550,000 ft^3 each. The average detention time in the activated sludge system is 4 hours which further reduces the BOD_5.

Final sedimentation consists of 12 final clarifiers, each having a diameter of 150 ft. This provides an overflow rate of 707 Gal/ft^2-day at average flow and 1180 Gal/ft^2-day at peak flow.

---

*Most wastewater treatment facilities throughout the nation were upgraded to secondary treatment by the mid or late 1980s. However, inadequate funding from district voters delayed improvements to this facility for several years.

The cost for expanding the facility from primary treatment to secondary treatment was approximately $250,000,000. This included design and construction of the packed towers, aeration tanks, final clarifiers, sludge handling and treatment, piping, pumping, and automated control systems.

INDUSTRIAL WASTEWATER TREATMENT

Industrial facilities must either treat wastewater prior to discharge into a receiving water, or send it to a POTW, or in some cases, do both. Industries in areas not served by a POTW must obtain an NPDES permit from the EPA. Industries in areas served by a POTW may discharge to the public sewer system, but they must pretreat the wastewater if it contains toxic materials that will pass through the system or that will interfere with the treatment process. Industrial dischargers must also pretreat any wastewater that has a sufficiently high organic or solids content that it will interfere with the POTW operation. Thus, pretreatment is not necessary if the industry is discharging wastewater comparable in strength to typical municipal wastewater and which does not contain appreciable amounts of toxic materials.

Industrial Pretreatment

Since industrial wastewater composition is highly variable, its treatment is very site- or industry-specific. Treatment methods can include ion exchange or precipitation of toxic metals, carbon adsorption of toxic organics, or simply biological processes to reduce the organic content of a high-strength industrial wastewater prior to discharge. Individual municipalities are responsible for enforcing the requirements for industries using their sewer systems.

Direct Discharge with an NPDES Permit

Industries not discharging to a municipal sewer system must have an NPDES permit issued by the state or the EPA. This permit will list the allowable concentrations and total mass of discharges allowed. (Such permits will often have an effluent concentration limit and a maximum mass discharge allowed per day, week, or month.) Treatment systems for direct industrial dischargers include systems similar to those already discussed, but because their water goes directly to the receiving water without benefit of additional treatment at a POTW, these industries must attain lower discharge limits for suspended solids and organics (measured as BOD or COD) than industries that pretreat.

Review Questions

1. Describe the major provisions of the current water quality regulations as they relate to municipal wastewater treatment.

2. Describe the major provisions of the current water quality regulations as they relate to industrial wastewater treatment.

3. Define or explain the following:
   - **a.** activated sludge
   - **b.** sedimentation
   - **c.** attached growth
   - **d.** suspended growth
   - **e.** preliminary treatment
   - **f.** secondary treatment
   - **g.** advanced treatment
   - **h.** disinfection

4. What are the purposes of grit removal?

5. Select a local industry or POTW. Contact the nearest federal or state EPA office (in writing or by telephone, as your instructor directs) and obtain the discharge reports of the facility for the last 12 months. Estimate the amount of BOD and SS that are removed. Did the facility comply with its report for the entire period? List any non-compliance periods, the parameter(s) not met, and any reason for the non-compliance.

6. Draw a schematic of the following wastewater treatment systems. Include all preliminary operations, show where and how oxygenation occurs, and describe the hydraulic flow regime where indicated.
   - **a.** Conventional activated sludge process
   - **b.** High-rate activated sludge process
   - **c.** Trickling filter process
   - **d.** RBC process

7. A city of 50,000 people has an average wastewater flow of 7 MGD. Assuming two units are to be constructed, estimate the area and diameter of the primary clarifiers and the depth of the basins needed. What would the overflow rate and detention times be at a peak flow of 2.5 times average?

8. What primary clarifier area is required per person for a typical average wastewater flow? What volume?

9. As a consulting engineer, you are asked to estimate the size of a packed tower for a city of 50,000 population. The average wastewater flow is 7 MGD with a BOD_5 of 170 MG/L. The laboratory at your firm has constructed a pilot 20-ft packed tower to estimate the k value for the Schultz–Germain equation. It was found to be $0.85 \, ft^{1/2}\text{-} day^{1/2}$. Your supervisor has determined that there will be two packed towers with heights of either 20 or 30 ft. You should calculate the areas of the towers

for both heights if the effluent BOD_5 must be 10 MG/L. The constant k varies with height according to this relationship:

$$k_2 = k_1 \left(\frac{h_1}{h_2} \right)^{0.3}$$

where k_1, k_2 are the reaction constants at heights h_1, h_2.

10. A packed tower is to be used to treat the waste from a small municipality. The population is 800 and the wastewater flow is 100,000 gal/day. Estimate the size of the units if one packed tower and if two packed towers are to be used. The influent BOD_5 is 180 mg/L, and the required effluent BOD_5 is 25 mg/L. The Schultz–Germain k value was estimated to be 0.6 $ft^{1/2}$- $day^{1/2}$ for a unit 20-ft high.

11. Estimate the volume of sludge wasted per day for an activated sludge plant with a plant flow of 10 MGD. The hydraulic retention time in the activated sludge tank is 5 hr, the mixed liquor volatile suspended solids is 3000 mg/L, the waste sludge concentration is 9000 mg/L, and the mean cell residence time is 10 days. What percentage of the plant flow goes to sludge wasting?

12. Estimate the required reactor volume of an activated sludge plant if the flow to the plant is 10 MGD, the influent soluble BOD_5 is 250 mg/L and the required effluent soluble BOD_5 is 10 mg/L. You may assume that the plant will have a mean cell residence time of 10 days. The maximum yield coefficient is 0.6 mg cells/mg substrate, and the endogenous decay coefficient is 0.07/day. The reactor is to operate at an MLVSS of 3000 mg/L VSS.

13. A plant receives 25 MGD of wastewater with the characteristics shown in the accompanying table. Estimate the required hydraulic retention time and the volume of the activated sludge tank for a high-rate activated sludge plant. Also estimate the area required for the final clarifier.

PARAMETER	VALUE
Influent soluble BOD_5	150 mg/L
Required effluent soluble BOD_5	15 mg/L
Maximum yield coefficient, Y	0.55 mg cells/mg BOD
Endogeneous decay coefficient, k_d	0.006/day
Reactor MLVSS	2400 mg/L VSS
Estimated mean cell residence time, θ_c	10 days

14. A typical secondary clarifier overflow rate is $700 \, \text{gal}/\text{ft}^2$-day. If two circular clarifiers are used in a 10-MGD wastewater plant, what diameter would be required? (Since clarifiers are usually available only in 5-ft increments, round your answer up to the nearest 5 ft.)

15. A 10-MGD wastewater plant plans to use alum (aluminum sulfate) for phosphorus removal. Estimate the amount of alum required per year if the average phosphorus content of the water is $6 \, \text{mg}/\text{L}$ and a 50% excess of alum is required. How much aluminum phosphate precipitate will be produced each year? How much will the sulfate concentration increase in the effluent from the plant?

References

1. J. M. Kovalic, *The Clean Water Act of 1987* (Alexandria, VA: Water Pollution Control Federation, 1987). (An excellent discussion of U.S. water quality laws and their history).

2. Great Lakes–Upper Mississippi River Board of State Sanitary Engineers, *Recommended Standards for Sewage Treatment Works*, 1978 ed. (Albany, NY: Health Education Service, Inc., 1978).

3. L. Metcalf and H. P. Eddy, *American Sewerage Practice, Vol. III, Disposal of Sewage*, 2nd ed. (New York: McGraw-Hill, 1916).

4. K. L. Schultz, "Load and Efficiency of Trickling Filters," *Journal of the Water Pollution Control Federation*, vol. 32, no. 3 (March 1960), pp. 245–61.

5. J. E. Germain, "Economic Treatment of Domestic Waste by Plastic-Medium Trickling Filters," *Journal of the Water Pollution Control Federation*, vol. 38, no. 2 (February 1966), pp. 192–203.

6. P. A. Vesilind, P. Aarne, *Treatment and Disposal of Wastewater Sludges*, rev. ed. (Ann Arbor, MI: Ann Arbor Science, 1979).

7. *Environmental Regulations and Technology: Use and Disposal of Municipal Wastewater Sludge* EPA 625/10-84/003 (Washington, DC: U.S. Environmental Protection Agency, September 1984).

8. *Process Design Manual for Municipal Sludge Landfills*, EPA-625/1-78-010 (Washington: EPA, October 1978).

9. *Process Design Manual for Sludge Treatment and Disposal*, EPA 625/1-79-011 (Washington: EPA, September 1979).

10. *1988 Needs Survey Report to Congress: Assessment of Needed Publicly Owned Wastewater Treatment Facilities in the United States*, EPA 430/09-89-001 (Washington: EPA, February 1989).

11. *Process Design Manual for Nitrogen Control* (Washington: EPA, October 1975).

12. *Process Design Manual for Phosphorus Removal,* EPA 625-1-76-001a (Washington: EPA, April 1976).

Additional References

H. J. Glynn and G. W. Heinke, *Environmental Science and Engineering* (Englewood Cliffs, NJ: Prentice-Hall, 1989).

T. J. McGhee, *Water Supply and Sewerage,* 6th ed. (New York: McGraw-Hill, 1991).

Metcalf and Eddy, Inc., (revised by G. Tchobanoglous and F. L. Burton), *WASTE-WATER ENGINEERING, Treatment/Disposal/Reuse,* 3rd ed. (New York: McGraw-Hill, 1991).

H. S. Peavey, D. R. Rowe, and G. Tchobanoglous, *Environmental Engineering* (New York: McGraw-Hill, 1985).

S. R. Qasim, *Wastewater Treatment Plant Design* (New York: HRW, 1985).

Wastewater Treatment Plant Design, WPCF Manual of Practice No. 8 (Alexandria, VA: Water Pollution Control Federation, 1977).

11 Solid Waste Disposal

`` AMERICANS PRODUCE MORE AND MORE SOLID WASTE EACH YEAR; WE GENERATE MORE PER CAPITA THAN ANY OTHER NATION. BUT, AT THE SAME TIME THAT WE GENERATE MORE WASTE, WE ARE RUNNING OUT OF PLACES TO DISPOSE OF IT. LANDFILL CAPACITY IN SOME PLACES IS ALMOST FILLED TO THE SATURATION POINT, AND SOLID WASTE FACILITIES CONTINUE TO BE DIFFICULT TO SITE BECAUSE OF PUBLIC RESISTANCE, COMMONLY KNOWN AS THE 'NOT IN MY BACKYARD' (NIMBY) SYNDROME.''

J. WINSTON PORTER

Asssistant Administrator

U.S. Environmental Protection Agency
Solid Waste and Energy Response
February 1989

Gallatin National Landfill located near Fairview, Illinois. The landfill disposes of municipal solid waste and MSW incinerator residue. Photograph courtesy of SIS Consultants Ltd.

INTRODUCTION

Solid waste disposal is becoming more difficult with each passing year. Our "disposable society" is partly to blame. As someone who grew up in Missouri, I can recall a time when you could not buy Coca-Cola® in anything but refillable bottles. Today, you can hardly find a refillable bottle of anything. In fact, the last Coke that I purchased in a refillable bottle was as a souvenir at a hotel in Japan in 1988. In addition to our disposable society attitude, another problem is our increased population. Cities and states are finding the land necessary for traditional disposal, landfilling, less and less available. In addition, community opposition to landfills, or even incineration, is increasingly forceful. This leaves both the responsible government bodies, and the engineers involved, with an increasingly difficult task. The NIMBY* syndrome is alive and well. And, it is understandable. Would you want a landfill in *your* backyard? Probably not. Yet, because of our increased population density, almost everyplace is someone's backyard. Surprisingly, with all of the negatives, there are signs of improving attitudes. With improved attitudes comes other environmental progress. We, as a nation, are becoming more aware of recycling. Only a few years ago, everything used to go into the trash. Now many communities are adopting organized recycling programs. Even without community-sponsored plans, businesses and industries are supporting the effort. Other communities are beginning to compost yard wastes instead of landfilling them. There really are encouraging trends.

The hierarchy of solid waste disposal is shown in Figure 11.1. The best option is to avoid creation of the waste material. Obviously, we cannot eliminate the generation of solid waste. However, there is ample opportunity to significantly reduce the amount of waste created. Elimination of excess packaging and disposable products are two examples. The second best option is to recycle unwanted materials rather than disposing of them. In this category are recycling of plastic, glass, paper, and metals. The next option is to use the waste materials for energy recovery by use of a solid waste incinerator that produces usable energy. Landfilling is the least desirable option for solid waste disposal, yet our current disposal practices rely heavily on landfilling. Approximately 80% of all solid waste is currently landfilled. The remaining 20% is evenly divided between recycling and incineration. There is ample evidence that the trend is toward more recycling, composting, and incineration with energy recovery.

---

*NIMBY, not in my back yard!

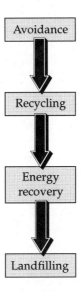

FIGURE 11.1 Solid waste disposal hierarchy.

SOLID WASTE SOURCES

The people of the United States dispose of more solid waste per person than the people of any other nation on earth. Sources of solid waste in the United States can be grouped into four broad categories: municipal, industrial, mineral extraction, and agricultural. The waste from municipalities, that is, the wastes from residences and commercial businesses, accounts for approximately 4 lb per person per day. Industrial sources account for an additional 6 lb per person per day. Agricultural wastes do not generally fall within the realm of environmental engineering and will not be discussed here. However, if all human sources are considered, including construction, minerals extraction (mining) and agriculture, the total solid waste produced in this country is an astounding 90 lb per person per day!

Municipal Waste Sources

Municipal solid waste (MSW) consists of the materials discarded primarily from residences but also includes business and commercial waste (not industrial waste). What's thrown away—the total quantity discarded as well as the amount discarded per person—has continued to increase. The EPA predicts, however, that new regulations will reduce the amount of waste produced per person, resulting in a plateau in total material discarded [1, 2]. The history of municipal solid waste disposal from 1960 to the present, and anticipated to the year 2000, is shown in Figure 11.2. (The figures for 1990 and before are actual quantities: those after 1990 are EPA predictions.) The EPA predicts that more recycling and energy recovery in the future will decrease the total amount of material that must be landfilled. Federal officials believe the amount of materials being landfilled peaked near the end of the last decade, or about 1990.

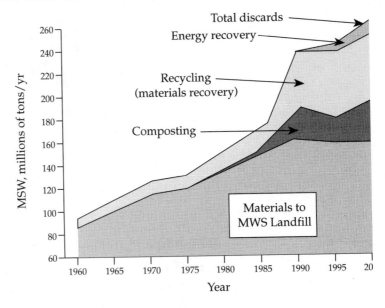

FIGURE 11.2   Municipal solid waste disposal in the United States from 1960 to 2000.

Many municipalities and/or states have implemented either voluntary or mandatory recycling and composting programs [3, 4]. A particularly easy source to eliminate from landfilling and incineration is yard waste, which can be composted and converted into a usable end product for landscaping or soil conditioning. Paper, glass, plastics, and metals can also be recycled. One problem with recycling, though, is that many times it is more costly than purchasing new materials. Only government regulations, public pressure, and financial incentives will bring about lasting changes in disposal and recycling practices.

Types of Municipal Solid Waste

Municipal solid waste consists of the materials discarded from residences and businesses. Figure 11.3 presents the percentages of different components in the municipal solid waste stream for 1990. We will focus on each of

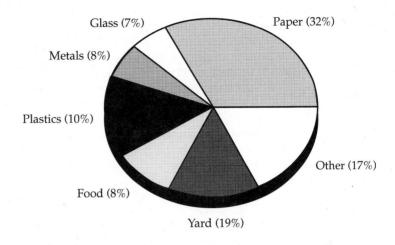

FIGURE 11.3   Composition of municipal solid waste in 1990.

these waste categories, discussing the possibility of recycling or reuse rather than landfilling or incineration.

Paper Wastes.   If they are not contaminated, newspapers, magazines, books, packaging, and other materials composed of paper or cardboard can be recycled. Waste paper is approximately 35% of municipal solid waste.

Paper is recycled by repulping the waste and then mixing it with some portion of new paper or pulp. The higher the recycle content, the smaller the amount of pulp that must come from newly cut trees. Thus recycling paper not only conserves landfill space but also reduces the demand on limited natural resources.

Yard Wastes.   The amount of leaves, grass clippings, brush, and limbs varies considerably with the season, as well as the area of the country. Grass clippings predominate in the summer, leaves in the fall and early winter. These materials represent approximately 20% of all municipal waste when averaged over a yearly period. Leaves alone, however, can represent as much as 50% of the municipal solid waste stream during peak periods in the fall.

Glass Wastes.   Bottles, glasses, jars, and other glass containers that are no longer used account for approximately 7% of all municipal wastes. Almost all glass wastes could be successfully recycled in a well-developed program. Many glass beverage bottles could be reused directly by returning them to their original bottling company where they could be washed, sterilized, and refilled. Instead, today, most bottles that are recycled are returned to a glass manufacturer, which grinds the glass, remelts it, and then reforms it into a new product. This requires much more energy than washing and refilling. However, both forms of recycling reduce the demand for natural resources and save limited landfill space.

Metal Wastes. Beverage and food containers, other small household containers, scrap appliances, and miscellaneous metal products represent another 8% or so of the total solid waste disposal load. Most communities now have recycling programs for aluminum cans, although these programs are often sponsored by private companies rather than the city government. However, most other metal products can be recycled in effective municipal programs. Recycled metal is normally returned to a foundry where it is melted and blended with other metal to obtain the desired alloy.

Plastic Wastes. Representing 10% of the total solid waste load, most plastic products now have the standardized plastic recycle code indicating the type of material they are made of. The codes are shown in Table 11.1. This system assists in the recycling process. Plastics are a large component of the fast-food industry. Some fast-food companies have begun to eliminate and/or recycle the plastics used in their restaurants.

Food Wastes. Wastes from food preparation and consumption have declined from 15% of the solid waste stream in 1960 to 8% as of 1985. This

TABLE 11.1 Plastic Recycling Codes

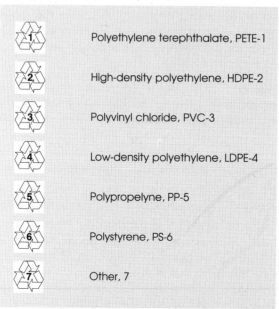

1	Polyethylene terephthalate, PETE-1
2	High-density polyethylene, HDPE-2
3	Polyvinyl chloride, PVC-3
4	Low-density polyethylene, LDPE-4
5	Polypropelyne, PP-5
6	Polystyrene, PS-6
7	Other, 7

change does not mean American consumers are producing less waste. Instead, it indicates the overall change in lifestyle—from meals prepared totally at home a few years ago to the dramatic increase in availability of ready to serve or "heat and eat" meals. Also, many households now have garbage grinders and compactors. Thus, whereas many other solid waste categories have increased markedly, food wastes have declined as a percentage of the total over the past 25 years. Food wastes offer little prospect for recycling or alternative disposal other than for citizens to waste less.

Alternatives and Trends. Of all those wastes, paper products are the largest portion in municipal load, 32%, with yard wastes comprising an additional 20%. Elimination of these two wastes alone, by recycling and composting, would reduce municipal solid waste disposal by more than half! Other materials that could be recycled include glass, metals, and plastics. Thus, comprehensive recycling and composting have the potential to reduce the solid waste burden to landfills and incinerators by more than 75%.

Table 11.2 summarizes some characteristics of solid waste generation and the implications for collection and disposal.

TABLE 11.2 Miscellaneous Generation and Density Figures for Solid Waste

PARAMETER	VALUE
MSW generation rate	4 lb/capita-day
Industrial solid waste generation rate	5 lb/capita-day
Uncompacted density of MSW	200 to 300 lb/yd^3
Compacted in collection vehicle	400 to 600 lb/yd^3
Compacted in MSW landfill	700 to 1000 lb/yd^3

Table 11.3 presents the quantities of solid wastes discarded and recycled from 1960 to the present [1, 2]. It also has estimates for future disposal quantities through the year 2000. This array indicates trends. For instance, there were almost no plastic wastes in 1960, where now some 16 million tons of plastic waste, or roughly 10% of the total discard stream, is produced each year. Paper wastes have also increased dramatically during the 30-year period from 1960 to 1990. They now are about 52 million tons per year, or one-third of the total discards! It is reasonable to expect changes in the waste stream over the next 30-years as well. Many states have started vigorous programs to limit the amount of landscape or yard waste placed in landfills. Some states have specific bans on this kind of

TABLE 11.3   Characterization of Municipal Solid Waste in the United States, 1960 to 2000

MATERIAL	1960	1965	1970	1975	1980	1985	1990	1995*	2000*
Paper	24.5	32.3	36.8	34.8	42.8	48.4	52.4	52.1	50.9
Glass	6.6	8.6	12.5	13.1	14.2	12.2	10.6	9.7	8.8
Metals	10.4	10.7	13.7	13.6	13.3	12.7	12.4	11.1	10.8
Plastics	0.4	1.4	3.1	4.5	7.8	11.5	15.9	19.1	22.3
Rubber/ leather	1.7	2.3	2.9	3.7	4.2	3.6	4.4	5.6	6.1
Textiles	1.7	1.9	2.0	2.2	2.6	2.8	5.3	5.5	6.1
Wood	3.0	3.5	4.0	4.4	6.7	8.2	11.9	12.6	14.4
Other	0.1	0.0	0.5	1.3	2.4	2.9	2.4	2.2	2.8
Food	12.2	12.7	12.8	13.4	13.2	13.2	13.2	13.2	13.2
Yard	20.0	21.6	23.2	25.2	27.5	30.0	30.8	22.6	17.1
Misc. inorganic	1.3	1.6	1.8	2.0	2.2	2.5	2.9	2.3	2.9
Discards to landfill	81.9	96.6	113.3	118.2	136.9	148.0	162.2	156.0	155.4
Energy recovery	0.0	0.2	0.4	0.7	2.7	7.6	29.7	22.5	32.0
Recycling	5.9	6.8	8.6	9.9	14.5	16.4	29.2	40.8	50.9
Composting	0.0	0.0	0.0	0.0	0.0	0.0	4.2	11.1	15.8
Total discards	87.8	103.6	122.3	128.8	154.1	172.0	225.3	230.4	254.1

*Data for 1995 and 2000 are estimates.

waste going to landfills. Recycling will also reduce the amount of metals, paper and plastics buried.

Industrial Waste Sources

Industrial solid waste includes the materials no longer of value that were used in the manufacture of industrial products. Since industries recycle many wastes in-plant, recycling figures for industry are difficult to determine. Specifically excluded, both from this discussion and from government figures on industrial solid waste generation, are hazardous wastes. We will take that subject up in the next chapter.

Standard Industrial Classification. There are many methods of classifying industries. The most-used system is the Standard Industrial Classification (SIC), originally developed by the U.S. Department of Commerce. This categorizes industries based on their activities. This SIC system is used for many purposes. We will use it to group industries according to the amounts of solid waste generated. The most common classifications relating to solid waste generators are shown in Table 11.4. Students interested in solid or hazardous waste disposal should be familiar with the system.

TABLE 11.4 Industrial Solid Waste Disposal Quantities

INDUSTRY	TOTAL, 1000 ton/yr
Organic chemicals	2,138
Ferrous metals	9,892
Agricultural chemicals	11,365
Electric power	54,612
Plastics and resins	4,270
Inorganic chemicals	44,651
Clay, glass, concrete	16,806
Pulp and paper	16,284
Nonferrous metals	10,512
Food	79,993
Water treatment	9,121
Petroleum refining	747
Rubber and miscellaneous	630
Transportation	880
Other chemicals	548
Textile manufacturing	159
Leather	20
Totals	262,628

SOURCE: EPA Report to Congress, *Solid Waste Disposal in the United States*, vol. 1, October 1988.

Industrial Solid Waste Generation. Most industries produce some amount of wastewater, emit some air pollutants, and generate conventional solid wastes as well as hazardous wastes. Of these various wastes, only solid wastes will be discussed here. We examined the wastewater components in Chapter 10. Hazardous wastes will be discussed in Chapter 12, and air pollutants will be the focus in Chapter 13. Solid wastes typically generated by industry are not greatly different from those produced by most households. They consist mostly of paper, glass, rubber, plastics, and metals. However, some food and yard wastes may also be generated. Industries also generate process wastes—unwanted by-products of industrial or manufacturing processes.

Unlike municipalities, the data available for industrial solid waste is less extensive. And several questions arise. Should wastes that are recycled in-plant be included? Should wastes disposed of on-site be included? Should wastes that are incinerated on-site be included? How can such information be obtained? It is particularly difficult to obtain information on in-plant recycling efforts because each company is competing with other similar companies. In general, many companies do not want to help competitors improve their operations. Any information, even about in-plant recycling efforts, could put a company at a disadvantage. Thus, many industries will not voluntarily disclose their solid waste practices unless required to by either the state or federal government. The EPA, however, has conducted studies of industrial disposal practices. The results of an EPA report to Congress are shown in Table 11.3. There are approximately the same amounts of wastes disposed of in landfills and in waste piles. Waste piles are wastes that are not buried, but simply piled on top of the ground. The total of these wastes is approximately the same as that for municipal solid waste.

Unlike municipal solid waste and hazardous waste, industrial solid waste is not tightly controlled. Differences in requirements between industrial and municipal solid waste disposal will be addressed in the section on regulations.

Mineral Extraction Sources

Mineral extraction, or mining, results in voluminous waste. The cast-offs include mining overburden—that is, soil and minerals that overlay the materials to be extracted. Removed to gain access to the valuable minerals, these layers of materials become waste.

Unlike the requirements for other forms of pollution, the control of these wastes has been minimal to date. Most mining activities are exempt from both the solid and hazardous waste requirements other industries must meet. However, Congress and the EPA will certainly control mining wastes more strictly in the future. Some regulations in place now require mining companies to restore the land surface after surface digging. It is reasonable to expect that other regulations dealing with solid waste from mining will be passed during your engineering career. Environmental engineers will be needed to help solve these problems.

SOLID WASTE DISPOSAL REGULATIONS

Today federal law regulates many facets of solid waste disposal. Even so, many states have taken the lead in encouraging recycling and limiting the materials that may be placed in landfills. In general, federal environmental regulations are written to allow the states to adopt more stringent standards. States are usually prohibited from having requirements that are not as strong as federal standards. Only the federal landfill requirements will be presented here.

The first significant step in federal regulation of solid wastes was the Solid Waste Disposal Act of 1965, PL 89-272. This law was intended to promote better management of solid wastes and to support resource recovery. Other parts of the act directed that the U.S. Public Health Service (PHS) promulgate and enforce regulations for solid waste collection, transportation, recycling, and disposal. In addition, the act provided financial assistance for states to study and develop solid waste management plans. And finally, it supported research and development to improve methods of solid waste management.

The next significant step in solid waste regulation was the Resource Recovery Act of 1970, PL 95-512. It redirected the nation's emphasis from solid waste disposal to recycling and energy recovery. In addition, the Resource Recovery Act required the U.S. PHS to investigate and report on the disposal of hazardous waste in the nation. The U.S. Environmental Protection Agency was formed in the interim, and in 1973 its staff issued the final *Report to Congress: Disposal of Hazardous Wastes* [5]. This was an important guide for solid and hazardous waste management in its early stages.

The Resource Conservation and Recovery Act of 1976 (RCRA) included significant requirements for the control of hazardous waste storage, treatment and disposal. It also directed the EPA to establish regulations to control solid waste disposal. The Hazardous and Solid Waste Amendments of 1984 (HSWA) called upon the agency to revise criteria for landfills that receive hazardous household waste or small quantities of industrial hazardous waste. Additional requirements on the quality of the surfacewater running off landfills and the methods of disposing of wastewater sewage sludge at landfills were included in the Clean Water Act as amended.

In October 1991 the EPA promulgated new regulations for municipal solid waste landfills [6]. These require a minimum performance for groundwater protection, restrict landfill locations, establish minimum standards for landfill design, and set forth minimum operating procedures. The technical requirements will be discussed later in this chapter.

RECYCLING

Recycling is the recovery and reuse of products that would otherwise be thrown away. Industries have been practicing recycling for many years. In many cases it is simply separating such items as scrap metals or plastics and then reusing them or returning them to their suppliers for processing and reuse. States and municipalities are beginning to encourage recycling as well. Almost everyone is familiar with one or more aluminum can or wastepaper return programs. However, much more can be done to reduce the amount of material finding its way to landfills. From Figure 11.3 you can see that there are many good candidates for recycling, including paper, glass, metals, and plastics. As we have noted, programs for these materials, combined with composting of yard wastes, could reduce solid waste disposal by more than 75%. And recycling could reduce natural resource consumption significantly. Preliminary steps to create a recycling program are [4]:

1. Accurate analysis of the sources and content of the solid waste stream.

2. Evaluation of existing recycling programs, which must be integrated into new or expanded efforts.

3. Identification of public attitudes.

4. Finding markets for the potential recycled materials. (It does no good to collect materials for recycling if there is no place to sell them!)

5. Choosing the best options.

Recycling efforts vary from voluntary programs, in which people must take recyclables to specific drop-off points, to mandatory programs with curbside pickup of separated materials. A few of the recycling options are discussed below.

Drop-off Centers

In this method specific places are designated for drop-off of recyclable materials. These may be unstaffed with only bins for specific types of wastes, such as old newspapers, or they can be staffed with workers collecting several different types of materials. Drop-off centers should be in convenient locations such as shopping malls or on heavily traveled city routes. Drop-off facilities are cheaper to operate than curbside collection, but fewer materials will be collected.

Curbside Collection

Curbside collection frequently involves special containers. These may be color-coded for different materials, or all materials may be placed in a single recycle container, separated by bagging. Collection can be in conjunction with normal garbage pickup, or it may be a separate service. Special collection vehicles, with multiple receptacles, are available for cities with curbside recycling programs.

Voluntary Versus Mandatory Recycling

Voluntary programs will reduce the amount of waste going to landfills, but not to the extent that mandatory programs will. As an example, Austin, Texas, has a voluntary curbside recycling program. Its initial participation was 25% of the population served. With improvements in curbside collection (primarily allowing all containers, metal or glass, to be mixed) and better public awareness, the participation rate has risen to 60% [7]. Hamburg, New York, has a mandatory curbside recycling program, and its participation rate exceeds 95%. The primary enforcement mechanism there consists of not collecting the trash from residences who do not separate for recycling [4]. A significant incentive!

COMPOSTING

Comprising a fifth of the total municipal solid waste stream, yard wastes have been targeted by the EPA for special emphasis in coming years [8]. If they were composted and put to good uses elsewhere, leaves, grass clippings, limbs, and other wood wastes could lighten the load on landfills and incinerators significantly.

In nature, leaves fall to the forest floor and are gradually decomposed by a variety of microorganisms including fungi, bacteria, and protozoa. This degradation process returns the nutrients contained in the leaves to the soil where they become available again to the trees and other vegetation. In contrast, leaves falling in cities become a solid waste. Prior to the 1970s many people burned their leaves. However, many states now have air pollution laws prohibiting open burning of yard wastes. As a result, yard wastes are generally landfilled.

Composting, which is the controlled aerobic partial degradation of organic wastes, produces a material that can be used for landscaping, landfill cover, or soil conditioning. Leaves are an ideal material for composting. Grass, which has a higher nitrogen content and is often more moist, can cause composting problems. The moisture and nitrogen result in a very rapid degradation, depleting the oxygen. This anaerobic condition produces unpleasant odors. The EPA recommends that grass clippings be left on the lawn for their nutrient value where possible. Brush and limbs do not compost well, either. They are best shredded and used as landscaping material.

The decomposition that takes place during composting typically reduces the waste volume by 40 to 75%. Weight reduction, which is less, depends to a great extent on moisture levels. Composting is accomplished by placing the wastes in long rows, called windrows, several feet high and wide. To control temperature, maintain aerobic conditions, and reduce odor, the windrows are mixed and turned periodically. After several months the microbial activity will have slowed sufficiently for the material to be utilized.

Collection and Preprocessing

Materials for composting can be gathered at curbside in bags or bulk containers, or they may be collected at preselected drop-off locations. Residents must be informed of materials acceptable for composting and encouraged to not place other items in their yard waste. Preprocessing steps may include grinding to produce a material of uniform size, separation, wetting, and screening. Where nondegradable plastic bags are used, the bags must be emptied and removed.

Composting Parameters

Decomposition depends on several important factors, including oxygen level, nitrogen content, temperature, and moisture. Oxygen is supplied to most composting piles by natural processes. Smaller windrows provide better oxygen penetration but retain less heat. Since little oxygen may penetrate into the center of large windrows, they must be turned and mixed frequently to increase oxygen transfer. If there is insufficient oxygen, anaerobic conditions will develop, creating offensive odors. Composting near residential areas, therefore can be a major concern.

Sufficient nitrogen must be present for the microbes. This is usually judged based on the carbon/nitrogen ratio in the composting material. Since the decomposition process can be expected to form carbon dioxide as an end product (but no gaseous nitrogen products), the carbon/nitrogen ratio can be expected to decrease with time. The composition of yard wastes varies from season to season, and nitrogen may be insufficient in some instances. Woody materials have very low carbon/nitrogen ratios — several hundred to one. Leaves also have low carbon/nitrogen ratios, ranging from $60:1$ to $80:1$. Grass clippings, with ratios of about $20:1$, have much more nitrogen. Other materials with a higher nitrogen content can be mixed with the nitrogen-deficient material, or nitrogen can be added in the form of fertilizer. However, the addition of fertilizer often raises microbial activity and can cause anaerobic conditions to occur if the compost is not turned and mixed often enough.

The temperature of the compost is maintained by the heat released as the microoganisms oxidize the organic matter to carbon dioxide and water. This heat release can be substantial. The temperature can even rise to the point that biological activity is depressed. On the other hand, during cold weather in northern climates, maintaining an adequate temperature can require the use of larger windrows to provide better insulation for processes going on in the interior.

Microorganisms require moisture for growth. Moisture within the compost materials must be adequate to allow biological activity. About 50% moisture by weight appears to be a minimum [9]. Where extremely dry materials are being processed, addition of water may be necessary. The material can be sprayed prior to composting.

EPA officials recommend that wood materials such as branches and brush be shredded for mulch or bedding material and not composted if possible. They also encourage people to leave grass clippings on the lawn for nutrient value [4]. Such an approach would make leaves the primary constituent in composting.

Minimal-Level Composting

Current composting technology divides into four different levels, depending on the resources available, labor required, and the land area. Minimal-level composting requires the least equipment or labor investment, but it takes more land than more active methods. Yard wastes are collected, placed in windrows 12 feet high by 24 feet wide, and turned only annually. Windrows can be formed, turned, and mixed using a front end loader attached to a tractor. The center of the windrow becomes anaerobic. Active composting only occurs in the outer layers where oxygen can penetrate. Naturally, an adequate buffer is required between the composting area and any nearby residences or businesses since foul-smelling anaerobic gases are produced by this process. Typical requirements are 1 acre/4000 yd^3 or, using 5 yd^3/ton, 800 tons of compost per acre [9]. This process takes about 3 years to obtain a stabilized, usable end product.

Low-Level Composting

Low-level composting is the most common approach at present. In this approach windrows are approximately 6 feet high by 12 feet wide (slightly larger in extremely cold climates). A minimum moisture content of 50% is needed. Materials are initially sprayed with water if necessary to attain this. After 1 month two windrows are mixed and combined. Additional turning takes place about three times per year. The compost will be ready for use after about a year and a half.

Intermediate-Level Composting

Intermediate-level composting utilizes special machinery to turn the windrows, which are limited in size by the equipment—a maximum height of 5 to 7 feet and a width of 10 to 14 feet. When the windrows have been turned each week for 4 to 6 months, the compost is ready to use.

High-Level Composting

In high-level composting, forced air ventilation provides oxygen. Large windrows, at least 10 feet high by 200 feet wide, are formed, and water and nitrogen are added if required. Windrows are allowed to decompose under these near-optimal conditions for 2 to 10 weeks, after which the ventilation is stopped. The windrows are then turned periodically. The process is complete in 3 months to 1 year depending on turning frequency [3].

Compost Uses

Finished compost can be used by municipalities for landscaping in parks, around buildings, and along roadways. Home owners can use it for gardening or landscaping. Commercial landscapers and nurseries can use the material, as may farmers in the area. Where practical, a city may sell the compost to raise revenue to offset composting costs; in other cases city work crews may use the material themselves and/or give it away. All of these options are better than burying the material.

COLLECTION

Efficient solid waste collection is a challenge. For the municipality or private contractor involved, collection represents from 60 to 80% of the total solid waste expenditure [10]. Although this figure can be expected to decrease in response to increasing landfill disposal costs, it will remain a major financial concern. The types of vehicles, frequency and type of collection, and length of routes are all variables that must be considered. Each can have a significant impact on the cost, convenience, and efficiency of the collection operation.

Collection Services

Collection services can vary considerably. For single-family residential areas, the minimum is curbside pickup, where residents are responsible for carrying the containers to the curbside in the morning and returning them in the afternoon after they have been emptied by collection crews. Optimal service is backyard carry, where the collection crew is responsible for entering a resident's property, carrying the containers to the curbside, emptying them into the collection vehicle, and then returning the empty containers to their storage location. Table 11.5 describes three common collection methods. These methods, which have several variations, are complicated by recycling efforts. It is much more expensive for collection services to enter a resident's property and gather several recyclable containers than it is simply to collect one or two large ones. Thus, recycling efforts tend to make curbside or alley collection more favorable.

TABLE 11.5 Solid Waste Collection Services

COLLECTION SERVICE	DESCRIPTION	COST
Curbside	Residents responsible for placing trash containers at curbside and returning them after collection.	Low
Backyard carry	Collection crew responsible for entering residents' property, transporting containers to collection vehicle and returning them to storage location.	High
Alley	Residents responsible for placing trash containers by alley and returning them after collection.	Low

Apartment complexes often utilize larger containers. Apartment dwellers are responsible for taking their trash to a central location and placing it in the containers. Collection vehicles can pick up and dump the container contents using an automatic hydraulic system. Many of these automated collection vehicles require only a driver. Businesses and commercial establishments often use similar systems. Some larger establishments may use a hauled container system. The collection vehicle brings out an empty container. The empty container is left, and the full container is then removed.

Collection Vehicles

Collection vehicles vary from the trash compactors used in many residential areas, to larger trucks that collect and haul only one large container from apartment complexes or commercial establishments, to vehicles fitted with multiple containers for various recyclable items.

Collection of nonrecyclable single-family residential trash is usually accomplished by collection vehicles with hydraulic compactors. (See Figure 11.4.) The compactor reduces the waste volume, thus increasing the waste density as well as the amount of trash that can be carried on one trip. Such vehicles usually carry one to four workers, including the driver. The

FIGURE 11.4   Rear-load municipal solid waste collection vehicle. The vehicle hydraulically compacts the waste to allow larger volumes to be collected. Photograph courtesy of the Leach Company, Oshkosh, Wisconsin.

additional persons collect the containers and empty them into the vehicle.

In areas where recyclables are collected at curbside, collection vehicles must have multiple bins for containing the recyclable materials. This is usually accomplished using one vehicle to collect the trash and a separate vehicle with multiple bins to collect recyclable materials. Several manufacturers produce collection vehicles for recyclables. Both open-top and closed-top vehicles are available. There are also multiple-bin-trailers designed for recycling [11]. A typical recycle collection trailer is shown in Figure 11.5.

Vehicles used to collect larger trash containers such as those from apartments and businesses usually have a hydraulic system to lift and empty the container. In some, drivers can operate the equipment without leaving the cab. Others require the driver or a second person to position the container for lifting.

The largest vehicles involved in collection are those that lift and carry a single container. These usually drop off an empty container and then pick up the full container. Such containers are used only by larger apartment complexes and businesses in which the volume of trash is sufficient to warrant the larger containers.

Typical sizes of various collection vehicles are given in Table 11.6. A typical compactor truck will compress the waste to half its original volume.

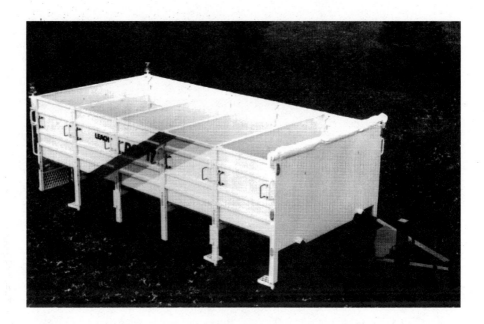

FIGURE 11.5   Multiple-bin recycle collection vehicle. The trailer can be pulled behind a typical solid waste collection vehicle. Photograph courtesy of the Leach Company, Oshkosh, Wisconsin.

TABLE 11.6   Capacities of Selected MSW Collection Vehicles

VEHICLE TYPE	CAPACITY, yd^3
Rear-loaded compactor	20 to 25
Front-loaded compactor	30 to 40
Side-loaded compactor	25 to 35
Multiple-bin recycle	20 to 30
Hauled container bins	20 to 40

Routing

Determining the optimal routes for collection is a real challenge. There are many factors to consider: avoiding rush-hour traffic; not traveling through the same area twice; beginning and ending routes near major streets; traveling downhill, not uphill, during collection; and minimizing left turns across traffic. Knowing the capacity of a collection vehicle and the number of trips to the landfill it can make each day, we can estimate the number of residences the vehicle can serve. This is illustrated in the following example.

EXAMPLE 11.1	ESTIMATION OF THE NUMBER OF RESIDENCES SERVED BY A SINGLE COLLECTION VEHICLE

A typical side-loaded compactor truck has a capacity of 30 yd³. Estimate the maximum number of residences it can serve per week. It can make 3 trips per day, 4 days per week. The fifth day is reserved for special collections, holidays, and the like. Assume each residence contains 3.5 persons.

SOLUTION: A typical residence produces approximately 4 pounds of solid waste per person per day (from Table 11.4). So, the generation rate per residence is

$$G_{\text{MSW}} = 3.5 \ \frac{\text{persons}}{\text{residence}} \times 4 \ \frac{\text{pounds}}{\text{person/day}} \times 7 \ \frac{\text{days}}{\text{week}}$$

$$= 98 \ \frac{\text{pounds}}{\text{residence/week}}$$

The vehicle has a capacity of 30 yd³. It can make 3 trips per day. The approximate density of the refuse is 500 lb/yd³ (also from Table 11.4). The truck's capacity per week is then

$$C_{\text{truck}} = 3 \ \frac{\text{trips}}{\text{day}} \times 30 \ \frac{\text{yd}^3}{\text{trip}} \times 4 \ \frac{\text{days}}{\text{week}} \times 500 \ \frac{\text{lb}}{\text{yd}^3}$$

$$= 180{,}000 \ \frac{\text{lb}}{\text{week}}$$

Thus, the maximum number of residences served will be

$$N = \frac{180{,}000 \ \dfrac{\text{lb}}{\text{week}}}{98 \ \dfrac{\text{lb}}{\text{residence/week}}}$$

$$N \simeq 1836 \ \text{residences}$$

SOLID WASTE PROCESSING

Processing of solid waste includes sorting for recycling, compaction to reduce volume, and shredding to reduce particle size. The intent of processing is to recover usable materials or to improve the disposal process. Brief discussions of these topics follow.

Sorting

Sorting can be done at the point of generation, at a transfer station prior to shipment to a landfill, or at the landfill. The ideal method is for the material to be sorted at the point of generation, by the residents or businesses. Source separation increases the amount of materials that can be recycled, and it has been shown to be effective. However, recycling rates are improved where a minimum of source separation is required.

Sorting after collection can be either mechanical or manual. Manual separation usually involves a wide conveyor belt. The materials are mechanically delivered to the belt, and people pick out the recoverable items by hand. In mechanical separation large magnets remove ferrous metals from the mass, and compressed air or inertia-driven separation processes remove the remaining items. Several manufacturers produce automated solid waste separation equipment.

Compaction

Compaction often begins with the collection vehicle. Many such vehicles have low-pressure compactors to reduce the volume of the collected wastes. This improves the efficiency of the vehicle, enabling it to collect more waste before traveling to the landfill or transfer station. Transfer stations and landfills sometimes have mechanical compactors of their own. Low-pressure compaction results in finished densities of 700 to 1000 lb/ft^3. High-pressure compaction often involves baling the waste and can produce final densities of 1600 to 1800 lb/ft^3. In either method, the waste occupies less space for transit and landfilling. Baled waste is also easier to transport.

Shredding

Shredders reduce the size of the materials to be landfilled but not necessarily the volume. There are several types, but all use brute force to hammer, tear, grind, and pound the wastes into smaller, more uniform sizes. A typical shredder is depicted in Figure 11.6. Shredders are also sometimes used prior to compacting solid wastes.

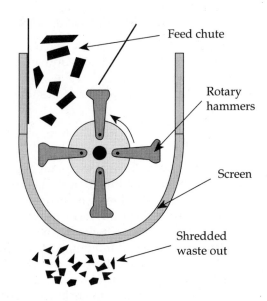

FIGURE 11.6   Solid waste shredder.

LANDFILLING

Landfilling is the legal and controlled placement of wastes in the ground. In contrast, **dumping**, is putting wastes onto or into the ground illegally or in an unregulated manner. Even though landfills pose some risk to the surrounding air, land, and water environment (and even the best-designed ones do), controlled landfilling is superior to leaving the material in the backyard or dumping it on a nearby roadside.

When material is placed in a landfill, it undergoes gradual chemical and biochemical changes. Liquids are usually present in small volumes in some solid wastes, and water enters during the landfilling process from runoff and infiltration associated with rain or snow. This water both promotes biologic activity and acts as a transport mechanism for contaminants that dissolve in the water. If this water, now termed **leachate**, is allowed to enter the groundwater below, contamination of a valuable natural resource results. Landfill leachate can be quite high in both organic and inorganic constituents. Landfills often collect the leachate from the bottom of the pile and pump it to the top. Recirculation of leachate in this way solubilizes additional organics within the landfill. And because microbial processes occur only in a wet environment, leachate recirculation speeds the degradation process. Typical concentration ranges of some leachate constituents are listed in Table 11.7 [12].

TABLE 11.7   Landfill Leachate Characteristics

CONSTITUENT	CONCENTRATION RANGE, mg/L
BOD$_5$	9–55,000
COD	0–90,000
Total solids	6–45,000
TDS	0–42,000
TSS	6–2,700
Chloride	34–2,800
TKN	0–1,400
Sulfate	1–1,800
Phosphate	0–154
Lead	0–5
Copper	0–10
pH	3.7–8.5

The organic portion of wastes placed in a landfill will biochemically degrade over time. In many cases microbial degradation also results in a volume reduction and settling. Since regulations require that wastes placed in a landfill be covered daily with 6 to 12 inches of soil, oxygen does not penetrate into the wastes. The existing oxygen is rapidly consumed, and the landfill section or "cell" rapidly becomes anaerobic. In anaerobic degradation, discussed in Chapter 6, Microbiology and Microbial Growth, the complex organics convert first to simpler organics, then to organic acids, and finally to methane and carbon dioxide. Acetic acid serves as a primary pathway to the end products. The process can be depicted in this way:

$$\text{Complex organics} \rightarrow \text{Simple organics} \rightarrow \text{Acetic acid} \qquad \textbf{11.1}$$

and then

$$CH_3COOH \rightarrow CO_2 + CH_4 \qquad \textbf{11.2}$$

As degradation continues, the acid content of the leachate begins to diminish, as does the organic content. As this occurs, the leachate pH increases to near neutral or above. At the same time, landfill gas production diminishes. This is a sign that the landfill degradation process is nearing completion.

In an effort to force landfill owners and operators to minimize the effects landfills have on the surrounding environment, the EPA promulgated the Solid Waste Disposal Facility Criteria in 1991 [6]. These regulations place substantial restrictions on the location, operation, and design of landfills.

Siting Restrictions

Location is important to minimize the impact of a landfill on adjacent areas and to ensure that the environmental conditions of the area do not present an undue hazard to successful landfill operation. The first consideration should be to attempt to locate a landfill in an area with as low a population density as possible. This will minimize exposure and risk to the population. Such a task is more difficult to accomplish in the populous Northeast and West Coast areas than in the less populated Midwest. Other siting considerations include potential flooding problems and wetlands destruction. Low-lying, flood-prone areas often have lower land costs than other areas. It is thus tempting for a landfill operation to locate in such areas. However, floodplains pose the problem of possible environmental damage from erosion during flooding. And wetlands, although often considered of low value, provide valuable habitat to wildlife. Another factor is that because scavenging birds are often present in large numbers at landfills, it is important to keep landfills and airports separated by some distance. This minimizes the possibility of low-flying aircraft colliding with birds.

Landfill regulations include the following restrictions on location [6]:

1. Prohibit the placement of a landfill facility near an airport because of dangers from scavenging birds.

2. Require the landfill to be located outside the 100-year floodplain, or that the design protect against the washout of solid waste during a major or catastrophic flood.

3. Prohibit the placement of a new landfill or expansion of an existing landfill into or on a wetland.

4. Prohibit the placement of a landfill within 200 feet of an earthquake fault.

5. Prohibit the placement of a landfill in an area with a high probability of a strong earthquake.

6. Prohibit the placement of a landfill in an area with unstable soil.

7. Require landfills that cannot meet the airport, flood plain, or un-stable-area requirements above to close within 5 years. The state may grant a maximum of a 2 year extension.

Operational Requirements

Landfill operations have improved much during the past two decades. Where open dumps were commonplace in the 1950s and 1960s, they are now illegal. But even with the improvements, landfills operating in the 1980s often had problems with groundwater pollution. The 1990 regulations require a higher standard of operation than many landfills have previously attained. These requirements are designed to reduce the opportunity for contamination of the groundwater and minimize the disagreeable effects of the landfill on nearby areas. According to the regulations [6], landfill operators are to

1. Exclude hazardous waste.

2. Provide at least 6 inches of soil cover daily over newly placed waste.

3. Control disease vectors such as rodents and insects.

4. Monitor methane concentrations in the landfill and associated buildings. (Methane is explosive when combined with the oxygen in air.)

5. Eliminate most open burning.

6. Control public access.

7. Construct run-on and run-off controls for water.

8. Meet water quality discharge requirements (NPDES) for surfacewater.

9. Prohibit all liquid wastes except small quantities of household liquid wastes.

10. Maintain records indicating compliance.

Design Requirements

The Solid Waste Disposal Facility Criteria promulgated on October 9, 1991, set new standards for municipal solid waste landfills. More restrictive than any previous federal requirements [6], they offer two basic options for landfill design (see Figure 11.7). One is to include a synthetic membrane liner at least 30 mils thick, with a compacted soil liner at least 2 feet thick. The soil liner must have a maximum hydraulic conductivity of 1×10^{-7} cm/s. (This means that any contaminants passing through the synthetic liner will travel very slowly through the underlying soil.)

Other designs can be different from this, but they must be state approved and certified not to allow contaminants above the limits shown in Table 11.8. The state environmental agency will certify to EPA that the design is acceptable. A photograph of a lined solid waste landfill is shown in Figure 11.8.

TABLE 11.8   Maximum Contaminant Levels for Groundwater at
Landfill Boundaries

CONSTITUENT	MCL, mg/L
Arsenic	0.05
Barium	1.0
Benzene	0.005
Cadmium	0.01
Carbon tetrachloride	0.005
Chromium(VI)	0.05
2,4-dichlorophenoxy acetic acid	0.1
1,4-dichlorobenzene	0.075
1,2-dichloroethane	0.005
1,1-dichloroethylene	0.007
Endrin	0.0002
Fluoride	4
Lindane	0.004
Lead	0.05
Mercury	0.002
Methoxychlor	0.1
Nitrate	10
Selenium	0.01
Silver	0.05
Toxaphene	0.005
1,1,1-trichloroethane	0.2
Trichloroethylene	0.005
2,4,5-trichlorophenoxy acetic acid	0.002
Vinyl chloride	0.002

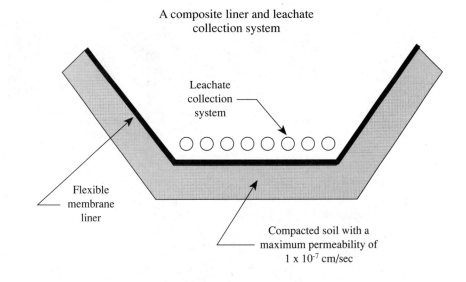

A composite liner and leachate
collection system

Leachate
collection
system

Flexible
membrane
liner

Compacted soil with a
maximum permeability of
1×10^{-7} cm/sec

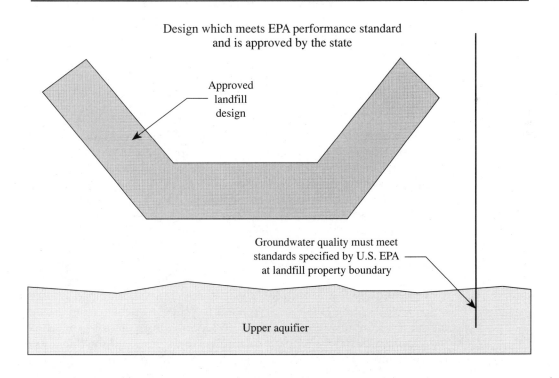

Design which meets EPA performance standard
and is approved by the state

Approved
landfill
design

Groundwater quality must meet
standards specified by U.S. EPA
at landfill property boundary

Upper aquifer

FIGURE 11.7 Landfill designs acceptable to EPA.

FIGURE 11.8   Installation of a flexible membrane and porous drainage material at a solid waste landfill facility. Photograph courtesy of Gundle Lining Systems, Inc., Houston, Texas.

Leachate Collection and Treatment

In landfills that do not have an intact liner, leachate eventually migrates to the groundwater. This has been a common problem with older landfills. To prevent groundwater contamination, current regulations require that a leachate collection system be installed in all landfills. The collected leachate cannot be discharged to surface waters unless it meets the applicable federal and state water quality (NPDES) standards. Leachate is usually recycled through the landfill, speeding the process of degradation of organic matter.

Groundwater Monitoring

Low-permeability soil and the flexible membrane, combined with the leachate collection system, are intended to prevent leachate from contaminating the groundwater. Nevertheless, landfills must install groundwater monitoring wells in the uppermost aquifer to determine if the liner system is functioning. This is accomplished by monitoring both the upgradient (and uncontaminated) groundwater and the downgradient groundwater at the landfill boundary. The difference in quality is assumed to be a result of landfill contamination. Wells must be sampled at least semiannually. The maximum allowable level of contamination for various chemicals is listed in Table 11.8. Should contamination occur, landfill operators are required to take corrective action.

A groundwater monitoring well is installed by boring a hole (usually 8 to 10 inches in diameter) to the needed depth. A well casing (typically 4-inch diameter) is installed with a screen at the desired sampling region. Well casings may be made of PVC plastic, stainless steel, or teflon pipe, depending on the contaminants that might be detected. The bore hole around the casing is filled with a permeable material such as gravel at the sampling region. An impermeable material such as bentonite clay is used for the remainder of the hole up to about three feet from the top. The top part of the casing is filled with concrete. The clay and concrete prevent migration of surface water or groundwater into the sample from heights other than the one being sampled. A typical well is depicted in Figure 11.9.

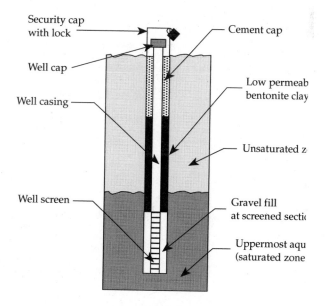

FIGURE 11.9 Cross section of a groundwater monitoring well.

Landfill Types

Landfill construction varies with terrain. In some cases an area is excavated, filled, and covered. In others, a natural depression is filled and the waste covered. Traditionally, the wastes are disgorged from collection or transport vehicles at appropriate locations within the landfill. Bulldozers or other earth-moving equipment are then used to compact the wastes. There are several methods of doing this.

Trench Method.   In this approach a long trench is initially dug to a depth and width sufficient to contain the anticipated volume of wastes for some time period, typically months or years. The soil is placed along one side of the trench. Low-permeability soil is then compacted to a minimum depth of 2 feet. This layer of soil must be purchased elsewhere and hauled to the site if local soil is unacceptable for this purpose. The synthetic liner is placed on top of the compacted soil. Next, the leachate collection system is installed. The trench is then filled with solid waste in thin layers (1 to 2 ft) and compacted. The **cell** size should be planned such that the layers reach the desired height at the end of the day. Before operations cease at day's end, the entire contents of the cell must be covered with 6 to 12 inches of soil to prevent scatter by wind and reduce rodent activity [10]. A cross-sectional view of the trench method is shown at the top in Figure 11.10.

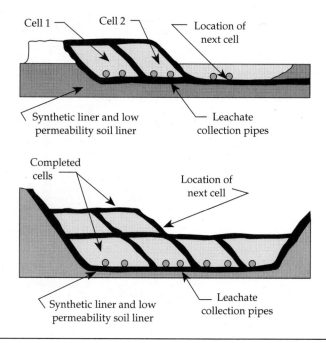

FIGURE 11.10   Cross section of the trench method, above, and the ravine method of landfilling, below.

Ravine or Depression Method.   This approach, which utilizes natural contours of the land, can also make good use of abandoned surface mines and quarries. A 2-foot thick, low-permeability soil must be put in place to contain the leachate. This must be covered with a flexible membrane liner, and a leachate collection system must be constructed. The waste is discharged into cells, just as in the trench method, but multiple cells are constructed upon each other. The material is compacted as it is placed in the landfill [10]. Fill material is either excavated from the site or brought in to cover each day's waste. The ravine method is depicted at the bottom in Figure 11.10.

Bale Disposal

Rather than disposing of loose solid waste from collection or transport vehicles, some landfills accept only baled waste, which is much more dense [13]. Baling machines, located at transfer stations or at the landfill, compact and strap the waste. The resulting parcels are easier to handle during placement in the landfill, and their higher density provides for better utilization of landfill volume.

CASE STUDY

The Gallatin National Landfill, Fairview, Illinois

A new landfill has been constructed on 2700 acres of abandoned strip-mining property. Located in central Illinois, Gallatin National landfill receives wastes from six counties in the Chicago area, and six counties near the landfill. Wastes from the Chicago area are shipped by truck and rail to the facility [13]. The initial landfill site is 80 acres. The landfill extends 50 feet below ground and will extend 100 feet above ground when complete. Landfill capacity is estimated to be approximately 10 million cubic yards. Wastes are accepted at a rate of approximately 1000 tons per day. The landfill life is estimated to be 13 years. The landfill will also have a yard-waste composting facility at the site.

Groundwater protection includes a 10-ft-thick compacted clay liner above a second liner. A bentonite–clay slurry wall is used to prevent excessive groundwater from entering the landfill during construction and operation and to prevent leachate from coming into contact with outside runoff. A leachate collection, storage, and treatment system is also used to prevent groundwater contamination. A system will be installed to collect and remove landfill gas. The gas will initially be burned, but there are plans to collect and use the gas as the landfill size (and gas production) increases.

Closure Requirements

When the landfill stops accepting waste, the EPA requires that a cover (often called a landfill cap) be placed over it to minimize erosion and infiltration of surface water. The operator must start this activity within 30 days of the last receipt of wastes, and the project must be completed within 180 days. The cover consists of at least two soil layers. The lower layer must be at least 18 inches of low-permeability soil. To prevent buildup of liquids in the landfill, this layer must be less permeable than the bottom layer. The top layer must be a minimum of 6 inches of earthen material to allow native plant growth. Synthetic membranes similar to those used at the bottom are often placed between the two soil layers. Each unit of a landfill must be closed as it is completed. These covers are usually sloped to prevent standing water.

Finally, the property deed must have a notation that the land contains solid waste. This notation remains with the deed in perpetuity. Any person purchasing the property must agree to the closure and postclosure requirements.

Postclosure Care

After closure, the EPA requires that a landfill be monitored and maintained for 30 years. The cap must be maintained and repaired as necessary to correct for settling, subsidence, and erosion — the objective being to prevent water from infiltrating the landfill. The groundwater must be monitored twice per year, just as when the landfill was operating. Landfill gas must also be checked. In general, oxygen must be kept out of the landfill to prevent the possibility of an explosion.

CASE STUDY

Landfill Closure, Cambridge, Massachusetts

In the 1980s the City of Cambridge contracted with two consulting engineering firms to study the possibility of converting a 50-acre former site within the city into a park. Mayor Thomas W. Danehy Park now contains three softball and three soccer fields, one multisport field, two children's play areas, more than 2.5 miles of trails for hiking, jogging, and biking, a 300-car parking area, a small wetland for drainage control, and other recreational attractions [14].

In and around the park are monitoring wells to ensure that landfill leachate does not contaminate the groundwater. A public water supply reservoir is only 2000 feet from the landfill. There is also gas monitoring to ensure that the landfill gas is neither toxic nor explosive. To date, monitoring indicates that the groundwater is protected and gaseous emissions are safe.

EXAMPLE 11.2	ESTIMATION OF REQUIRED LANDFILL VOLUME AND AREA

Estimate the landfill volume and area required for a city of 25,000 persons. Also estimate the volume to be landfilled during each year. The facility is to operate for 10 years, beginning 3 years from the present, allowing for permitting and initial construction. The maximum height is to be 30 ft. Assume an additional area of 20% for buffer zones, offices, scales, and slopes. It is estimated that an additional 10% of solid waste (by volume) will come from business and nonhazardous industrial sources. You may assume a population growth of 2% per year.

SOLUTION: The first step is to estimate the population during each of the next 13 years. Then use the generation rate of 4 lb/person-day and the density of the landfilled refuse to calculate the actual yearly volumes. This is presented in the accompanying table, generated from a spreadsheet.

YEAR	POPULATION	GENERATION RATE lb/yr	REQUIRED VOLUME yd³/yr
0	25,000		
1	25,250		
2	25,503		
3	25,758		
4	26,015	41,780,251	46,423
5	26,275	42,198,054	46,887
6	26,538	42,620,034	47,356
7	26,803	43,046,234	47,829
8	27,071	43,476,697	48,307
9	27,342	43,911,464	48,791
10	27,616	4,350,578	49,278
11	27,892	44,794,084	49,771
12	28,171	45,242,025	50,269
13	28,452	45,694,445	50,772
			Sum = 85,682

The generation rate for each year is the population times the per capita generation rate plus 10% for business and nonhazardous industrial solid wastes. The generation rate in the first year of landfill operation (year 4) is thus calculated as follows

$$G_r = (56{,}015 \text{ capita})\left(4 \ \frac{\text{lb}}{\text{capita-day}} \times 365 \ \frac{\text{days}}{\text{year}}\right) \quad (1.1)$$

$$= 41{,}780{,}000 \ \frac{\text{lb}}{\text{year}}$$

So the required landfill volume for the 10-year period is 530,000 yd^3.

Using the total refuse volume, we calculate the landfill area as follows

$$A = \frac{485{,}682 \ \text{yd}^3 \times 27 \ \dfrac{\text{ft}^3}{\text{yd}^3}}{30 \ \text{ft} \times 43{,}560 \ \dfrac{\text{ft}^2}{\text{acre}}}$$

$$= 10 \text{ acres}$$

With the additional 20% for a buffer zone, offices, scales, and other facilities, the total area required is 12 acres.

INCINERATION

Incineration is used for approximately 10 to 15% of all municipal solid waste. Industry and government have accepted burning as a preferred disposal method for many solid and hazardous wastes — that is, compared to landfilling. Incineration destroys the toxic organics in waste in a matter of minutes or seconds, whereas those chemicals might lie for decades in a landfill or, worse, migrate to groundwater. Another advantage of incineration is the reduction in long-term liability. A landfilled waste represents legal exposure as long as there is a potential for hazardous components in the waste to spread to the surrounding environment. Should that occur, not only the landfill operators are liable, but also the original producer of the waste if it is an industry or municipality. Incineration reduces that extended liability because the only waste remaining is ash. In addition to destroying hazardous materials and reducing risk, incineration presents other advantages. It uses an otherwise worthless material to produce energy and it can vastly reduce the volume required for landfilling. Incinerators produce ash that must be used or disposed of in some manner of course. However, the ash has much less mass and volume than the original solid waste. It is anticipated that incineration of solid waste will increase during next few decades [1, 16]. The biggest problem in solid waste incineration now is public opposition. Because incinerators produce small amounts of air pollutants, a segment of the public invariably opposes them [15, 16]. Table 11.9 indicates the current status of solid waste incineration in the United States.

TABLE 11.9 Municipal Solid Waste
Incineration

Number of units [15]	200 to 300
Capacity, 10^6 tons/yr [15]	31 to 47

Most municipal solid waste incinerators do not have air pollution control equipment. That is, they do not have scrubbers or electrostatic precipitators to remove air pollutants escaping from the combustion process. However, the EPA is now requiring that these incinerators meet more stringent standards. This will probably mean adding air pollution control equipment to existing units, and building new units with such equipment.

Several incineration options are available. In all cases, recyclables should be removed prior to burning. In one method, wastes are mixed with other fuel, depending on the energy content of the waste, and then fired to produce usable energy. Another option is to process the waste, including shredding, chemical addition, and pelletizing. This refuse-derived fuel (RDF) can then be marketed for mixing with coal in conventional power generation plants. A third method is to combust the entire waste stream without energy recovery. Clearly, the first two methods are preferable.

Energy Content of Solid Wastes

The energy content of the different classes of solid waste varies considerably. Table 11.10 presents the energy content of several waste classifications as well as common fuels. As we recycle more and more solid waste, the fuel characteristics of the waste will change. The following example shows how the fuel content of solid waste has altered over the past three decades.

TABLE 11.10   Energy Content of Solid Wastes

SOLID WASTE OR FUEL	Btu/lb	kJ/Kg
Solid waste Paper	6,800	15,800
Plastics	14,000	32,600
Rubber/leather	8,000	18,600
Textiles	7,500	17,400
Wood	8,000	18,600
Food wastes	2,200	5,100
Other organics	2,500	5,800
Fuel: Coal	12,000	27,900
Natural gas	21,000	48,900

Note: Glass and metal solid wastes may have 100 to 500 Btu/lb due to organic residues and coatings.

EXAMPLE 11.3	COMPARISON OF THE FUEL CONTENT OF SOLID WASTE IN 1960 AND 1990

There has been a change in the composition of solid waste in the United States since 1960 due both to changes in the materials we consume and to changes in disposal and recycling practices. Estimate the average energy content of solid waste in both 1960 and 1990.

SOLUTION: The first step is to determine the composition of solid waste in both 1960 and 1990. From this we can estimate characteristic energy content. The composition data are available in Table 11.2. The energy values appear in Table 11.10. A simple table or spreadsheet can be used to calculate the values. For example, the energy content of paper in a pound of solid waste in 1960 was

$$E = \text{(Fraction of paper)(Energy content of paper)}$$

$$= \frac{\text{Weight of paper discarded per year}}{\text{Total weight of discarded material per year}}$$

$$\times \text{ Energy content of paper}$$

$$= \left(\frac{24.5 \times 10^6 \text{ lb/yr}}{81.9 \times 10^6 \text{ lb/yr}}\right)\left(6800 \frac{\text{Btu}}{\text{lb}}\right) = 2034 \frac{\text{Btu}}{\text{lb}}$$

MATERIAL	VOLUME, tons/yr	% OF TOTAL	Btu/lb	Btu contrib.
		1960		
Paper	24.5	29.9	6,800	2,034
Glass	6.6	8.1	250	20
Metals	10.4	12.7	250	32
Plastics	0.4	0.5	14,000	68
Rubber/leather	1.7	2.1	8,000	166
Textiles	1.7	2.1	7,500	156
Wood	3.0	3.7	8,000	293
Other	0.1	0.1	2,500	3
Food	12.2	14.9	2,200	328
Yard	20.0	24.4	8,000	1,954
Misc. inorganic	1.3	1.6	250	4
Discards to landfill	81.9			
Average fuel value, Btu/lb				5,058

Other assumptions required: that yard-waste's energy content is the same as wood's, and that the miscellaneous inorganic compounds have an energy value of 250 Btu/lb.

		1990		
MATERIAL	**VOLUME, tons/yr**	**% OF TOTAL**	**Btu/lb**	**Btu contrib.**
Paper	52.4	32.3	6,800	2,197
Glass	10.6	6.5	250	16
Metals	12.4	7.6	250	19
Plastics	15.9	9.8	14,000	1,372
Rubber/leather	4.4	2.7	8,000	217
Textiles	5.3	3.3	7,500	245
Wood	11.9	7.3	8,000	587
Other	2.4	1.5	2,500	37
Food	13.2	8.1	2,200	179
Yard	30.8	19.0	8,000	1,519
Misc. inorganic	2.9	1.8	250	4
Discards to landfill	162.2			
Average fuel value, Btu/lb				6393

Thus, the energy content of solid waste has increased from 5058 Btu/lb in 1960 to 6393 Btu/lb in 1990. Why has this occurred? Primarily because the volume of two high-energy components of waste, discarded paper and plastics, has increased.

TRENDS

With suitable space diminishing and a public quick to oppose, landfill real estate is at a premium. Similarly, incinerators, whether designed only to incinerate waste or to convert it to energy, often face stiff local public opposition. The answers are not easy, but avoiding waste in the first place, recycling, and composting offer reasonable means to reduce the load on landfills. And incinerating solid waste to produce electricity is superior to landfilling. As engineers, we cannot simply present a plan and walk away. More and more, engineers are called upon to convince a doubting public that there really is a better way. In the future this may well be a greater challenge than any facility design.

Review Questions

1. Describe the recycling program(s) in your town or city. What materials are recycled? What others could be? Are the programs voluntary or mandatory?

2. Describe the recyling program(s) on your campus. What materials are recycled? What others could be? Is the situation voluntary or mandatory?

3. For one week keep a listing of every piece of solid waste material you throw away, including its weight. You may collect the nonperishable items and weigh them at the school laboratory. Estimate the perishable wastes as best you can and record their weight and type. What portion can be recycled? What portion could be incinerated for energy? What portion offers no possibility of recycling or recovery? Construct a pie chart similar to the one in Figure 11.3 for your personal waste. How does yours differ from the typical? Based on this 1 week period, how much waste do you generate per year?

4. Survey your apartment or dormitory room. List all toxic materials that could end up in the local landfill.

5. Choose five common products or have your instructor choose five. Discuss their packaging. How could the packaging be changed to minimize amount of solid waste?

6. Typical leachate characteristics are given in Table 11.7. Determine the mean for those concentration ranges $(C_{min} + C_{max})/2$. Table 10.1 presents typical wastewater characteristics. Determine the ratio of each component given in both tables. That is (mean $BOD_{5,leachate}$)/(typical $BOD_{5,wastewater}$) = ? What is your impression of leachate strength to that of domestic wastewater?

7. Obtain a copy of your state's solid waste landfilling regulations (either from your instructor or the appropriate state agency). Compare those regulations to EPA regulations. In what areas is your state stricter?

8. Estimate the energy content of solid waste in the year 2000. You may assume a reasonable increase in the recycle rate and in composting. You will be required to use engineering judgment to solve this problem. There is not a single correct answer. Example 11.3 will be a guide.

9. Estimate the change in energy content (in Btu/lb or kJ/kg) from 1975 to 2000. Assume a reasonable increase in the recycling rate and in composting. You will be required to use engineering judgment to solve this problem. There is no single correct answer, and again, Example

11.3 is a reference point.

10. Changes in our lifestyles, product packaging, and recycling have a direct effect not only in the amount of solid waste generated, but also in its energy content. Estimate the effect that elimination of all yard wastes, a 50% increase in paper recycling, and a 90% increase in plastic recycling (over those values shown in Table 11.2) would have on the energy content of solid waste in the year 2000.

11. Estimate the amount of solid waste generated in your community at present.

12. Estimate the amount of solid waste that will be generated in your community for the next 15 years, including a reasonable increase in population growth. Your instructor may provide guidance about population growth.

13. Estimate the amount of land required for a landfill for your community. The facility is to commence operation in 5 years and operate for the following 15 years. Assume there will be a reasonable increase in the recycling rate in your community.

14. Compare the energy content of municipal solid waste (MSW) in 1995 with that of coal.

15. Calculate the energy content of a refuse-derived fuel composed of all the combustible materials found in MSW. That is, the glass, metals, and other inorganics have been removed.

16. Estimate the energy content of a fuel composed of 10% MSW and 90% coal. Assume that only the combustible materials are used in the fuel. That is, the metals, glass, and other inorganic materials have been removed in a preprocessing step.

References

1. *Characterization of Municipal Solid Waste in the United States 1960–2000, Update,* EPA 530-SW-88-033 (Washington, DC: U.S. Environmental Protection Agency, Office of Solid Waste Management, 1988).

2. *Characterization of Municipal Solid Waste in the United States, 1992 Update,* EPA/530-R-92-019 (Washington: EPA, Office of Solid Waste, 1992).

3. *Yard Waste Composting, A Study of Eight Programs,* EPA/530-SW-89-038 (Washington: EPA, April 1989).

4. *Recycling Works! State and Local Solutions to Solid Waste Management Problems,* EPA/530-SW-89-014 (Washington: EPA, Office of Solid Waste, January 1989).

5. *Report to Congress: Disposal of Hazardous Wastes* (Washington: EPA, Office of

Solid Waste, 1973).

6. *Solid Waste Disposal Facility Criteria*: *Final Rule*, Federal Register, 40 CFR Parts 257 and 258 (October 9, 1991).

7. J. Glenn, "Improving Collection Efficiency for Curbside Recyclables," *Biocycle* (August 1991), pp. 30–32.

8. *The Solid Waste Dilemma*: *An Agenda for Action*, EPA/530-SW-89-019 (Washington: EPA, Office of Solid Waste and Emergency Response, February 1989).

9. P. F. Strom and M. S. Finstein, *Leaf Composting Manual for New Jersey Municipalities* (New Brunswick, NJ: Cook College, 1989).

10. G. H. Tchobanoglous, H. Theisen, and R. Eliassen, *Solid Wastes*: *Engineering Principles and Management Issues* (New York: McGraw-Hill, 1977).

11. "Recycling Collection Vehicles," *Biocycle* (May/June 1988), pp. 39–41.

12. *Procedures Manual for Ground Water Monitoring at Solid Waste Disposal Facilities*, EPA/530/SW-611 (Washington: EPA, August 1977).

13. L. M. Burke, A. E. Haubert, "Burying by the Bale," *Civil Engineering* (August 1991), pp. 58–60.

14. J. Kissida and N. K. Beaton, "Landfill Park: From Eyesore to Asset," *Civil Engineering* (August 1991), pp. 49–51.

15. E. M. Steverson, "Provoking a Firestorm: Waste Incineration," *Environmental Science & Technology*, vol. 25, no. 11 (November 1991), pp. 1808–14.

16. T. Austin, "Waste to Energy? The Burning Question," *Civil Engineering* (October 1991), pp. 35–38.

12 Hazardous Waste Treatment and Disposal

``... THAT THE PREMISES ABOVE DESCRIBED HAVE BEEN FILLED, IN WHOLE OR IN PART, TO THE PRESENT GRADE LEVEL THEREOF WITH WASTE PRODUCTS RESULTING FROM THE MANUFACTURING OF CHEMICALS BY THE GRANTOR (HOOKER ELECTRO-CHEMICAL COMPANY) AT ITS PLANT IN THE CITY OF NIAGARA FALLS, NEW YORK, AND THE GRANTEE (BOARD OF EDUCATION OF THE SCHOOL DISTRICT OF THE CITY OF NIAGARA FALLS, NEW YORK) ASSUMES ALL RISK AND LIABILITY INCIDENT TO THE USE....''

DEED FILED JULY 6, 1953
Niagara County Clerk's Office
Lockport, New York

Organic solvents have contaminated the groundwater around Dayton Ohio's water supply. The white towers shown above are used to strip the volatile organic solvents.

INTRODUCTION

Hazardous wastes are materials no longer of value to their owner or producer and which represent a threat to human health or the environment. (A more technical definition used by the Environmental Protection Agency will be presented later.) As an example, many older exterior paints contained lead, a known neurotoxin. Therefore, old lead paint sitting in a trash can at an industrial facility is a hazardous waste. But if the industry plans to use paint and keeps it in a place that is clearly a storage area, it is not considered a hazardous waste because it still has value. (Wastes from private residences are exempt from federal regulations, no matter how hazardous.) Most hazardous wastes are an unfortunate by-product of twentieth century industrial and technological innovations exemplified by the chemical and petrochemical industries and a wide spectrum of new manufacturing processes.

The development of the synthetic organic chemical industry preceding, during, and after World War II caused a massive increase in the amount of hazardous wastes generated. Often, the detrimental effects of such wastes were not apparent until many years after their introduction to the environment, so the need for special disposal techniques and regulations went unrecognized for several decades. Large quantities of wastes were disposed of without regard to the risks. Therefore, today's scientists and engineers must assume responsibility for remediation of past handling mistakes. And to avoid similar errors, we must appraise the possible negative social and environmental ramifications of planned activities before we proceed. About 1 ton of hazardous waste is produced in this country each year for each person in the nation. That's right. Collectively, we generate approximately 275 million tons per year! Clearly, proper handling of such wastes in the United States and around the world is critical to the future of life on this planet.

IMPETUS FOR CHANGE

In the 1960s we began to discover that better living through chemistry had a hidden price. The harmful effects of residuals of the pesticide DDT on wildlife was among the first problems to surface concerning the use of synthetic organic chemicals. The pesticide problem was highly publicized by environmental groups and, through the publication of Rachel Carson's *Silent Spring* in 1962 [1], DDT was ultimately banned in the United States. Before 1962, this poison was a widely used general purpose insecticide. A few years later discoveries at Love Canal became the motivation for regulations controlling the production, storage, and disposal of hazardous wastes.

CASE STUDY

Love Canal

The highly publicized incident that connected improper hazardous waste disposal to the deterioration of human health hit the news in 1978. The disposal site, Love Canal, was originally built in the late 1800s by developer and businessman William T. Love [2]. It was to be part of a hydroelectric plant that would supply inexpensive electric power to lure industry to northwest New York State. See Figure 12.1. The Niagara River flows north, connecting Lake Erie with Lake Ontario—New York State being on the east and Ontario, Canada, on the west. The canal was to bring water to the plant from above Niagara Falls and return the water to the river below the famous falls. Love intended to name the industrial community Model City. However, economic conditions in the 1890s were poor, and the only part of the development to be completed was a small segment of the canal, a small factory, and a few homes [3].

The canal and adjoining land, located in a sparsely populated area, lay abandoned for 50 years. Then, in the 1940s, Hooker Chemical and Plastics Company leased the property for use as a waste disposal site. Later, having purchased the site, Hooker lined the canal (estimated to be 20 m wide, 3 m

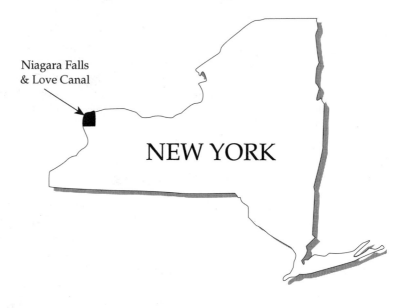

FIGURE 12.1   Location of Love Canal.

deep, by 1000 m long) with clay to reduce the leaching of wastes. The company has acknowledged that it disposed of more than 20,000 tons of chemical wastes between 1942 and 1953 at Love Canal.

On April 28, 1953, the Niagara Falls School Board acquired title to the property for the fee of $1. The deed contained this disclaimer:

> "...Prior to the delivery (sale)...the grantee herein (school board) has been advised by the grantor (Hooker) that the premises above described have been filled, in whole or in part, to the present grade level thereof with waste products resulting from the manufacturing of chemicals by the grantor at its plant in the City of Niagara Falls, New York, and that the grantee assumes all risk and liability incident to the use thereof...."
>
> (from the deed recorded July 6, 1953, City of Niagara Falls)

In addition to this, the school board received advice from its legal counsel that to purchase the land would place undue financial risk upon the city and the board. The construction company hired to build the school planned for the site also advised against construction, but school officials were not to be dissuaded. A school was built and opened in 1955—with some buildings directly atop the waste-filled canal. (See Figure 12.2.) The school board sold the remaining property to developers, who built and sold homes sought after by families with small children.

As the area developed, evidence of the earlier excavations was eliminated. By 1972 most of the "Ring One" homes, those with backyards abutting the school ground property, had been built.

Construction of the LaSalle Expressway in the late 1960s filled in the drainage ditch between the canal and the Niagara River, which aggravated the site's environmental problems. Previously, natural subsurface drainage from the canal flowed into the nearby river, but construction of the expressway impeded this natural drainage. During the fall of 1975 and spring of 1976, heavy rainfall in the area caused the groundwater to rise much higher than normal. This caused subsidence in the landfilled area and contaminated ponded surface water with high levels of toxic chemicals. Chemicals were also transported by the groundwater into nearby residences, many of which had basements. At that time, the contamination was not discovered.

The first indication of problems at the site was the detection of low levels of organic chemicals in the Niagara River and Lake Ontario in 1976. Then, in April 1977, the Calspan Corporation, a private consulting company, found that 21 of 188 homes adjacent to the canal (most near the south end of it) had chemical residues in their basements. Calspan also found PCBs in nearby sewers and unidentified organics buried in drums in the canal. These and more problems continued to be uncovered during

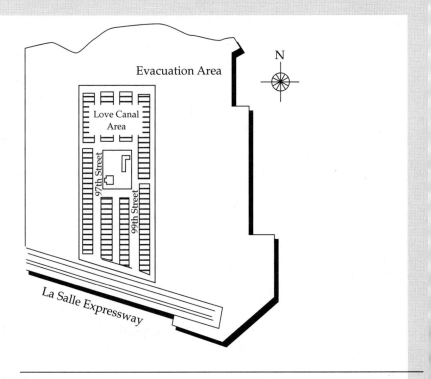

FIGURE 12.2 Niagara Falls, New York, and Love Canal.

the remainder of 1977 and into 1978. The contamination remained in place while local and state officials discussed the issue, which received extensive news coverage at the local level but little or none at the national level.

On August 2, 1978, the New York State Department of Health declared that a state of emergency existed for those persons living in the "Ring One and Ring Two" homes. These residents were evacuated at state expense. Later that same month President Jimmy Carter declared that a national emergency existed, and 237 families were evacuated.

Occidental Chemical Corp. (OxyChem) had purchased Hooker Chemical in 1968. A federal court has since found OxyChem to be financially liable for the cleanup at Love Canal. Although OxyChem still denies any liability, it has agreed to store and destroy the wastes from area. The estimated expenditures up to June 1989 were $140 million [4]. Total costs are expected to be in the range of $250 to $300 million [5].

SOURCES

In 1985 the EPA surveyed facilities permitted to treat, store, and dispose of hazardous wastes, finding that they handled a total of 275 million metric tons of waste each year [6]. Examples of hazardous wastes that result from the manufacturing and processing of common consumer products are shown in Table 12.1.

TABLE 12.1   Common Sources of Hazardous Wastes

CONSUMER PRODUCT	INDUSTRIAL PROCESSES	WASTES
Leather belt with brass plated buckle	Leather tanning Electroplating	Chromium, caustics Toxic metals and cyanides in sludges
Gasoline	Refining	Leaded sludges from tank bottoms, other toxic sludges
Electric motor	Copper refining (for wire) Steel finishing (for the case and internal parts Electroplating (for electrical connections, switches) Manufacturing and assembly	Copper sludges Acidic metal wastes, metallic sludges Toxic metals and cyanides in sludges Organic solvents and paints, chlorinated solvents
Bicycle	Electroplating Painting Assembly	Toxic metals and cyanides in sludges Oganic solvents and paints Chlorinated solvents

HEALTH EFFECTS

The EPA uses human health effects and environmental or ecological indicators to determine if a particular waste is harmful. That is, if improperly managed, could the waste be injurious to human health or damage the environment?

A material can be hazardous for a variety of reasons. Hazardous compounds can affect human health because they are infectious or contain acute or chronic toxins. Infectious materials are those that contain disease-producing microorganisms. A few possible sources of infectious wastes are hospitals, pharmaceutical manufacturers, and military biological weapons installations.

Acutely toxic compounds cause death or serious illness within a short time of exposure. Acute toxicity is usually measured as the LD_{50}. This is the exposure level, under precisely described conditions, in which 50% of the exposed population will die. Since scientists cannot use humans for this testing, the LD_{50} for humans is estimated from exposure data on animals. The results are then extrapolated for humans.

In contrast, a chronic toxin will not exhibit its negative effects until after an extended period of exposure. The effects vary: Carcinogenic materials produce cancer; teratogenic materials cause birth defects; mutagenic materials change the DNA of the exposed organism. Since DNA contains the entire "construction information" for the organism, alteration of DNA can result in cell mutations. These can cause cells to malfunction, resulting in a variety of problems from cancer to reproductive failure. Thus, a mutagen can also be a teratogen or a carcinogen.

FEDERAL REGULATIONS

Three major pieces of federal legislation deal with hazardous wastes and toxic substances. The first to become law was the Resource Conservation and Recovery Act of 1976 (or RCRA, pronounced *rĕk' rà*). RCRA is intended to control hazardous waste from creation to ultimate disposal and at all steps in between—an approach that is often referred to as "cradle to grave" management. RCRA was followed shortly (in legislative time!) by the Comprehensive Environmental Response, Compensation and Liability Act of 1980 (CERCLA, or Superfund). The intent of Superfund is to clean-up abandoned hazardous waste sites. The other significant federal law is the Toxic Substances Control Act of 1976 (TOSCA). This deals primarily with the manufacture of new chemicals and their effects on human health and the environment.

Detailed requirements involved in these laws are contained in the Code of Federal Regulations. They can be obtained by contacting the U.S. Environmental Protection Agency in Washington, DC. These regulations are also available at any library that is a federal repository. Most university libraries are federal repositories.

Resource Conservation and Recovery Act

The Resource Conservation and Recovery Act (RCRA) instructed the EPA to develop regulations to control the handling of hazardous waste from "cradle to grave." To accomplish this mission, one of the agency's first tasks was defining a hazardous waste. Although this might seem like a simple matter, it is actually quite difficult. There are so many different types of wastes from so many sources, any or all of which may be harmful to human health or the environment, that to develop a single definition is almost impossible. Therefore, the EPA decided to use two definitions. One declares a waste hazardous if it possesses one or more hazardous characteristics: ignitability, corrosivity, reactivity, or toxicity. The other definition identifies a waste as hazardous if it is on one of two lists of processes that generate hazardous waste, or if it is on a list of discarded or spilled chemicals. The EPA's hazardous waste definition is depicted in Figure 12.3. Classification is discussed in turn in the remainder of this section.

Characteristics: Listing:

▶ Ignitable   ▶ Nonspecific sources

▶ Corrosive   ▶ Specific sources

▶ Reactive   ▶ Discarded or spilled

▶ Toxic   chemicals

FIGURE 12.3 Hazardous waste classifications.

Ignitable wastes are those which may cause a fire during their storage, transportation, or disposal. The EPA defines ignitability in terms of the material's **flash point** and other characteristics. An example of an ignitable waste is a product containing any of the more volatile fuels such as gasoline.

Corrosive wastes have pH measurements outside the range of 2 to 12.5 or, corrode steel at a rate greater than 6.35 mm (0.25 in) per year. An example of wastes that might meet the corrosivity criteria are highly acidic or basic metal cleaning solutions.

Reactive wastes include those that react violently with water, form potentially explosive mixtures with water, are normally unstable, contain cyanide or sulfide in sufficient quantity to evolve toxic fumes at high or low pH, can explode if heated under pressure, or are listed as explosive compounds in Department of Transportation (DoT) regulations.

Toxic wastes are defined by a rigorous test procedure, the Toxicity Characteristic Leaching Procedure (or TCLP). This simulates the leaching conditions found in landfills and identifies materials that will **leach** toxic materials if placed in a landfill. In the procedure, a waste is extracted (or leached) for 24 hours with an acetic acid solution. The extraction is simply the dissolving of any constituents that are soluble in acetic acid solution. The residual acidic liquid, or **leachate** (not the solid material) is analyzed for several different substances listed in the EPA procedure. It is assumed that if the acetic acid will not solubilize the contaminants, neither will the action of rain water infiltrating throughth the landfill. Table 12.2 contains a partial listing of contaminants that the EPA has determined are a threat to groundwater because they can be leached from wastes. If the leachate

TABLE 12.2   Toxicity Concentration Limits for the TCLP

EPA NUMBER	CONTAMINANT	MAXIMUM CONCENTRATION, mg/L
D004	Arsenic	5.0
D005	Barium	100.0
D018	Benzene	0.5
D006	Cadmium	0.1
D019	Carbon tetrachloride	0.5
D020	Chlordane	0.03
D021	Chlorobenzene	100.0
D022	Chloroform	6.0
D007	Chromium	5.0
D008	Lead	5.0
D009	Mercury	0.2
D010	Selenium	1.0
D011	Silver	5.0
D012	Endrin	0.02
D013	Lindane	0.4
D014	Methoxychlor	10.0
D015	Toxaphene	0.5
D016	2,4-D	10.0
D017	2,4,5-TP Silvex	1.0

SOURCE: From 40 Code of Federal Register Part 261.24, Characteristics of TCLP Toxicity, May 19, 1980.

contains any of the listed compounds in excess of the specified amounts, it is considered to be "TCLP toxic" and is assigned the corresponding hazardous waste number. An example of wastes that might meet the TCLP toxicity criterion are industrial process sludges that contain metals or organic contaminants listed in the table.

Three other definitions of hazardous substances are in the form of lists that either name wastes or identify processes that produce such wastes. The three EPA lists identify nonspecific source wastes, specific source wastes, and discarded or spilled commercial products. (See Figure 12.3.)

Hazardous Waste from Nonspecific Sources. Table 12.3 describes generic wastes commonly produced in a variety of industrial processes. The EPA has included in this list several common halogenated degreasing solvents, electroplating wastes, and wastes from metal heat treating and recovery. An example of such a waste is trichloroethylene (TCE), which is used in removing oils and other contaminants during many metal manufacturing operations. Because electroplating wastewater typically contains cyanides, metals, and chromates, any sludge generated by treating such wastewaters is also included.

TABLE 12.3   Hazardous Waste from Nonspecific Sources.

HAZARDOUS WASTE NUMBER	WASTE DESCRIPTION	HAZARD CODE*
F001	Spent halogenated solvents used in degreasing, tetrachloroethylene, trichloroethylene, methylene chloride, 1,1,1-trichloroethane, carbon tetrachloride, and the chlorinated fluorocarbons; and sludges from the recovery of these solvents in degreasing operations.	T
F006	Wastewater treatment sludges from electroplating.	T
F007	Spent plating bath sludges from electroplating operations.	R, T
F010	Quenching bath sludge from metal heat treating operations.	R, T

*The four hazardous waste classifications: ignitability (I), corrosivity (C), reactivity (R), and toxicity (T). Partial listing from 40 CFR Part 261.31, Hazardous Waste from Nonspecific Sources, May 19, 1980.

Hazardous Waste from Specific Sources.   Wastes from specific industrial processes identified by the EPA are divided into 11 categories: wood preservation; production of inorganic pigments, organic chemicals, pesticides, and explosives; petroleum refining, leather tanning, and iron and steel manufacturing; primary copper, lead, and zinc production and secondary lead production. Hazardous wastes from specific sources are shown in Table 12.4.

TABLE 12.4   Examples of Hazardous Waste from Specific Sources.

HAZARDOUS WASTE NUMBER	WASTE DESCRIPTION	HAZARD CODE*
Wood preservation: K001	Bottom sediment sludge from the treatment of wastewater from wood-preserving processes that use creosote and/or pentachlorophenol	T
Inorganic pigments: K005	Wastewater treatment sludge from the production of chrome (green) pigments	T
Organic chemicals: K020	Heavy ends from the distillation of vinyl chloride in vinyl chloride monomer production	T
Pesticides: K032	Wastewater treatment sludge from the production of chlordane	T
Explosives: K044	Wastewater treatment sludges from the processing of explosives	R
Petroleum refining: K052	Tank bottoms (leaded) from the petroleum refining industry	T
Leather tanning, finishing: K053	Chrome (blue) trimmings generated by the following subcategories of the leather tanning and finishing industry:...	T
Iron and steel: K060	Ammonia still lime sludge from coking production	T

*The four hazardous waste classifications; ignitability (I), corrosivity (C), reactivity (R), and toxicity (T). Partial listing from 40 CFR Part 261.32, Hazardous Waste from Specific Sources, May 19, 1980.

Discarded or Spilled Commercial Chemical Products.   The EPA has determined that certain cast-off chemical products are hazardous. This classification includes only commercial chemical products that are discarded, their containers if not triple rinsed, and any spills or spill residues of these products. The grouping, which includes off-specification species as well, is further divided into acute hazardous wastes and toxic wastes. Acute hazardous wastes are those the EPA has determined represent a greater risk to human health and the environment than others. Examples are given in Table 12.5.

TABLE 12.5   Discarded Commercial Chemical Products, Off-specification Species, Containers, and Spill Residues Thereof

HAZARDOUS WASTE NUMBER	DESCRIPTION
P012	Arsenic trioxide
P089	Parathion
P110	Tetraethyl lead
U051	Creosote
U151	Mercury
U226	1,1,1-trichloroethane

Partial listing from 40 CFR Part 261.33, May 19, 1980.

Thus, to summarize, a waste can be a hazardous waste if it meets any of the four characteristics of a hazardous waste, is from a listed nonspecific source, is from a listed specific source, or is a discarded or spilled commercial product listed by the Environmental Protection Agency. In all cases, it is the responsibility of the waste producer to determine if a waste is hazardous as defined by the government, and if it is, to follow the RCRA regulations concerning its treatment, storage, transportation, and disposal. It is important to note that a waste material so defined as hazardous is, nevertheless, not considered hazardous if the material has value. Thus, scrap or by-products from industrial processes, if they are still of value (that is, they can be reused by the facility or sold, but not given away), are exempt from hazardous waste regulations.

RCRA Program Requirements.   RCRA regulations for the management of hazardous wastes are divided into four interrelated program areas. This is shown in Figure 12.4. Subtitle C comprises the basic regulations for managing hazardous wastes. However, the underground storge tank (UST) regulations are also an integral part of the overall hazardous waste management program.

Subtitle C of RCRA places hazardous waste in three management classifications: generators; transporters; and treatment, storage, and disposal facilities (TSDFs). There are different control methods and regulations for each of the three areas. These classifications are depicted in Figure 12.5.

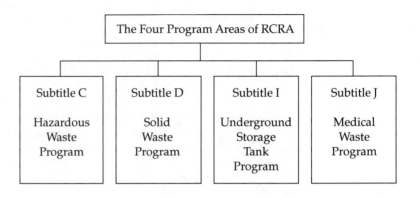

FIGURE 12.4 RCRA's four interrelated program areas.

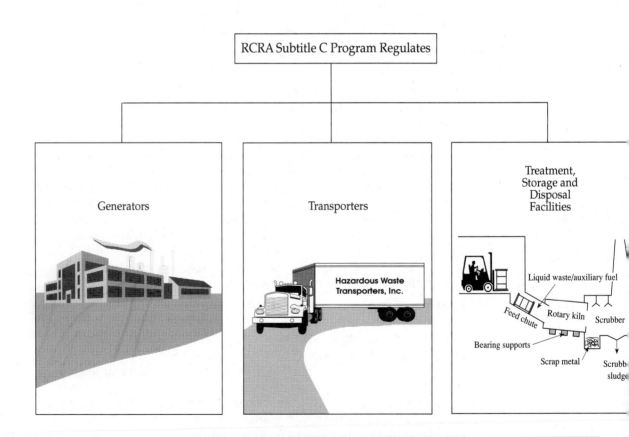

FIGURE 12.5 Subtitle C regulatory coverage.

Generators. All generators are required to obtain an EPA identification number. No waste may be transported by or sent to any facility that does not also have an EPA ID number. Prior to transportation all hazardous waste must be properly packaged so it will not leak or spill. Containers must be identified as containing hazardous waste and have the name and telephone number of an emergency contact familiar with the waste's characteristics and dangers. The EPA adopted the U.S. Department of Transportation requirements for this purpose.

As indicated in Figure 12.6, federal officials divide generators into three categories, depending on quantity of waste generated. Large-quantity generators are those that generate more than 1000 kg of hazardous waste per month or more than 1 kg of acutely hazardous waste per month. The EPA has estimated that these generators produce more than 99% of all regulated hazardous waste in the United States. Small-quantity generators, or SQGs (often called squeegees), are those that produce more than 100 kg but less than 1000 kg of hazardous waste, and less than 1 kg of acutely hazardous waste per month. There are separate requirements for these categories.

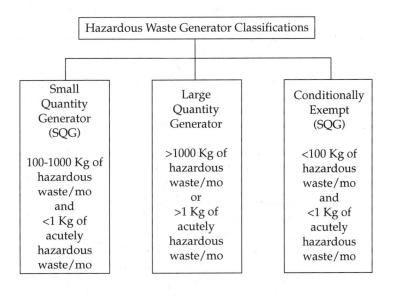

FIGURE 12.6 Hazardous waste generator classifications.

Depending on size, generators are allowed to accumulate different amounts of waste prior to shipment. The length of time a waste may be stored also varies. For large-quantity generators, the limit is 90 days. For SQGs it is 180 to 270 days. If a facility plans to exceed these limits, it is required to obtain a TSDF permit. Many large-quantity generators obtain a TSDF permit to store wastes for a longer time period prior to shipment, and in some cases to treat or partially treat the wastes themselves.

And there are many other requirements. All materials must be clearly marked "Hazardous Waste" with the date accumulation first began. Facilities must have emergency plans for accidents. People handling the wastes must be properly trained. Generators must also determine if their wastes are banned from land disposal and, if so, ship them only to appropriate disposal facilities. In addition, rules state that a generator of hazardous waste cannot sell or transfer liability for wastes they may generate. And no matter how irresponsibly a contract transporter or TSDF facility may act, the original generator is financially liable for mishaps. This is one of the biggest dangers an industrial generator faces in shipping wastes to another facility for disposal.

A final requirement of generators is that they initiate a manifest, or written record, of all materials shipped off site. The manifest is used to track the hazardous wastes. One copy is retained by the generator after shipment. The transporter retains a copy. The remaining copies are given to the TSDF receiving the waste. The TSDF must then mail a copy back to the generator. Thus, the generator knows that the wastes have in fact arrived at the intended destination. If the generator does not receive a return copy within 45 days, the EPA must be notified.

Transporters. Any organization transporting hazardous waste is required to obtain an ID number from the EPA. Transporters must also comply with the manifest requirements, and they have certain designated responsibilities if an accident or spill occurs.

Treatment, Storage, and Disposal Facilities. Treatment is any method, technique, or process intended to improve the character of a waste. This could make the waste nonhazardous or less hazardous, or it could improve its handling characteristics — making a sludge into a solid, for example, or concentrating a waste by removing a portion of the water. Storage is the temporary holding of hazardous waste (excepting the short-term storage by generators discussed earlier). Disposal is the discharge or placement of any waste material into or on the land, or into air or water. Any facility engaging in any of these activities must obtain an RCRA TSDF permit. One example would be a commercial hazardous waste incineration facility that stores incoming waste from generators, sorts and blends the liquid wastes to provide optimal fuel values, and then incinerates the wastes and places the residues in a landfill at the site. Another example would be a manufacturing plant that stores wastes more than the 90 days allowed for generators and treats some of them to make them less hazardous before shipment to a commercial disposal facility.

TSDFs have the most extensive regulations. Requirements are divided into administrative standards, permit general standards, and permit specific standards. The administrative standards include rules for waste analysis, site security, inspections, training, emergency preparedness, and manifests and other record keeping.

Administrative and permit general standards are required of all TSDFs. The general standards include requirements for record keeping, groundwater monitoring, closure and postclosure activities, and financial considerations. The objective of the groundwater monitoring is to detect any leaks of wastes into the groundwater. Facilities must install monitoring wells and take corrective action should leaks occur.

Closure is the period after a facility stops accepting waste and completes final treatment, storage, and disposal operations. During this period equipment, buildings, and soil are decontaminated, and the cover or cap is applied to landfills. Postclosure monitoring and maintenance is required for a minimum of 30 years for landfill facilities.

Financial requirements are intended to ensure that TSDFs can afford to close, conduct monitoring afterward, and to pay any liabilities that may arise due to injuries to other parties. These requirements include an estimate of all closure and postclosure costs. Proof of financial responsibility can be in the form of a trust fund, a financial test, a corporate guarantee, or other measures taken. The requirements must be met prior to the issuance of the TSDF permit.

Permit specific standards regulate the various treatment, storage, and disposal operations at a given facility. Containers, tanks, surface impoundments (ponds), landfills, incinerators, and other equipment are all addressed in these standards. For example, landfills are required to have at least two liners and two leachate collection systems, one above the first liner, and one between the liners, as shown in Figure 12.7. Amendments to the RCRA regulations required that the placement of liquids in landfills be minimized, and in 1984 Congress fine-tuned the RCRA with the Hazardous and Solid Waste Amendments (HSWA). A significant part of HSWA is the "land ban" of many wastes formerly landfilled. This includes solvents and dioxins, chlorinated organics, and most liquids containing dissolved metals.

Incinerators are the preferred disposal method for hazardous wastes whenever possible. Many industries have chosen incineration over less expensive methods because it effectively eliminates future liability. The RCRA requires that incinerators remove 99.99%, or "four nines," of each principal organic hazardous constituent. (For dioxins, due to their extreme toxicity, the removal requirement is 99.9999%, or "six nines.")

Subtitle C management is a large EPA program. As of June 1989, the agency maintained records on 180,000 generators, 17,000 transporters, and 7000 treatment, storage, and disposal facilities [6].

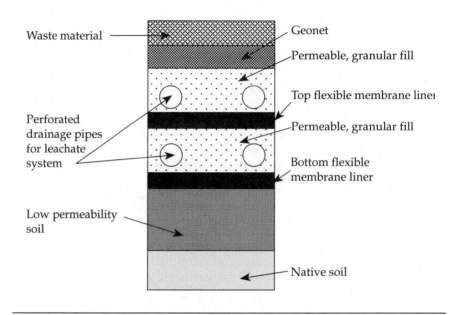

Waste material —

Geonet

Permeable, granular fill

Top flexible membrane liner

Perforated
drainage pipes
for leachate
system

Permeable, granular fill

Bottom flexible
membrane liner

Low permeability
soil

Native soil

FIGURE 12.7   Cross section of a hazardous waste landfill with two liners and
two leachate collection systems.

Comprehensive Environmental Response, Compensation and Liability Act

When RCRA legislation was passed, its only purpose was to prevent improper waste disposal by dictating proper handling methods; it proposed no remedy for sites at which hazardous materials had been previously disposed of by inadequate methods. Site operators were expected to clean up their past mistakes, but many refused and others were financially unable to do so. Since there was no regulation or law directly stating that operators were responsible for decontamination of such sites, the only means of forcing them to assume responsibility was the judicial system—a process that could take many years for each site. Likewise, there was no regulation requiring cleanup of sites contaminated by continued mismanagement of disposal facilities or by accidental spills.

To close this legal "gap," Congress passed the Comprehensive Environmental Response, Compensation and Liability Act (CERCLA, or Superfund) in 1980. **Site remediation** under Superfund is the process of cleaning up disposal and accidental spill sites that have been abandoned by the responsible parties or at which the responsible parties refuse to clean up or cannot afford to. The severity of problems at such sites ranges from minor spills requiring emergency cleanup to long-term remedial actions necessitating the expenditure of hundreds of millions of dollars over a number of years.

Superfund has two levels of activity: emergency response and long-term remediation. Emergency responses are initiated at sites where there is an immediate threat to human health or the environment. Conditions such as explosion hazards, the threat of groundwater contamination, and significant leaks threatening human health would be included in this category. Cleanup of such crisis conditions is to be completed within six months.

Long-term activity is begun only after a complete site evaluation determines that cleanup is required. This determination is based on a Hazard Ranking System (HRS) developed by the EPA. A site must score over 27.5 (out of a possible 100) on the HRS to be added to the National Priorities List (NPL) for eventual Superfund cleanup. In 1990 there were approximately 1300 sites on the NPL. Estimates of the total number of sites eventually to be added to the NPL vary from over 2000 to 10,000.

Before remedial action is undertaken with Superfund monies, the EPA tries to get the responsible parties to do the work. If this fails, the site is cleaned up under EPA authority (usually by private contractors). The agency then attempts to recover the costs from the responsible parties. Superfund, originally authorized for $1.6 billion, acquired this name because of the large amounts of money it raises from taxes on various industrial products. This money was obtained primarily from a tax on petrochemicals, with a small portion coming from general revenues. Superfund was reauthorized by the Superfund Amendments and Reauthorization Act (SARA) in 1986, increasing the funding level to $8.5 billion. SARA also changed the method of financing to allow for a much wider industrial base. Currently, monies are obtained from petrochemical and chemical taxes, a broad-based corporate environmental tax, responsible party funds, Superfund penalties, and general tax revenues.

Toxic Substance Regulations

In addition to RCRA and Superfund, which deal with hazardous wastes, federal legislation during the 1970s dealt with toxic substances other than wastes. The Toxic Substances Control Act of 1976 (TOSCA) empowered the EPA to evaluate new chemicals prior to their manufacture and introduction into the market. TOSCA empowers EPA to restrict or prohibit chemicals that cannot be used and disposed of without undue danger to human health and the environment. It did not prohibit companies from researching and testing chemicals before beginning commercial production, however. TOSCA also allowed the EPA to evaluate existing chemicals for health or environmental hazards. Chemicals determined to represent a significant threat to human health or the environment, even when managed properly, may be banned or restricted.

The Federal Insecticide, Fungicide, and Rodenticide Act of 1972 (FIFRA), which was amended in 1978, controls the manufacture and use of pesticides in a manner similar to that used by TOSCA for other toxic substances.

WASTE MINIMIZATION

Hazardous **waste minimization** is the elimination or reduction of a waste prior to its generation. It thus reduces the amount of waste for treatment, storage, or disposal. This is the ideal approach for any potential waste, but it has limits. Through RCRA and its arm of implementation, Congress encourages waste minimization. Oddly, the EPA has spent little money to develop minimization programs, yet its programs for hazardous waste treatment, storage, and disposal are so costly that, indirectly, the agency has encouraged waste minimization. Industry is profit motivated, and in a free market economy such as ours, it must be. The hazardous waste regulations have increased the cost of hazardous waste disposal several-fold in the past 15 years. This, combined with the potential liability for any harm caused by such wastes, has provided industry with the incentive to decrease or minimize the quantity of hazardous waste produced.

Waste minimization methods often require an intimate knowledge of the industry involved. However, several general concepts concerning minimization can be presented here. Waste minimization can be accomplished by one of two methods: recycling or source reduction.

Recycling

Recycling is the reuse of a waste product, either by directly reusing the material or by reclaiming it. In some cases a material may require additional processing for reuse, and in other cases materials can be reused directly.

A good example of the EPA's indirect encouragement of waste minimization or recycling occurs in the electroplating* industry. Electroplated materials are rinsed to remove the plating solution. This rinsewater contains the same toxic metals and salts (often cyanide) as the solutions themselves, but in concentrations too low to use. Prior to RCRA, most electroplaters used chemical treatment to remove contaminants. This resulted in a clean water for discharge, but it also produced a hazardous waste sludge that contained high levels of heavy metals. Because the RCRA defines these sludges as hazardous, the combination of rinsewater treatment and sludge disposal is so expensive that many electroplaters have switched to rinsewater treatment processes that recover the metals for recycling. Ion exchange, reverse osmosis, or evaporation is used to concentrate a rinse solution without chemically changing its constituents. The concentrated solution then returns to the electroplating bath. These processes were available prior to RCRA but were considered too costly because it was cheaper simply to place the sludges in a landfill or lagoon — practi-

---

*Electroplating is used on many common consumer products, including jewelry, belt buckles, watches, watchbands, many electronic components, automobiles, furniture, and house wares.

ces that are now prohibited. Expensive procedures such as chemical fixation are the only currently approved disposal methods for heavy metals, so reuse is a financially attractive alternative to disposal.

Source Reduction

Source reduction is the elimination or reduction of a waste at its source by modifying the process that produces it. Thus, it requires substitution of the process feedstock, modification of the process, improved inventory control, or better process control.

The electroplating industry has also been able to practice a significant amount of source reduction. Prior to RCRA, most electroplating facilities used cyanide-based metallic plating for metals such as zinc, copper, silver, brass (zinc–copper), and gold. Although other processes have been available for many years, they were seldom used because they were more costly than the cyanide processes and required better control to achieve the same product quality level. As noted, disposal costs for cyanide electroplating wastes increased several-fold after RCRA regulations came into effect, so alternative methods became more cost-effective. Many electroplaters have switched to these processes, thereby eliminating the production of cyanide-bearing rinsewaters.

TREATMENT AND DISPOSAL METHODS

Hazardous waste treatment is intended to decrease or eliminate the threatening properties of the materials, to reduce their volume, or to destroy the wastes. From an environmental standpoint, it means fewer dangerous materials enter the environment. From a regulatory standpoint, a waste that is less toxic or on hand in smaller amounts is easier to dispose of. (A waste that does not exist is even easier to dispose of!) And from a corporate liability point of view, less toxic waste should be less of a risk while a waste that is destroyed no longer represents any liability. Clearly, when treatment procedures eliminate the hazardous characteristics of wastes, special disposal methods are not necessary for those wastes. Not all wastes can be treated, but many can, thus reducing the ultimate disposal costs to the industry and the risk to human health and the environment.

Physical and Chemical Treatment

Physical and chemical treatment processes include precipitation, coagulation, filtration, neutralization, and oxidation and reduction.

Precipitation and Coagulation. The precipitation and coagulation process is used primarily to remove metals from wastewater. Many metals are readily precipitated when the pH is raised to the 8 to 10 range, usually by the addition of lime (calcium hydroxide). Coagulation is used in conjunction with the precipitation to agglomerate the material and remove it from suspension.

A representative reaction for the removal of copper is

$$Cu^{2+} + Ca(OH)_2 \rightarrow Cu(OH)_2\downarrow + Ca^{2+}$$ **12.1**

For chromium(III) the reaction would be

$$2Cr^{3+} + 3Ca(OH)_2 \rightarrow 2Cr(OH)_3\downarrow + 3Ca^{2+}$$ **12.2**

The calcium ions produced in the reaction will precipitate with carbonate, if it is present in sufficient quantity, and will be removed in coagulation. However, calcium is not toxic to humans or aquatic life, so it may safely remain in the effluent. These processes, often carried out in the wastewater treatment facilities of many industries, may or may not be considered treatment of hazardous wastes, depending on the characteristics of the wastewater. Hazardous materials in some instances may be so diluted that the wastewater is not considered hazardous under RCRA regulations. Even so, the toxic materials must be removed to comply with the federal wastewater treatment regulations. (See Chapter 10, Wastewater Treatment.) Since the toxic or hazardous components are concentrated in the sludges generated from these processes, the sludges are usually hazardous wastes by the RCRA definitions. So for such wastewater, precipitation and coagulation greatly reduce the amount of waste that is considered hazardous. However, these processes do not eliminate the waste, which remains in the sludge needing to be treated further or disposed of.

EXAMPLE 12.1	ESTIMATION OF LIME REQUIRED TO PRECIPITATE COPPER

Use the precipitation reaction shown in Equation 12.1 to estimate the amount of lime required per day to precipitate 20 mg/L of aqueous copper from an industrial process that has a flow of 15 GPM. Assume an additional 25% of lime is consumed in reactions with other species in the water.

SOLUTION: To determine the ratios in which the materials combine, use this approach:

$$Cu^{2+} + Ca(OH)_2 \rightarrow Cu(OH)_2 + Ca^{2+}$$
$$63.6 \quad [40+2(16+1)] \quad [63.55+2(16+1)] \quad 40$$

There are 20 mg/L of Cu^{2+} in the water. So the amount of lime required to precipitate the copper is

$$m_{Cu} = \left(20\,\frac{mg\ Cu^{2+}}{L}\right)\left(\frac{74\ mg\ Ca(OH)_2}{63.6\ mg\ Cu^{2+}}\right)\left(15\,\frac{gal}{min} \times \frac{3.78\ L}{gal}\right)$$

$$\times \left(1440\,\frac{min}{day} \times \frac{kg}{10^6\ mg} \times 1.25\right)$$

$$= 2.4\ kg\ lime\ per\ day$$

Filtration. Filtration can decrease the amount of water (or other liquid) in liquid or sludge wastes, thereby reducing the volume. Mechanical filtration devices include belt filter presses, rotary vacuum filters, and other types of pressure filter presses. These devices are similar to those described in Chapter 10.

Neutralization. Neutralization of acidic and basic liquids can eliminate the characteristic of corrosivity by bringing the wastes inside the 2 to 12.5 pH range. (Recall that a corrosive waste has a pH outside this range.) Since neutralization is simply the mixing of an acid and a base, this can sometimes be accomplished by mixing two or more wastes from the same facility, without purchasing additional chemicals. When self-neutralizing waste combinations are not available, commercial acids such as sulfuric or hydrochloric acid or bases such as sodium hydroxide or lime can be used.

Oxidation and Reduction. Oxidation and reduction processes (redox) can treat some hazardous wastes. One common application is the destruction of toxic electroplating wastes. Cyanide, used in large quantities in electroplating and mining, can be oxidized to carbon dioxide and nitrogen gas. Another electroplating application is the reduction of chromium(VI) to the less toxic form chromium(III). An additional benefit is that Cr(III) can be precipitated, while Cr(VI) is very soluble.

The cyanide oxidation process, illustrated in Figure 12.8 uses Cl_2 gas, calcium hypochlorite, or sodium hypochlorite to oxidize cyanide to carbon dioxide and nitrogen gas. When sodium hypochlorite is used,

$$Ca(OCl)_2 \rightarrow Ca^{2+} + 2OCl^- \qquad\qquad \textbf{12.3}$$

When chlorine gas is used, it forms hypochlorite as it dissolves into the water:

$$Cl_2 + H_2O \rightarrow HOCl + H^+ + Cl^- \qquad\qquad \textbf{12.4}$$

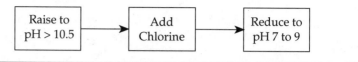

FIGURE 12.8 Cyanide destruction using alkaline chlorination.

Prior to the addition of the oxidizer, a base such as sodium hydroxide is used to increase the pH to about 10.5. At the higher pH, the oxidation reaction proceeds much faster. The oxidation–reduction reactions are then

$$CN^- + OCl^- + H_2O \rightarrow CNCl + 2OH^- \qquad \textbf{12.5}$$

$$CNCl^- + 2OH^- \rightarrow CNO^- + Cl^- + H_2O \qquad \textbf{12.6}$$

$$2CNO^- + 3HOCl + H_2O \rightarrow 2HCO_3^- + N_2 + 3Cl^- + 3H^+ \qquad \textbf{12.7}$$

The hexavalent chromium [Cr(VI)] reduction process is analogous to the cyanide oxidation process. However, in chromate reduction the pH is lowered, a reducing agent such as sulfur dioxide [SO_2] or sodium bisulfite [$NaHSO_3$] is added, and then the pH is raised to precipitate the resulting Cr(III).

Adsorption Processes.   Carbon adsorption and ion exchange can be used to concentrate a contaminant at the surface of a solid. Carbon adsorption uses activated carbon to bring organics to the surface of the carbon. After the surface is filled, the carbon can be regenerated and reused. During the regeneration process, the adsorbed contaminants are incinerated.

As described in Chapter 4, Physical Chemistry, ion exchange processes exploit chemical reactions in which one ion is exchanged for another. Their most common use is for water softening, but ion exchange is also used to remove toxic metal ions from industrial wastewaters. Figure 12.9 depicts the removal of zinc and cyanide from an industrial process with ion exchange. The metallic ions can be recovered for reuse.

Landfill Disposal

Landfilling is the placement of wastes into the land under controlled conditions to minimize their migration or effect on the surrounding environment. During the 1980s, approximately 20% of all hazardous wastes were disposed of by landfill [6]. Because of problems associated with landfilling in the past, Congress has passed legislation discouraging landfilling of many hazardous wastes. However, in the foreseeable future, landfilling will remain a necessary part of our nation's hazardous waste program. Landfilling can be used successfully for certain kinds of wastes. Important criteria for determining the suitability of a waste for landfill

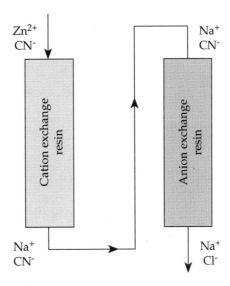

FIGURE 12.9 Ion exchange removal of zinc cyanide solution.

disposal are its leachability, state (solid or liquid), toxicity, and volatility. Essentially, will the waste stay in the landfill or does it have a significant potential for migration into the surrounding soil, air, and/or groundwater? Good candidates for landfill disposal are dewatered sludges that do not contain leachable metals or organics, incinerator ash, and other nonleachable hazardous solids. Congress has specifically banned the landfill disposal of organic solvents, dioxins, and certain other toxic or hazardous wastes.

A hazardous waste landfill must contain the wastes and collect the leachate generated. It must also collect any gas produced in anaerobic degradative processes. Thus, the landfill must have a liner system, as discussed previously, but also a leachate treatment system and a gas collection or venting system to prevent flammable gases from migrating underground and causing an explosion in basements of nearby buildings. After wastes are placed in a landfill, the top must be covered to prevent precipitation from entering the landfill, creating leachate. A typical landfill cross section is shown in Figure 12.10.

Groundwater monitoring is an essential part of any landfill operation. It is used to establish the ambient or background level of groundwater quality as well as detect any contamination. Figure 12.11 depicts the detection of landfill leachate using groundwater monitoring wells. Where it is known that groundwater contamination has occurred, monitoring wells are used to determine the extent.

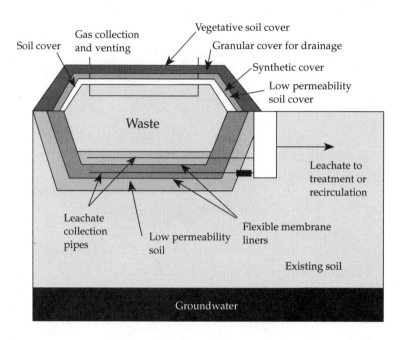

FIGURE 12.10   Cross section of a typical hazardous waste landfill.

The EPA has two levels of groundwater protection rules. Currently, permitted and constructed hazardous waste landfill facilities are required to have two liners with one leachate collection system above the top liner and one between the liners. The second liner and leachate collection system will be used only after the first liner fails due to degradation or puncture. The landfill depicted in Figure 12.10 has two liners and two leachate collection systems. Facilities constructed and permitted prior to 1985 are required to have only one liner and one leachate collection system. However, most of these facilities should be out of service in the next few years.

Incineration

Incineration is the preferred option for the disposal of many hazardous wastes because it reduces relatively large volumes of many wastes to only a small amount of residual ash and air-scrubber sludge. The EPA requires that incinerators have an air pollution control system, which normally consists of a wet scrubber. This is a device that sprays water with an acid-neutralizing agent (such as lime or sodium hydroxide) into the exhaust gases. The spray removes the particulates, and the alkaline compound neutralizes any acids produced by the combustion.

Many corporations are choosing incineration over other disposal

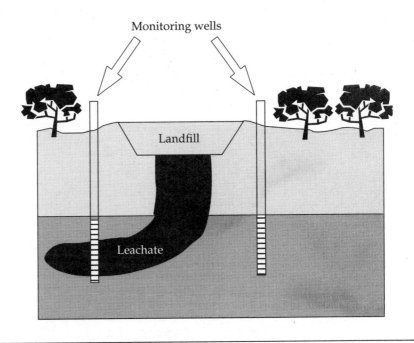

Monitoring wells

Landfill

Leachate

FIGURE 12.11 Landfill construction and groundwater monitoring wells.

methods even when burning costs are significantly greater because the process leaves no appreciable waste and no liability. The scrubber sludge can be dewatered and placed in a **secured** hazardous waste **landfill** if it is found to be hazardous. Congress and the EPA have designated incineration as the only acceptable method of disposal for chlorinated organics.

Combustion of most organic wastes results in end products of water, carbon dioxide, and other inorganics. As an example, the combustion of the PCB 2,3,3′,4′-tetrachlorobiphenyl [$C_{12}H_6Cl_4$], depicted in Figure 12.12, yields

$$C_{12}H_6Cl_4 + 12\tfrac{1}{2}O_2 \rightarrow 12CO_2 + H_2O + 4HCl \qquad \textbf{12.8}$$

Thus, each mole of PCB requires 12.5 mol of oxygen for complete combustion. The resulting products are 12 mol of carbon dioxide, 1 mol of water, and 4 mol of hydrochloric acid. This acid production is common for organic compounds containing chlorine, bromine, and in many cases sulfur or nitrogen as well. The acid production is one of the reasons an air scrubber is required for hazardous waste incinerators. The other is that such devices remove particulates resulting from ash formation and incomplete combustion.

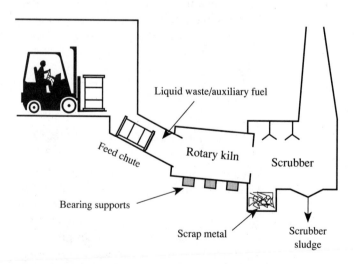

FIGURE 12.12   PCB structure.

The 4 mol of acid produced from the combustion of the PCB could be neutralized in the scrubber with lime or sodium hydroxide. The reaction for sodium hydroxide would be

$$4HCl + 4NaOH \rightarrow 4NaCl + 4H_2O$$ **12.9**

Several types of incinerators are used for hazardous waste incineration. The most versatile design is the rotary kiln incinerator, shown in Figure 12.13. These incinerators can burn injected liquid wastes, liquid wastes in drums, sludges, and a variety of solid wastes. They have relatively large capacities, from 9 to 15 tons/hr [7] and are used at commercial TSDFs. Some versions of the rotary kiln furnace are transportable on two or more tractor trailers, which allows them to be moved to a site and set up for temporary use. However, setup requires considerable time. This method of

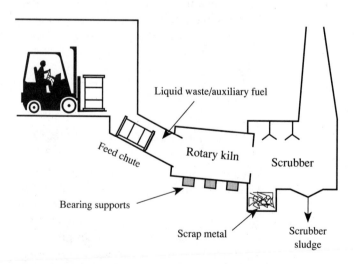

FIGURE 12.13   Schematic of a rotary kiln incinerator.

disposal is effective for organic contaminants but will not remove heavy metals or some other inorganic contaminants.

Liquid injection furnaces are used to incinerate pumpable liquids. In fact, the only materials they can burn are those that can be pumped into the combustion chamber. The advantages of liquid injection furnaces include lower capital costs and simpler operation and maintenance compared to the rotary kiln furnace. They are not nearly as versatile, however, in that they cannot incinerate solids, sludges, or very viscous liquids.

Let us examine the incineration of a typical chlorinated organic. The reaction for destruction of 1,1,1-trichloroethane (TCA) is

$$Cl_3CCH_3 + 2O_2 \rightarrow 2CO_2 + 3HCl \qquad\qquad \textbf{12.10}$$

Thus, each mole of TCA produces 3 mol of acidic HCl gas. The EPA requires that hazardous waste incinerators meet stringent air quality standards, so furnaces must have some method of neutralizing the acidic HCl gas produced from the incineration of chlorinated organics. Furnaces must also remove particulates resulting from the combustion process. These tasks are usually accomplished by a wet air scrubber. (See Chapter 13, Air Pollution.)

Chemical Fixation

Chemical fixation* is a term for several different methods of chemically immobilizing hazardous materials into a cement, plastic, or other matrix. Other terms commonly used for this process are **stabilization** and **solidification**. The end product should be a solid that is resistant to natural and manufactured processes that could leach contaminants into the surrounding soil or groundwater. The objective of chemical fixation is to render the waste harmless according to RCRA definition.

During the past 15 to 20 years several techniques have been developed to stabilize hazardous waste. These processes detoxify the materials, prevent toxic chemical migration, improve the physical properties (convert a sludge into a solid for better handling, for example), and decrease the surface area across which chemical migration can occur. Thus, stabilization converts an unstable (or leachable) solid, liquid or sludge into a solid or semisolid waste that does not exude hazardous materials under leaching action in the ground. Table 12.6 shows several methods of stabilization [8]. Most processes developed to date utilize solidification agent(s) as well as silicates or other chemical binding materials. Several of the processes are proprietary.

---

*Chemical fixation, stabilization, and solidification clearly represent different concepts. Unfortunately, the terms are used interchangeably in the environmental enginering field.

TABLE 12.6   Methods of Stabilization and Solidification

SOLIDIFICATION METHOD	APPLICABLE WASTES
Portland cement	Sludges, contaminated soil, metal salts, low-level radioactive waste
Lime or quick lime	Sludges containing metals and oils, contaminated soils, flue gas desulfurization wastes, other inorganic wastes
Thermoplastics	Strong oxidizers, inorganic salts, low-molecular weight volatiles, radioactive wastes
Self-cementation	Flue gas desulfurization wastes, other wastes with large proportions of calcium sulfate or calcium sulfite

Cement stabilization uses common cement as the solidifying agent, along with silicates or other chemical binding agents to prevent the chemicals from migrating after land disposal. The waste materials are mixed with the cement and silicates and allowed to solidify. The process is commonly used for sludges containing toxic metals. At the resulting high pH, the metals are bound as insoluble carbonates or hydroxides that resist the leaching action of soil and groundwater, particularly when placed in **secure landfills** — that is, those that are lined and capped, thereby limiting the intrusion of liquids. The raw materials for this process are widely available and inexpensive. Extensive dewatering is not required because water is largely used up in the solidification reactions. Also, the process is tolerant of varying waste characteristics, and the technology of cement mixing is widely known. The disadvantages are that large quantities of cement are often required and cement requires much energy to produce [9].

Lime-based stabilization combines the waste with either lime [Ca(OH)$_2$] or quick lime [CaO] and materials containing silicates. The process has been used to treat metals, flue gas desulfurization wastes, and fly ash. The materials are relatively low cost and available, little or no specialized equipment is required, and dewatering is not necessary since the setting reactions require water.

Self-cementing techniques are possible for some hazardous wastes such as flue gas desulfurization (FGD) wastes, since these contain large amounts of calcium sulfate and calcium sulfite. A portion of the waste is dewatered and processed to produce a solidifying agent. It is then reintroduced into the FGD sludge along with fly ash to reduce the moisture content.

Thermoplastic encapsulation uses asphalt materials, paraffin, or polyethylene plastic materials to provide a protective wrap for some hazardous materials. This process has been used for strong oxidizing wastes, inorganic salts, and low–molecular weight volatile wastes as well as some radioactive materials [10]. The process is effective but expensive.

CASE STUDY

The DuPont Chambers Works, A Multiprocess TSDF

E. I. duPont de Nemours & Company is a large chemical company with an established environmental and safety record that most companies envy. DuPont has several large chemical production facilities in the Delaware–New Jersey area and others throughout the United States and the world. DuPont makes everything from the Teflon coatings on cookware to textiles for clothing and carpet, to specialty chemicals for a variety of other industries.

The company's Chambers Works, located in Deepwater, New Jersey, and pictured in Figure 12.14, has worked to reduce the quantity of wastes

FIGURE 12.14 The DuPont Chambers Works in Tidewater, NJ. Photograph compliments of the E. I. duPont de Nemours Company.

produced while at the same time developing a multiprocess TSD facility. The TSD facility has an advanced wastewater treatment plant, a secured hazardous waste landfill that accepts only solids and dewatered sludges, and an incineration facility [11]. A schematic of the facility appears in Figure 12.15. In addition to wastes produced at the Chambers Works, the facility accepts hazardous materials from other DuPont facilities in the Delaware–New Jersey area, and from a few other local industries.

The Chambers Works is a large plant. The environmental facilities alone are enormous. The wastewater treatment operation processes 20,000 to 25,000 gallons of water per minute (1.2 to 1.6 m^3/s). That flow rate is equivalent to that of an average city of 250,000 population.

Incineration.   The incineration facility has a large liquid injection furnace equipped with a wet scrubber to remove the acids produced in the combustion of halogenated organics. It also has an electrostatic precipitator to remove particulates. Incineration wastes are sent to the wastewater treatment facility for neutralization and removal.

Secured Landfill.   The Chambers Works landfill accepts solid chemical

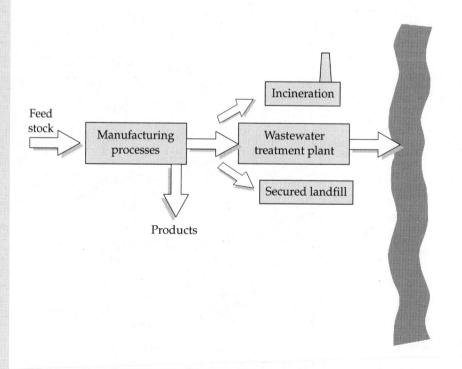

FIGURE 12.15   The DuPont Chambers Works integrated TSD facility.

wastes that can neither be incinerated nor processed in the wastewater treatment facility. For handling ease these are containered in 55-gal drums and then buried. The landfill has a double liner with two leachate collection systems. (It is interesting to note that DuPont installed these devices before the EPA made them a requirement.)

Wastewater Treatment Plant. Acidic wastewater is first neutralized. Then solids produced in primary sedimentation are dewatered and placed in the landfill. The resulting dewatered sludge is used as packing material for the drums containing the other chemical wastes. The wastewater is then treated with an advanced process, combining activated sludge with activated carbon to adsorb organic contaminants that would otherwise upset the biological process.

SITE REMEDIATION PROCEDURES

The technologies used to make abandoned hazardous waste sites safe range from simply removing contamination and transporting it to a secured disposal site (TSDF) to extensive cleanup of groundwater to on-site incineration of wastes or contaminated soils. The strategies can be divided into three broad categories: containment techniques to prevent migration and further contamination of surrounding areas; conventional ways of removing contaminated soil, water, or other material, followed by treatment and disposal, and special procedures for removing or destroying contaminants in place — without removing the soil or groundwater. Many of the procedures are the same as those used by TSDF facilities. Some, which have been presented earlier, will only be mentioned here. Others we will examine in more depth.

Containment

The first steps in any remediation project are identification of the contaminant(s) and the extent of the problem, followed by containment of the wastes. The purpose of containment is to prevent the contamination from moving and thus adversely affecting other areas. Wind, surfacewater runoff, and groundwater flow can all cause contaminant migration.

Containment is usually a temporary measure intended to "buy time" while other strategies are developed and implemented. Some of the more common techniques are pumping, capping, draining and employment of "slurry walls." The choice of techniques varies with the type of materials present and the geology of the site. Often, more than one containment method is used in a given location.

Pumped Containment. Where groundwater is contaminated and moving off site, pumping intercepts the contaminant plume and keeps it from expanding. In this process, wells are placed at or near the advancing edge of a plume as shown in Figure 12.16. As the contaminated water approaches the well or wells, it is pumped out of the ground and removed for treatment. The wells act as a barricade to further advancement of the plume. The disadvantage of this method is that the collected groundwater must then be treated to remove the contaminants prior to discharge to surfacewater or reinjection into the ground.

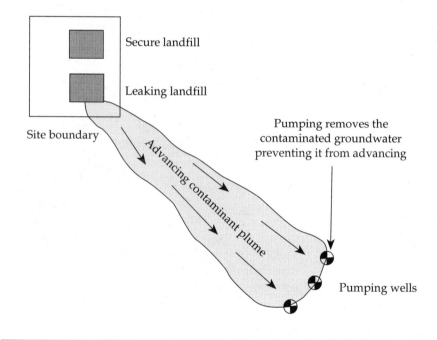

FIGURE 12.16 Pumped containment of contaminated groundwater.

Capping. Leaking landfills and disposal sites can be given a protective cover. This prevents additional surfacewater and precipitation from entering the ground in the area and producing more leachate laden with contaminants. Capping is usually accomplished by placing a 1- to 3-foot-thick clay layer or a synthetic membrane over the area. Runoff is thereby directed away from the site.

Slurry Walls. To prevent the movement of groundwater out of a contamination area, remediation workers sometimes use strategically placed slurry walls. They do this by digging a trench to an impermeable layer, then filling it with a clay or grout mixture that has a very low permeability. This prevents further underground movement of contaminated water. Slurry walls are usually combined with some form of pumping to prevent water from building up adjacent to the wall and then flowing around it. Slurry walls cannot be used effectively where there is not an impermeable rock or clay barrier in the upper geologic strata. Usually, the impermeable layer must be in the upper 30 feet. (See the Case Study on page 423.)

Conventional Cleanup Methods

These include removal and incineration of contaminated soil, direct pumping and treating of contaminated groundwater, chemical treatment, fixation, or solidification, and other approaches to the problem. Contaminated vegetation is often incinerated and buildings decontaminated, incinerated, or landfilled as required. Soil and contaminant incineration can be accomplished by literally digging up the affected soil and combusting it in a rotary kiln furnace. And chemical fixation (or stabilization) can be used for some contaminants found at hazardous waste sites.

In Situ Cleanup Methods

In situ (or in place) cleanup methods involve removal or destruction of contaminants without removal of the subsurface water or soil involved. These methods are far less expensive than removing the contaminated soil and burning it or pumping large quantities of water to the surface for treatment. The EPA, the Department of Energy, and the Department of Defense have funded several demonstration projects in attempts to prove these technologies. We will discuss several of the more promising techniques.

Vacuum Extraction. Where subsurface hazardous materials are sufficiently volatile, vacuum extraction is becoming a common method of removing contamination above the saturated zone. In this procedure wells are drilled into the contaminated soil. Screened pipe sections are placed only in the contaminated area. Suction is applied, drawing contaminants out of the soil and sweeping them away with the air flow, as shown in Figure 12.17. A single vacuum well will remove contaminants for a radius of 20 to 150 feet, depending on soil conditions.

The EPA requires that an air discharge permit be obtained if the quantities released above ground are to be significant. In cases where the discharge of the contaminant is too great, some form of air pollution abatement is required. This may be a carbon adsorption column to trap the organics, or it could be a furnace to burn the contaminants, the combustion being followed by scrubbing to remove any acidic emissions.

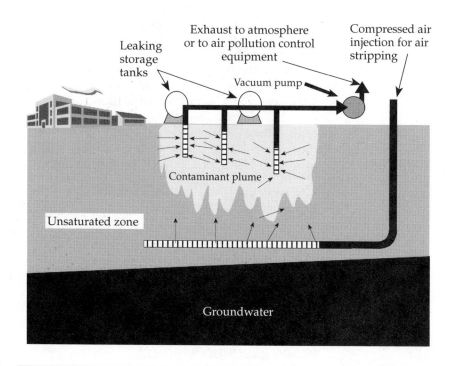

FIGURE 12.17   Vacuum extraction and air stripping of soil contaminants.

Air Stripping.   Volatile organics can be used to strip contaminants from the groundwater and from soil above the water table. The process, "air stripping," can augment vacuum extraction. It involves using compressed air to volatilize the subsurface organics. One or more wells with a screened pipe section are drilled into or under the contamination zone and compressed air is forced into the ground. New technologies allow the placement of horizontal wells for this purpose, as shown in Figure 12.17. Vacuum extraction wells are usually mounted above the contamination zone to capture and remove the contaminants. Air stripping has the advantage of being effective not only in the unsaturated zone, but also in the saturated or groundwater zone.

In Situ Biological Treatment.   Supplying nutrients and/or an electron acceptor (hydrogen peroxide or oxygen) to soil bacteria already present can also reduce contamination. In general, one or more supply wells and one or more withdrawal wells are drilled into or on the boundaries of the contamination zone. A rich supply of nutrients is pumped into the ground and recirculated through the contamination zone. This spurs bacterial growth and reproduction, which creates a demand for even more nutrients.

However, no additional substrate is supplied in this case, so the bacteria utilize the organic contaminants as substrate for energy or synthesis, effectively removing and destroying the contaminants. In some instances, special bacterial populations able to degrade particular contaminants have been injected into the ground. However, it is often difficult to obtain EPA approval for the application of such microorganisms.

CASE STUDY

Containment and Treatment at the Rocky Mountain Arsenal

The Rocky Mountain Arsenal is just outside Denver, Colorado, adjacent to Stapleton International, which was the city's principal airport until 1994 when the larger Denver International Airport opened on a prairie site farther east. Near Stapleton the upper aquifer lies only a few feet below the surface and extends to an impermeable clay layer 20 to 30 feet below the surface. Weapons production during the past 40 years has contaminated this upper aquifer with chlorinated organic solvents. So remediation specialists are faced with the challenge of three separate contaminant plumes moving off site, as shown in Figure 12.18.

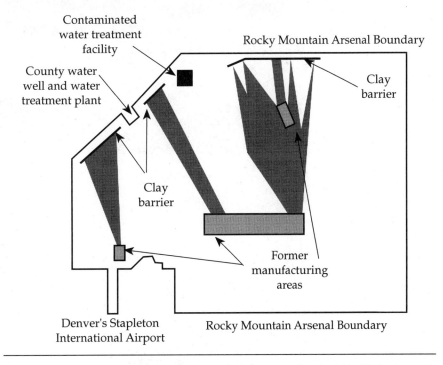

FIGURE 12.18 Groundwater contamination at the Rocky Mountain Arsenal.

In an effort to prevent contamination from reaching nearby drinking water supply wells, engineers used a combination of slurry walls and pumping. Three slurry walls were constructed to intercept the plumes, and water entering the wells on the contaminant side of the walls is pumped to a treatment facility. After treatment with filtration and activated carbon adsorption, a portion of the water is returned to the down-gradient side of the slurry walls, as shown in Figure 12.19. Returning water to the opposite side of the slurry wall keeps the hydraulic gradient greater on the "clean" side of the slurry wall, preventing contaminant leaks.

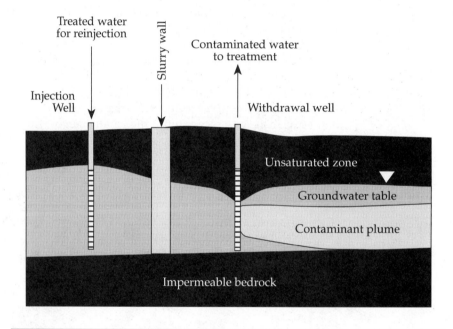

FIGURE 12.19 Slurry wall at the Rocky Mountain Arsenal.

Review Questions

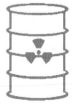

1. Define the following terms:
   a. air stripping
   b. vacuum extraction
   c. pumped extraction
   d. secured disposal site
   e. dump

2. What is the main objective of the hazardous waste regulations in RCRA?

3. What is the EPA definition of a hazardous waste? Hint: Briefly describe the four characteristics of a hazardous waste and lists the EPA uses to identify waste as hazardous.

4. Briefly describe the RCRA's four program areas.

5. What is the purpose of CERCLA, or Superfund?

6. How is Superfund funded?

7. What are the advantages of hazardous waste incineration versus other methods of treatment or disposal?

8. Choose a local industry. List the hazardous wastes it generates and describe the methods it uses to treat and dispose of them.

9. List the chemicals present in your dorm room (or apartment). Which would be a hazardous waste if disposed of improperly (in larger quantities)?

10. An industrial plant produces approximately 1000 kg per month of waste pentachlorophenol [C_6Cl_5OH]. How much oxygen is required to combust the material? How much acid is produced? What quantity of NaOH is required to neutralize the acid?

11. What are the advantages and disadvantages of landfill disposal versus incineration?

12. List and discuss *in situ* methods of hazardous waste remediation.

13. List and describe the hazardous waste treatment and disposal methods that a TSDF might use.

14. Using your campus library, find and describe a new method of hazardous waste treatment or disposal — one not listed in this text. Good information sources might be *Chemical and Engineering News* or *Environmental Science and Technology*. Both are published by the American Chemical Society.

15. Describe the methods used in a secured hazardous waste landfill to protect groundwater.

16. What is the difference between a dump and a landfill?

17. Choose a product in your apartment or room. Use the library to determine what hazardous wastes are produced during its manufacture.

References

1. R. Carson, *Silent Spring* (Boston: Houghton Mifflin, 1962).

2. J. Deegan, Jr., "Looking Back at Love Canal," *Environmental Science & Technology*, vol. 21, no. 4 (1987), pp. 328–31.

3. A. G. Levine, *Love Canal: Science, Politics, and People* (Lexington, MA: Lexington Books, 1982).

4. L. Ember, "Occidental Agrees to Store, Treat Love Canal Wastes," *Chemical and Engineering News* (June 19, 1989).

5. L. Ember, "OxyChem Told to Pay for Love Canal Cleanup," *Chemical and Engineering News* (Feb. 29, 1988).

6. *RCRA Orientation Manual,* 1990 ed., EPA/530-SW-90-036, Washington, DC: U.S. Environmental Protection Agency, Office of Solid Waste, 1990).

7. N. P. Johnson and M. G. Cosmos, "Thermal Treatment Technologies for Haz Waste Remediation," *Pollution Engineering* (October 1989), pp. 66–85.

8. S. P. Tucker and G. A. Carson, "Deactivation of Hazardous Chemical Wastes," *Environmental Science and Technology*, vol. 19, no. 3 (1985), p. 219.

9. *Survey of Solidification/Stabilization Technology for Hazardous Industrial Wastes*, EPA-600/2-79-056, Municipal Environmental Research Laboratory, Cincinnati, Ohio (Washington: EPA, 1979).

10. *A Compendium of Technologies Used in the Treatment of Hazardous Waste*, EPA/625/8-87-014 (Washington: EPA, September 1987).

11. T. E. Lewis, "DuPont Chambers Works—Integrated Site Waste Disposal System," in *Hazardous Waste Disposal*, J. P. Lehman, ed. (New York: Plenum Press, 1983), pp. 345–356.

Additional References

Y. Kiang and A. M. Metry, *Hazardous Waste Processing Technology* (Boston: Ann Arbor Science, 1982).

H. M. Freeeman, ed., *Standard Handbook of Hazardous Waste Treatment and Disposal* (New York: McGraw-Hill, 1989).

C. A. Wentz, *Hazardous Waste Management* (New York: McGraw-Hill, 1989).

13 🏭

Air Pollution and Control

`` . . . WHOSOEVER SHALL BE FOUND GUILTY OF BURNING COAL SHALL SUFFER THE LOSS OF HIS HEAD.''

King Edward II, circa A.D. 1300

Kaoshiung, Taiwan shrouded in smog. Air pollution problems persist not only in the U.S. but in most heavily populated industrial areas throughout the world.

INTRODUCTION

Air pollution, a difficulty in selected areas of the world for several hundred years, has become a serious and widespread problem during this century. There have been several documented episodes of air pollution in which loss of human life occurred. In 1931 in the heavily industrialized area of Manchester, England, more than 500 people lost their lives by being exposed to particulates and acids in the air. In 1948 in the steel town of Donora, Pennsylvania, about 20 people died and several thousand became ill from the effects of airborne contaminants. Today, an estimated 3000 cancer deaths yearly are attributable to secondhand tobacco smoke alone.

Before beginning a discussion of air pollution, we should ask what it is. We earlier defined pollution as something that poses a potential risk to human health or the environment. Then air pollution is the presence, in the outdoor air or the air inside a structure, of substances that pose a threat to human health and/or the environment. Airborne particles that decrease visibility are also considered a form of pollution now, as evidenced by EPA's negotiated agreement with a major western power company to improve visibility in the Grand Canyon National Park area. The EPA estimates that half of the American population is exposed to air pollutants exceeding current federal standards. Air pollutants also degrade or destroy architectural structures, such as buildings, bridges, and statues — a particularly serious problem in some parts of the United States and Europe.

Air pollution control is complicated by the many disparate sources and types of pollutants, each of which may require a unique treatment system. While some pollutants are produced by the earth naturally, such as radon gas, air pollution is generally created by human activities. These include energy production (coal- and oil-fired electric plants), transportation, the burning of oil or natural gas to heat living and work spaces, the use of home and automotive air conditioning, and many industrial activities. Chemicals implicated in indoor air pollution include formaldehyde from home insulation, residues from pesticide applications, and secondhand tobacco smoke.

Air pollutants can be classified in different ways: gaseous and particulate; conventional and toxic; conventional, acidic, and toxic; or greenhouse, acidic, toxic, and particulate. Some pollutants may fall into more than one category. Others may not fit well within any of these categories. Greenhouse gases include carbon dioxide [CO_2], nitrous oxide [N_2O], chlorofluorocarbons, and some other organic compounds. Acidic pollutants include sulfur oxides [SO_x], nitrous oxides [NO_x], hydrochloric acid [HCl], and other acids. Toxic pollutants include carbon monoxide [CO], many organic compounds such as benzene, PCBs, dioxins, and furans, and inorganics like lead, arsenic, beryllium, mercury, and asbestos. Particulate pollutants include materials formed from combustion and mechanical processes. The major sources of air pollution are summarized in Table 13.1.

TABLE 13.1   Major Sources of Air Pollution

ACTIVITY	AIR POLLUTANTS CREATED	DETRIMENTAL EFFECT
Energy production from fossil fuels	Carbon dioxide, sulfur oxides, particulates	Increase in greenhouse gases, acidic precipitation
Automobiles, other transportation sources	Carbon dioxide, nitrogen and sulfur oxides, products of incomplete combustion	Increases in greenhouse gases, acidic precipitation
Refrigeration devices including home, commercial, and those in vehicles	Chlorofluorocarbons	Destruction of the stratospheric ozone layer
Industrial manufacturing	Various depending on the industry and process, including toxic materials	Destruction of the stratospheric ozone layer, toxic emissions

Within these categories, contaminants are classified as either primary or secondary. Primary pollutants are substances that can cause harmful effects in the form in which they are produced and emitted to the atmosphere. Secondary pollutants are those formed in the atmosphere by the reaction of primary air pollutants. For example, sulfur dioxide, created in the combustion of coal, is a primary air pollutant. In the atmosphere, one molecule of sulfur dioxide reacts with one molecule of water to form sulfurous acid, a secondary air pollutant.

REGULATIONS

The first air pollution regulations were passed in England to reduce the emissions from the burning of coal. They were decreed in A.D. 1273. That's right, 1273. So the problem is not a completely modern phenomenon. In this country the first regulations to control air pollution were passed in California, primarily to control air pollution in the Los Angeles basin.* Because air pollution has many causes, a broad approach has been necessary to reduce and control pollutants. These controls today include restric-

---

*The Los Angeles basin is the area encompassing the metropolitan Los Angeles area. Air flow from the ocean is restricted by mountains surrounding the city, preventing the movement of pollutants out of the area. Thus, in certain weather conditions, the air becomes stagnant and air quality deteriorates. Los Angeles continues to have the worst air quality in the nation.

tions on motor vehicles, motor fuels, industries, and power plants. Initial attempts at regulating air quality suffered from the same problems that water quality regulations faced. The U.S. Congress was hesitant to force the states into any form of compliance. It was the general feeling in Congress that if a state desired to reduce pollution, it could do so, but the federal government would not infringe upon "states rights." Thus, the initial legislation passed by Congress in the 1950s and 1960s did little to solve the problem. It was simply an admission that problems existed.

The first legislation aimed at requiring compliance, with or without state concurrence, were auto emission standards passed in 1965. But these were placed on automotive manufacturers, not the states. It was not until the Clean Air Act Amendments of 1970 that Congress finally felt confident in requiring state compliance with air quality regulations. The major federal air pollution laws are summarized in Table 13.2.

The latest round of federal air pollution control regulations, the Clean Air Act of 1990 (CAA), has taken a broad approach. Some of its provisions are regulation of toxic air emissions (at present 189 different chemicals), significant reductions in acid rain, ozone protection by the phaseout of chlorofluorocarbons, better air emission control systems for motor vehicles, altered fuels for nonattainment areas, and a national permitting system for air emissions similar to the NPDES for water discharges. And there are more penalties for violations.

The CAA recognizes two regulatory categories: **primary** and **secondary ambient air quality standards**. Primary standards are intended to protect human health. Secondary standards are to protect "human welfare." Generally, secondary standards protect the environment and infrastructure from damage. Examples include defense of buildings, bridges, and statues from acid deposition, protection of forests from particulates, acid deposition and lead, and protection of lakes and rivers from acid rain.

The major sections of the CAA are presented in Table 13.3. The "new" clean air act requires additional removal of sulfur dioxides and nitrogen oxides to reduce acid precipitation, as well as phaseout of CFCs, which are implicated in stratospheric ozone depletion. Motor vehicle emissions are to be further reduced. And there are special requirements limiting development for nonattainment areas, or those that have not met the national ambient air quality standards.

It is reasonable to expect that current regulations will become more strict during the next few decades. We are beginning to understand the adverse environmental and health effects of various pollutants. As this occurs, such pollutants will most probably be limited to lower levels.

Twenty years ago, an important topic in the field was lead in the air. Tetraethyllead was used for years as an octane boosting additive in gasolines.* Lead is known to lower intelligence and cause behavior prob-

---

*This additive prevents gasoline from igniting prematurely as the fuel and air mixture in the cylinder is compressed. Fuels that do not contain "ethyl" must be refined to a greater extent, increasing the cost.

TABLE 13.2   Federal Air Pollution Control Regulations

ACT	SUMMARY
Air Pollution Control Act of 1955, PL 84-159	Established federal funding for air pollution research, federal technical assistance and training.
Air Pollution Control Act Amendments of 1960, PL 87-761	Continued the APCA of 1955 and began a study of human health effects caused by motor vehicle emissions.
The Clean Air Act of 1963, PL 88-206	Through matching grants to state and local government (federal share of 66 to 75%), increased research and training, efforts to control air pollution from federal facilities.
Motor Vehicle Air Pollution Control Act of 1965, PL 89-272	Required automobile exhaust emission standards to be met in 1968.
The Air Quality Act of 1967, PL 90-148	Set timetables for establishment of air quality criteria for various pollutants. Provided for state or federal enforcement of air quality limits. Understaffed and underfunded, the program was unsuccessful.
The Clean Air Act Amendments of 1970, PL 91-604	Established national ambient air quality standards for particulates, carbon monoxide, sulfur oxides, hydrocarbons, and other pollutants. Also set national emission standards for existing and new facilities, fines and criminal penalties for intentional violation, new stricter automobile emission standards, additional research funding.
The Clean Air Amendments Act of 1977, PL 95-95	Continued the 1970 requirements. Added restrictions for nonattainment areas.
The Clean Air Act of 1990, PL 101-549	Completely revamped air pollution control regulations, including compliance timetables (3 to 20 years) for major noncompliance areas. Mandated tighter standards for vehicle emissions, reformulated gasolines, air toxics, and acid rain controls. Began a new permitting program with stiffer civil and criminal penalties.

lems in children and high blood pressure in adults. It has also been suspected of causing defects in developing fetuses. Today, little is heard about lead in the atmosphere. Why? Because, atmospheric lead has been almost eliminated. Beginning with the 1965 Motor Vehicle Air Pollution

TABLE 13.3   Major Sections of the Clean Air Act of 1990

SECTION	COVERAGE
Subchapter I	Programs and activities
Part A	Air quality and emission limitations, including primary and secondary air quality standards
Part C	Prevention of significant deterioration of air quality (including clean air and visibility protection)
Part D	Plan requirements for nonattainment areas
Subchapter II	Emission standards for moving sources
Part A	Motor vehicle emission and fuel standards
Part B	Aircraft emission standards
Part C	Clean fuel vehicles
Subchapter III	General provisions, including administration, citizen law suits, air quality monitoring, protection of employees informing EPA of employer violations
Subchapter IV	Noise pollution
Subchapter IV-A	Acid deposition control, including phase in of SO_2 and NO_x requirements
Subchapter V	Permits
Subchapter VI	Stratospheric ozone protection

Control Act, the U.S. commenced a gradual phase-out of lead as a gasoline additive. Today only a fraction of the previous amount of lead is allowable in gasolines sold in this country. A complete phaseout of lead is scheduled for 1995. Lead is a regulatory success story. Many other industrialized nations have also eliminated lead. However, most developing nations are still using tetraethyllead as a gas additive. In 1988 Mexico inaugurated a program to reduce lead additives in gasoline. Almost certainly in the future we will witness an increased effort among developing and underdeveloped nations to reduce their levels of pollution, including air pollutants.

METEOROLOGY AND CLIMATOLOGY

Meteorology is the study of the lower atmosphere and particularly of weather. Meteorology is important to environmental engineers because of the way pollutants behave in the atmosphere. **Climatology** is the study of weather over long periods of time. Climatology is important because pollutants in the atmosphere are modifying the earth's climate. Even if stopped today, the pollutants already emitted into the atmosphere will continue to alter our climate for the next 100 or more years.

The Atmosphere

The earth's atmosphere is a mixture of several gases. The two primary gases are nitrogen and oxygen, which comprise about 99% of the atmosphere. The other gases, however, are of greater concern to environmental engineers. Table 13.4 provides the concentration of the major gases in the atmosphere, and some of the minor gases of interest to environmental engineers and scientists. There are myriad gases, some of which occur naturally, some of which result from human activity, and others of which have their concentration altered by human activity.

TABLE 13.4   Composition of the Earth's Atmosphere

GAS	CHEMICAL FORMULA	CONCENTRATION, % BY VOLUME
Nitrogen	N_2	78.1
Oxygen	O_2	21.0
Argon	Ar	0.9
Carbon dioxide*	CO_2	3.3×10^{-2}
Hydrogen	H_2	5×10^{-5}
Ozone*	O_3	1×10^{-6}
Methane*	CH_4	2×10^{-4}

*The concentration of these gases has been significantly modified by human activity over the past 100 years.

Meteorology

Air movement (or the lack of it) plays a major role in air pollution. Many localized severe air pollution episodes are related to particular weather patterns occurring at the time of the problem. The transport of acidic pollutants from the Ohio Valley into the northeastern United States and southeastern Canada is the result of the generally westerly flow of air across the continent.

Uneven heating of the earth's surface due to cloud cover and other factors causes pressure gradients to develop in the lower atmosphere. Where a pressure gradient exists, there will be a generalized air movement from a region of high pressure to a region of low pressure. This phenomenon is depicted in Figure 13.1a. However, an additional force, a result of the earth's rotation, acts perpendicular to the velocity vector for these air movements. This is the Coriolis force. In the Northern Hemisphere it acts

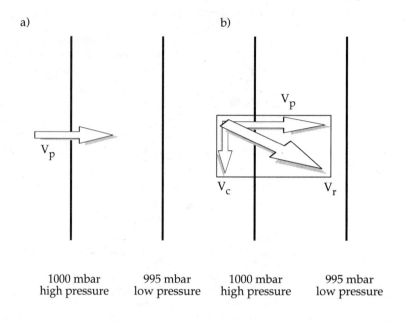

a)

b)

V_p

V_p

V_c

V_r

| 1000 mbar | 995 mbar | 1000 mbar | 995 mbar |
| high pressure | low pressure | high pressure | low pressure |

FIGURE 13.1 The Coriolis effect on air movement.

to the right of the velocity vector. In the Southern Hemisphere, it acts to the left of the velocity vector. The net velocity is shown in Figure 13.1b. Since air moves outward from a region of high pressure, the circulation is clockwise in the Northern Hemisphere. Similarly, the movement around a low-pressure region in the Northern Hemisphere is counterclockwise.

Localized air movements affect the dispersion of air pollutants. Low-pressure regions move air inward, and since the air must escape from the low-pressure center, it also moves upward. Thus, pollutants emitted in a low-pressure region tend to be dispersed and moved away from the ground.

In and around a high-pressure area, the air near the surface moves out and down from the center of the high-pressure region. In the upper atmosphere the air is moving into the region. This downward movement of the air at low altitudes tends to trap pollutants, intensifying air pollution problems. This is shown in Figure 13.2. The reverse is true of a low-pressure region. The air flows into the low-pressure region and upward. As shown in Figure 13.3, this tends to disperse the air pollutants.

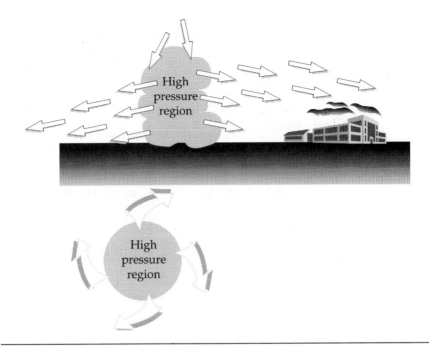

FIGURE 13.2 Air movement in a high-pressure region.

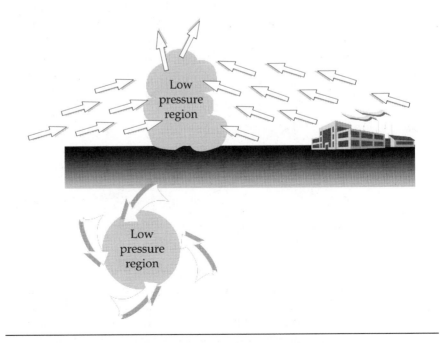

FIGURE 13.3 Air movement in a low-pressure region.

UNITS OF EXPRESSION

The units used to express concentration in air pollution are typically $\mu g/m^3$ (mass per volume) for gaseous, nongaseous, or particulate matter. However, parts per million (volume per volume) is frequently used for gases. Parts per million (or ppm) indicates the volume of contaminant gas per 1 million volumes of gas, or

$$\text{ppm} = \frac{\text{Volume of contaminant}}{10^6 \text{ volumes of air} + \text{contaminant}} \qquad \textbf{13.1}$$

For most cases the volume of contaminant is negligible compared to the volume of air. Thus, Equation 13.1 becomes

$$\text{ppm} = \frac{\text{Volume of contaminant}}{\text{Volume of air}} \qquad \textbf{13.2}$$

The conversion of volume per volume units of expression to mass per volume units of expression requires use of the gas density. At 20°C, each mole of an ideal gas is 22.4 L. The conversion proceeds as follows:

$$C_{\text{cont}} = \frac{V_{\text{cont}}}{10^6 \, V_{\text{air}}} = \frac{V_{\text{cont}} \times \dfrac{\text{mol}}{22.4\,\text{L}} \times \dfrac{10^3\,\text{L}}{\text{m}^3} \times \dfrac{\text{GMW g}}{\text{mol}} \times \dfrac{10^6\,\mu g}{\text{g}}}{10^6 \, V_{\text{air}}} \qquad \textbf{13.3}$$

The term $\dfrac{V_{\text{cont}}}{10^6 V_{\text{air}}}$ is ppm of contaminant. Thus,

$$C_{\text{cont}} \left[\frac{\mu g}{\text{m}^3} \right] = \frac{\text{ppm} \times 10^3\,\text{GMW}}{22.4} \qquad \textbf{13.4}$$

This conversion is only true for 20°C. It is different at different temperatures. The volume–temperature relationship from chemistry can be used to correct for different temperatures, or you can arrive at the same result by knowing the volume occupied by 1 mol of the gas. The following example illustrates.

EXAMPLE 13.1	CONVERSION OF ppm OF A GAS TO $\mu g/m^3$

Convert 20 ppm of NO_2 to $\mu g/m^3$ at 20°C.

SOLUTION:

$$C_{NO_2} = \left(20\, \frac{m^3\, NO_2}{10^6\, m^3\, air}\right)\left(\frac{46\, g\, NO_2}{22.4\, L\, NO_2} \times \frac{10^3\, L\, NO_2}{m^3\, NO_2} \times \frac{10^6\, \mu g\, NO_2}{g\, NO_2}\right)$$

$$= 41,000\, \frac{\mu g\, NO_2}{m^3\, air}$$

ACIDIC POLLUTANTS

Acidic pollutants in the atmosphere reduce the pH of natural precipitation. Acid precipitation was recognized as early as 1852 by Robert Smith in Manchester, England, where coal was burned extensively. Acidic precipitation reduces the buffer capacity of soils, lakes, and streams, and this eventually results in a depression of the soil or water pH. Acidity also increases the leaching of metals bound in the soil. Adding to the problem in some cases are toxic metals present in precipitation formed in the skies over the pollution source. And not all acidity reaches the earth's surface in rain or snow. Acid pollutants that have remained in the air uncombined with water can fall dry on soils, surfacewater, and plants. The fraction of acidity reaching the surface as dry deposition varies, but in some cases it can exceed 40% of the total acidity reaching the surface. Thus, acid deposition would be a better term than acid rain or acid precipitation, although the latter terms are more common.

Effects of Acidic Precipitation

Acid precipitation affects the environment in several ways. It consumes alkalinity in natural waters, eventually depressing the pH of the water. Most aquatic life are very sensitive to pH changes. In some cases, the acid precipitation causes the release of metals previously bound in the soil. Aluminum, for example, is thought to be a major factor in acid precipitation and aquatic toxicity [1].

A significant portion of the soils and lakes in the northeastern United States and southeastern Canada have already been damaged by acidic precipitation. The principal area affected is shown in Figure 13.4. Acidic precipitation is primarily the result of pollutant discharges from coal fired power plants such as those in the Ohio Valley. Emissions from motor

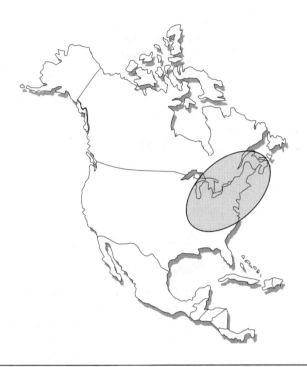

FIGURE 13.4 Areas of North America significantly affected by acidic precipitation.

vehicles exacerbate the problem. Midwest power plant emissions are transported by wind to the northeastern United States and southeastern Canada. However, other regions of this country are in imminent danger from acidic precipitation. These include northern Minnesota and several major metropolitan areas. A large portion of central and northern Europe has also been hurt by acidic precipitation. This includes much of Scandinavia and parts of Switzerland and Germany. Depressed pH levels in precipitation were recorded in central Europe as early as the 1870s. In recent years, the pH of precipitation in much of Europe has been 4.0 to 4.5 [2].

Acid precipitation varies in strength. Due to long atmospheric residence times, these pollutants are often deposited far away from their emission sources. The industrialized Ohio River Valley, for example, is literally hundreds of miles from some of the areas its pollutants reach.

Sources of Acidic Pollutants

The sources of acidic precipitation are sulfur oxides and nitrous oxides. Sulfur oxides are mainly attributable to stationary installations such as coal fired power plants. Sulfur in coal oxidizes during the combustion process to produce sulfur dioxide, a primary air pollutant:

$$S + O_2 \rightarrow SO_2 \qquad \qquad \textbf{13.5}$$

Sulfur dioxide reacts in the air with oxygen to form sulfur trioxide, a secondary air pollutant:

$$SO_2 + \tfrac{1}{2}O_2 \rightarrow SO_3 \qquad \qquad \textbf{13.6}$$

SO_2 and SO_3 are often paired for discussion purposes as SO_x. The formation of sulfurous acid from sulfur dioxide is

$$SO_2 + H_2O \rightarrow H_2SO_3 \qquad \qquad \textbf{13.7}$$

and for sulfuric acid

$$SO_3 + H_2O \rightarrow H_2SO_4 \qquad \qquad \textbf{13.8}$$

Eventually, these acids are swept from the air with precipitation or deposited dry.

Sulfur, a component of coal, varies from less than 1% sulfur to more than 3% sulfur by weight. About half of the sulfur in coal is in the form of pyritic sulfur; the remainder is organic sulfur. Some coal-cleaning processes, prior to burning, can remove much of the pyritic sulfur, but little of the organic sulfur can be removed by current precombustion cleaning technology. It has been estimated that 70% of all sulfur emissions in the United States come from coal-fired power plants [3].

Sulfur dioxide emissions can be reduced by switching to low-sulfur coal, by initiating flue-gas desulfurization processes, or possibly by coal desulfurization, an unproven but promising technology. Although alternative fuels might seem possible, the reserves of natural gas and fuel oil are not large enough to satisfy our energy-hungry society for an appreciable length of time. The primary reserves of low-sulfur coal are in Colorado, West Virginia, Utah, and Eastern Kentucky. The largest repositories of high-sulfur coal in the United States are Illinois, western Kentucky, and Ohio. Thus, fuel switching helps some regional economies while hurting others. Flue-gas desulfurization, the removal of sulfur from the combustion gases, is the other proven method. It requires a significant capital investment, however, and in many cases is more expensive than simply switching to low-sulfur coal. Several private and university research laboratories are working to develop practical methods of removing much of the sulfur in coal.

Nitrogen oxides are the other major group of acidic pollutants sent to the atmosphere. It has been estimated that stationary combustion sources emit 56% of these nitrogen oxides while transportation sources roughly

40% [4]. The remainder comes from industrial and other sources. The two major contributors to nitrogen oxides are nitric oxide [NO] and nitrogen dioxide [NO_2]. They are often considered together as NO_x. Most nitrogen oxides resulting from human activity occur as a result of combustion processes.

The source of most nitrogen oxide pollutants is the air itself. Combustion processes normally use air, which is 80% nitrogen, rather than pure oxygen. As the fuel is burned, atmospheric nitrogen reacts with the excess oxygen present in the combustor to form oxides of nitrogen. Higher combustion temperatures and residence times favor greater formation of NO_x, as does excess oxygen in the combustion chamber. Unfortunately, in power plants, the higher temperatures, longer residence times, and excess oxygen usually translate into improved combustion efficiencies. Thus, process optimization for fuel efficiency and NO_x reduction are often at odds. In addition, a low-oxygen excess contributes to other pollutants, particularly products of incomplete combustion, or PICs. In addition, coal and fuel oils contain some organic nitrogen. This reacts with the oxygen in the air as the oxidation of the carbon and hydrogen takes place. This puts more nitrogen oxides into the atmosphere, where they are converted to nitric acid, resulting in acidic deposition when the material is swept from the atmosphere by rain or snow.

Acid Emission Control—Electric Power Generation

The simplest way to limit SO_x emissions from coal-fired power plants is to convert the facility to low-sulfur coal. However, low-sulfur coal is more expensive than high-sulfur coal. Facilities using low-sulfur coal thus have higher supply costs. Many facilities already use low-sulfur coal. The new Clean Air Act will certainly force others to make this conversion as well. Another alternative is nuclear energy. There are no acidic emissions from nuclear power generation. Although new regulations allowing for faster permitting of nuclear power plants recently went into effect, the near-term outlook for nuclear power is not good. As of now, there is no permitted disposal facility for nuclear waste generated by such power plants. Wastes from existing nuclear plants are being stored on-site or at other places until proper disposal facilities are developed. In addition, the Three Mile Island accident and the Chernobyl disaster have left the public hesitant to accept additional nuclear power plants in the United States. As future energy needs increase, this public attitude may well change.

The remaining options are to remove the sulfur prior to combustion (coal cleaning or desulfurization), remove both SO_x and NO_x during combustion (fluidized bed combustion), or use scrubbers to remove both after combustion. There is currently much research into methods of cleaning coal prior to combustion. Pyritic sulfur, the inorganic form, is easier to remove because it is heavier than coal. Organic sulfur is very difficult to remove prior to combustion. The proportions of the two forms, and their total amounts, vary widely with different coal sources.

Postcombustion removal is usually accomplished by scrubbers, devices that spray water and/or basic solutions into the exhausting air to neutralize and remove the SO_x and NO_x from the gas stream. A schematic of a typical scrubber unit appears in Figure 13.5. These devices remove not only the gaseous pollutants associated with coal combustion, but also the particulates. The drawback of using them is that they produce a wet sludge that must usually be landfilled.

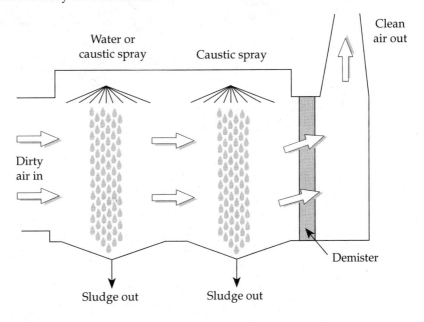

FIGURE 13.5 Wet scrubber.

In a wet scrubber an alkali is usually added to react with the acidic pollutants. This can be calcium oxide [CaO], or quicklime calcium carbonate ($CaCO_3$), or a magnesium or sodium alkali. An example is the neutralization of sulfur dioxide with calcium oxide. In a wet scrubber the sulfur dioxide produced from the combustion of iron pyrite and organic sulfur will react with water to form sulfurous acid:

$$SO_2 + H_2O \rightarrow H_2SO_3 \qquad\qquad \textbf{13.9}$$

The sulfurous acid will react with calcium oxide to form calcium sulfite:

$$H_2SO_3 + CaO \rightarrow CaSO_3 + H_2O \qquad\qquad \textbf{13.10}$$

Similarly, sodium hydroxide will produce sodium sulfite:

$$H_2SO_3 + 2NaOH \rightarrow Na_2SO_3 + 2H_2O$$ **13.11**

The following example illustrates the calculation of the quantity of chemical needed to neutralize a gas stream.

EXAMPLE 13.2	**CALCULATION OF THE CHEMICAL MASS REQUIRED TO NEUTRALIZE FLUE GAS**

Estimate the amount of calcium oxide required per 1000 ft³ of exhaust gas to neutralize 2000 ppm of sulfur dioxide at 20°C.

SOLUTION: First, calculate the mass (in moles) of SO_2 per 1000 ft³, noting that mass is concentration times volume:

$$m_{SO_x} = CV$$

or

$$m_{SO_x} = \left(2000 \ \frac{m^3 \ SO_2}{10^6 \ m^3 \ air}\right)\left(\frac{mol \ SO_2}{22.4 \ L \ SO_2}\right) \times \frac{10^3 \ L \ SO_2}{m^3 \ SO_2} \times \frac{m^3 \ air}{35.3 \ ft^3 \ air}$$

$$= 2.53 \ \frac{mol \ SO_2}{10^3 \ ft^3 \ air}$$

Thus, each 1000 ft³ of exhaust gas (air) contains 2.53 mol of SO_2. For neutralization, each mole of SO_2 (as H_2SO_3) combines with 1 mol CaO, or

$$H_2SO_3 + CaO = CaSO_3 + H_2O$$
$$1 \ mol \qquad 1 \ mol$$

The required amount of calcium oxide, or quick lime, is then

$$m_{CaO} = \left(2.53 \ \frac{mol \ SO_2}{10^3 \ ft^3 \ air}\right)\left(\frac{1 \ mol \ CaO}{1 \ mol \ SO_2} \times \frac{56 \ g \ CaO}{1 \ mol \ CaO}\right)$$

$$= 142 \ \frac{g \ CaO}{10^3 \ ft^3 \ air} = 0.312 \ \frac{lb \ CaO}{10^3 \ ft^3 \ air}$$

So it takes 142 g, or 0.312 lb, of quick lime to neutralize each 1000 ft³ of exhaust gas.

The waste produced in scrubbing must either be regenerated using additional energy, disposed of as a solid, or used in some manner. Today little of this waste material is used. However, there are projects under way to develop uses for the waste materials in such applications as construction materials for buildings and highways.

Fluidized bed combustion may become a less polluting process. In a fluidized bed combustor, coal, limestone, and ash are suspended by the incoming air, as shown in Figure 13.6. As the coal burns, the limestone absorbs and reacts with the acid produced by the combustion. Thus, the pollutants are neutralized during combustion, not as an additional treatment step. Additionally, because the fluidized bed combustor operates at lower temperatures, it produces less NO_x. Thus far, developmental, cost and operational problems have prevented widespread adoption of fluidized bed combustors. However, as air pollution regulations become more stringent, the process may become more advantageous.

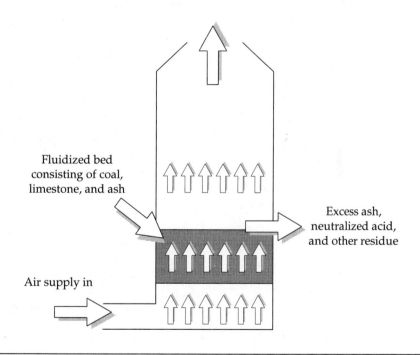

Fluidized bed consisting of coal, limestone, and ash

Excess ash, neutralized acid, and other residue

Air supply in

FIGURE 13.6 Fluidized bed combustor.

Acid Emission Control — Other Sources

The other major source of acidic pollutants is motor vehicles. Reductions in these emissions can be achieved by higher mileage standards, improved engine efficiency, and mass transportation. There are few, if any, emission control systems other than the catalytic converter that reduce PIC levels.

PARTICULATE POLLUTANTS

Anthropogenic sources of particulate pollutants include combustion processes, industrial facilities, power generation, and motor vehicles. Natural sources of particulates include ocean salt spray, volcanic ash, forest fires, pollen, and wind erosion. Fine particles are considered to be less than $1.0 \, \mu$m, large particles greater than $1.0 \, \mu$m. Most fine particles are created by condensation processes — often as a result of combustion. As a hot vapor cools upon entering the atmosphere, the particles condense. This is followed by coagulation. However, even the larger particles produced in this manner are usually less than $1 \, \mu$m. Yet larger particles are usually formed by mechanical processes: erosion or industrial machining processes, for instance [5]. **Primary particulate pollutants** are those produced in combustion before they have aggregated with other pollutants. **Secondary particulate pollutants** are aggregated masses of such particles, or pollutants otherwise transformed in the atmosphere.

Sources of Particulate Emissions

Particulate emission sources include power plants, waste incinerators, steel manufacturing plants, and motor vehicles. The major source of particulate emissions is the electric power industry, although most facilities have scrubbers, electrostatic precipitators, or other control devices to reduce their emissions.

Effects of Particulate Emissions

When we breathe, we trap large particles in the nasal passages and throat. Most of these larger particles do not penetrate into the lungs to do damage. However, smaller particles are less likely to be intercepted in the nasal passages, so they can get into the lungs. As shown in Figure 13.7 the terminal air passages to the alveolar sac, where actual gas transfer occurs, are typically only 0.6 mm in diameter. Particles entering the lungs are trapped in these passages. Soluble particles are dissolved into the blood stream. Insoluble particles remain for much longer times and may not ever be removed. In some instances the body builds scar tissue around the insoluble materials. The net result of airborne particles entering the lungs is that foreign pollutants are introduced into the human body. In metabolizing the foreign compounds, various human issues and organs are exposed to these pollutants with a concomitant increase in disease.

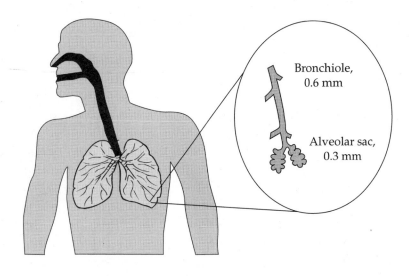

Bronchiole,
0.6 mm

Alveolar sac,
0.3 mm

FIGURE 13.7 The human respiratory system.

Particulate Emission Control

Particulate air pollutants can be kept from getting into the atmosphere by several processes, including filtration, electrostatic precipitation, and scrubbing.

Filtration of air particulates is similar in concept to the filtration used in water and wastewater treatment. The filtration medium is usually a heat-resistant porous fabric. Typical fabric filters can remove 99% or more of 0.5 μm particles [4]. A schematic of a fabric filter apparatus, often called a **baghouse filter**, is shown in Figure 13.8. As the particle layer builds up on the fabric, greater pressure is required to force the air through the filter. When this pressure becomes too high, the filter is cleaned. This is normally accomplished by vibrating the filters, which allows the particles to fall to a hopper at the bottom of the baghouse.

Electrostatic precipitators (ESPs) use a high-voltage electric field to move particles to a plate, as indicated in Figure 13.9. Since small particles possess a charge, they will migrate to oppositely charged surfaces. ESPs take advantage of this phenomenon. A typical ESP has many large parallel plates with small-wire electrodes spaced between them. As the particulate laden air traverses the field, the particles are attracted to the plate electrodes. The plates become covered with particles to a thickness of several centimeters. On a periodic basis one of the units will be removed from service by directing the air to the other units. The plates of the unit to be cleaned are then rapped (struck) automatically to break the particles loose. After the dust has settled, the unit is then ready to resume service.

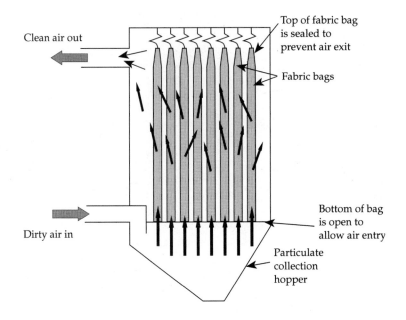

FIGURE 13.8   Baghouse filter.

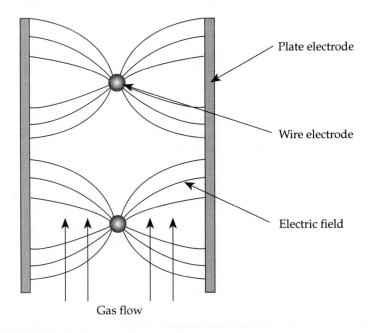

FIGURE 13.9   Electrostatic precipitator electric field.

STRATOSPHERIC OZONE DESTRUCTION

The **stratosphere** is the layer of atmosphere above the troposphere, from 12 to 70 km high. In this region, and above it, reactions involving atmospheric chemicals, pollutants, and light occur. These reactions are important in atmospheric studies because they are critical to life on earth, and they involve pollution because some chemicals emitted to the atmosphere rise to these altitudes and then affect the rate of formation or destruction of naturally occurring compounds. The oxygen in the upper atmosphere, above approximately 150 km, absorbs damaging shortwave—that is, high-energy—radiation, thus protecting life on earth. In doing so, the oxygen is split into atomic oxygen:

$$O_2 + Energy \rightarrow 2O \qquad \qquad \textbf{13.12}$$

This process is termed photodissociation. At 400 km, only 1% of the oxygen is combined as molecular oxygen [O_2]. The remainder exists as atomic oxygen [O]. At approximately 130 km, atomic and molecular oxygen occur in about equal proportions. At lower altitudes the amount of molecular oxygen increases. In the region from 15 to 30 km above the earth, shortwave radiation penetrates sufficiently to cause some dissociation of molecular oxygen, but not enough to reduce the level of molecular oxygen substantially. It is here that vital ozone is formed from the reaction of atomic oxygen [O] and molecular oxygen [O_2]:

$$O_2 + O \rightarrow O_3 + Energy \qquad \qquad \textbf{13.13}$$

In this intermediate altitude, 15 to 30 km, the ozone concentration is at a maximum of 0.03 ppm. The ozone absorbs ultraviolet solar radiation, protecting earth's inhabitants from injurious exposure.

Photochemical Dissociation

Solar radiation supplies the energy for the reaction forming atomic oxygen in the upper atmosphere. The O—O bond in O_2 has an energy of approximately 495 kJ/mol. To break this bond, 495 kJ/mol of energy must be supplied. In the upper atmosphere, light energy, in the form of photons, is that energy source. Similarly, ozone in the middle levels of the atmosphere absorbs ultraviolet radiation, breaking apart. Roughly 400 kJ/mol of energy is required to break the ozone bond, forming one atomic oxygen atom [O], and one oxygen molecule [O_2]. Absorption of this energy, carried by photons, is what protects us from damaging ultraviolet radiation. You may recall from chemistry and physics that

$$E = h\nu \qquad \qquad \textbf{13.14}$$

where

E = energy of photon [J]

h = Planck's constant [6.62×10^{-34} J·s]

v = frequency of photon [s^{-1}]

Also, the energy of a photon is inversely proportional to its wavelength, or

$$\lambda = \frac{c}{v}$$

13.15

where,

c = speed of light [3×10^8 m/s]

λ = wavelength [m]

Thus, the maximum wavelength that will supply sufficient energy, E, is

$$\lambda = \frac{ch}{E}$$

13.16

This energy can be absorbed by molecular bonds of molecules in the atmosphere to break the bond. For oxygen

$$O_2 + hv \rightarrow 2O$$

13.17

This free oxygen can then combine with an O_2 molecule to form ozone:

$$O + O_2 \rightarrow O_3$$

13.18

This ozone, which is present in the stratosphere, absorbs ultraviolet radiation before it reaches the earth's surface. As the ultraviolet radiation is absorbed, the ozone molecule is destroyed:

$$O_3 + hv \rightarrow O_2 + O$$

13.19

However, since there is a high concentration of molecular oxygen [O_2] the atomic oxygen [O] rapidly recombines, forming another O_3 molecule. It is primarily ultraviolet and shorter energy wavelengths that are of interest in these reactions. The electromagnetic spectrum, is displayed in Figure 13.10. The following example illustrates the energy levels and wavelengths needed for these photodissociation reactions.

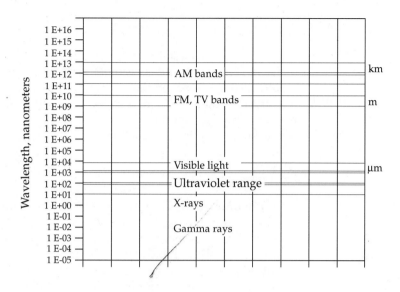

FIGURE 13.10 The electromagnetic spectrum.

<table>
<tr><td>**EXAMPLE 13.3**</td><td>**PHOTON ENERGY**</td></tr>
</table>

If the energy required to break the O_3 bond is approximately 400 kJ/mol, what maximum wavelength will be absorbed?

SOLUTION: First, recall that 1 mol is Avogadro's number of something, or 6.02×10^{23}. Thus, the energy to break one O_3 bond is

$$E = \frac{400 \text{ kJ/mol}}{6.02 \times 10^{23} \text{ (molecules/mol)}} \times \frac{10^3 \text{ J}}{\text{kJ}}$$

$$= 6.64 \times 10^{-19} \frac{\text{J}}{\text{molecule}}$$

Then,

$$\lambda = \frac{ch}{E} = \frac{[3.00 \times 10^8 \text{ (m/s)}](6.62 \times 10^{-34} \text{ J} \cdot \text{s})}{6.64 \times 10^{-19} \text{ J}}$$

$$= 2.99 \times 10^{-7} \text{ m} = 2400 \text{ Å}$$

Referring to Figure 13.10, we see that the light absorbed is in the ultraviolet range.

Risks in Losing the Ozone Layer

An increase in ultraviolet radiation reaching the earth would significantly increase skin cancer rates. Almost certainly, some other forms of cancer would also increase in the human population. Scientists have recently documented that the increased levels of ultraviolet radiation in sunlight reaching frog eggs in mountainous areas (less air to absorb the radiation), is causing a significant reduction in the percentage that are hatching.

Sources of CFCs

Chlorofluorocarbons, the same chemicals implicated in global warming, are also destroying the ozone layer. When emitted, they migrate to the upper layers of the atmosphere, where they are eventually broken down, releasing their chlorine. This chlorine then reacts with and destroys ozone. This is a recent problem. The world first used CFCs in the 1930s. Global atmospheric CFC concentrations did not increase substantially until the 1960s. Scientists became concerned about their impact on the ozone layer in the 1970s. Then, in 1985, researchers detected a "hole" in the ozone layer over the Antarctic region. More recently, a similar hole was detected over the Arctic region.

Control of CFC Emissions

CFC emissions can be controlled mainly by finding and using alternative chemicals for air conditioning, foam blowing, and solvent degreasing — and by developing refrigeration processes that do not require CFC gases. This process is being forced by a variety of national laws worldwide, and by the Montreal Protocol, an international agreement signed in Montreal, Canada, in 1987. In essence, the Montreal Protocol requires the phaseout of the most dangerous CFCs by 1997 for developed countries and by 2007 for developing nations. This agreement should lead to substantial reductions in global CFC emissions. This phaseout of CFCs will result in an improvement in stratospheric ozone. However, because CFCs have atmospheric lifetimes of 50 to 400 years, near-term expectations are poor [6]. Further reductions in CFC emissions in the United States, in excess of the Montreal Protocol, are required by the 1990 Clean Air Act Amendments. But, with CFCs implicated in both greenhouse warming and ozone depletion, President George Bush issued a directive banning CFCs by 1996, faster than either the Montreal Protocol or the Clean Air Act Amendments. Elimination of CFC emissions in this country is important because the United States is responsible for about one quarter to a third of world CFC emissions.

In an effort to accomplish the goal of CFC elimination, many industries have joined a global effort to find acceptable substitutes to CFCs. In the short term the substitute may be hydrofluorocarbons (HFCs). They also damage the ozone layer, but they have much shorter atmospheric lives than CFCs. Thus, although not ideal, they are much better than CFCs. In the longer term, it is expected that other substitute chemicals, and possibly different refrigeration processes, will be developed that have zero ozone depleting potential and are not greenhouse gases.

GREENHOUSE POLLUTANTS

Greenhouse gases impede the exit of reflected solar energy from the earth's atmosphere. Carbon dioxide (and other greenhouse gases) allow much incoming solar radiation to pass through the atmosphere and strike the earth. A portion of the light is reflected back upward, but at a different wavelength. This change in wavelength as the light energy travels toward space again is key to the greenhouse concept. The gases absorb a much higher fraction of the light radiating out, so they reduce the earth's energy loss. The result is a warming of the earth. The process of **global warming** is depicted in Figure 13.11.

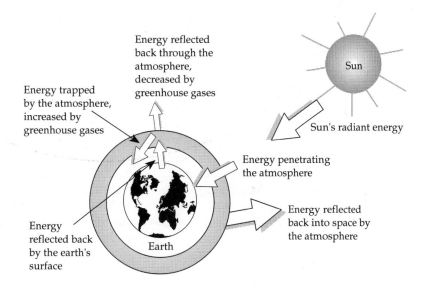

FIGURE 13.11 Greenhouse warming.

Effects of Global Warming

Table 13.5 shows the percentage of *additional* energy trapped in the earth by various gases [6]. Carbon dioxide is responsible for two-thirds of the additional solar radiation trapped in the atmosphere. And the atmospheric

TABLE 13.5 Major Greenhouse Gases and the Percentage of Additional Solar Energy Each is Trapping

GAS	SOURCES	FRACTION OF TRAPPED ENERGY ATTRIBUTABLE TO GAS	ANNUAL INCREASES IN GAS CONCENTRATION, %
Carbon dioxide	Fossil fuel combustion	0.66	0.5
Chlorofluoro-carbons	Vehicle and residential cooling systems, foams, aerosol propellants	0.10	4
Methane	Cattle, rice paddies	0.20	0.9
Nitrous oxide	Combustion processes	0.04	0.25

concentration of carbon dioxide is increasing at a rate of 0.5% per year. Although the atmospheric lifespan of CO_2 is not known precisely, it is probably on the order of 50 to 100 years. Although chlorofluorocarbons (CFCs) are responsible for only 10% of the excess solar energy trapped by the atmosphere, they are increasing at a rate of 4% per year. Even worse, CFCs have estimated atmospheric lifetimes of 50 to 400 years. So even if we stopped burning fossil fuels and stopped producing CFCs today, we could expect CFCs already in the atmosphere to affect the earth's temperature for at least the next one or two centuries. Estimates of how great these greenhouse gases will affect the earth vary. Temperature records are not of sufficient accuracy and number prior to the late 1800s to assist in forming a comprehensive global temperature record. However, since that period, we know there has been a gradual warming of the earth. The global average temperature has probably increased 0.3 to 0.6°C since about 1860 [6]. There are strong indications that the global temperature will increase more rapidly in the twenty-first century and that the increase will be in the range of 2 to 4°C (4 to 9°F). This corresponds roughly to a doubling of the atmospheric CO_2 level. Human activity has already increased global atmospheric CO_2 by 25%. The resulting increase in global temperature will have far-reaching effects on the planet.

Although the exact effects of global warming are not certain, experts believe that significant changes in the earth's weather patterns are possible in conjunction with alterations in the ocean currents. It has been predicted that increased CO_2 levels will also increase global precipitation and that the higher temperature will result in greater evaporation rates over the oceans. However, some areas may still have reduced rainfall.

In addition, the increased surface temperature will raise the sea level. During the peak of the last ice age, some 18,000 years ago, sea level was 100 m lower than it is today. Actually, the sea level has changed dramatically without human intervention. However, a rapid increase of only 0.3 to 0.7 m would devastate many coastal areas from wetlands to cities. This increase in sea level will be accompanied by decreases in polar sea ice.

It is also said that the polar caps will remain warmer in the winter. The warming in these areas of the globe is expected to be three times the global average. Most troubling, however, are the unknowns—eventualities that humans, with their limited insight, cannot foresee. Only time will answer these questions.

Sources of Greenhouse Gases

Since carbon dioxide is responsible for about two-thirds of the greenhouse warming, it is of major concern. In the Northern Hemisphere, the primary source of CO_2 is fossil fuel combustion. However, in the Southern Hemisphere, tropical deforestation is the principal source of CO_2 increases. On a global scale, tropical deforestation accounts for roughly 40% of all CO_2 emissions [7], and these emissions are increasing the atmospheric concentration at approximately 0.5% per year.

CFCs, which account for only 10% of the added trapped solar energy, are primarily used for refrigeration. They are also used as blowing agents for foam plastics, as cleaning solvents in the electronics industry, and as fire extinguishers for both civilian and military applications. Although CFCs account for only 10% of the additional trapped solar energy, the increase in their concentration in the atmosphere is phenomenal. As noted, due primarily to their ozone-depleting characteristics, they are expected to be phased out in the next 10 to 20 years. It is hoped that replacements will not increase the greenhouse effect as much as CFCs have.

Control of Global Warming

Two-thirds of all additional trapped energy is attributable to carbon dioxide emissions. To reduce them, we must decrease our energy usage or develop non–fossil fuel energy sources such as nuclear and solar. This seems unlikely to occur within the next few decades. On the other hand, CFC emissions account for a rapidly increasing portion of greenhouse warming. Whereas the level of CO_2 is expanding at 0.5% per year, CFC levels, again as noted, are expanding at the rate of 4% annually. So it is good that something is being done about them.

TROPOSPHERIC PHOTOCHEMICAL POLLUTANTS

Nitric oxide [NO] which is emitted in combustion processes, and volatile organic compounds (VOCs) are involved in photochemical reactions producing ozone in the lower atmosphere. The nitric oxide is oxidized to nitrogen dioxide [NO_2] in the atmosphere. Nitrogen dioxide can be split by solar radiation that reaches the lower atmosphere:

$$NO_2 + hv \rightarrow NO + O \qquad\qquad\qquad \textbf{13.20}$$

This free atomic oxygen [O] can then combine with molecular oxygen, producing ozone:

$$O_2 + O \rightarrow O_3 \qquad\qquad\qquad \textbf{13.21}$$

When nitrogen oxides and VOCs are present in the atmosphere, the problems are magnified. VOCs react with ozone produced by the nitrogen dioxide. This reaction is also photochemical:

$$\text{Alkene} + O \rightarrow R\cdot + RO\cdot \qquad\qquad\qquad \textbf{13.22}$$

Molecular oxygen in the atmosphere, in the presence of sunlight, forms highly reactive ozone near the earth's surface. This ozone is a secondary pollutant. It causes severe respiratory problems and is the principal component of smog* in many urban areas. It is ironic that the gas that protects us from ultraviolet solar radiation higher in the atmosphere is a strong chemical pollutant in the lower atmosphere.

Ozone affects the respiratory tract of humans. In higher concentrations, above 0.1 to 0.2 ppm, it reduces effective lung capacity and causes inflammation. With chronic (long-term) exposure there are indications that premature aging of the respiratory system occurs [8].

Smog Control

Since tropospheric ozone forms in the reaction of hydrocarbons and NO_x in the presence of sunlight, to reduce this ozone one or both of the precursors must be controlled. Motor vehicle hydrocarbon emissions have been reduced by 90% since the beginning of the clean air regulations in the

---

*The term *smog* originated as a combination or contraction of the two words *smoke* and *fog*.

late 1960s and early 1970s. However, during this same period, NO_x emissions have continued to rise. In many cases, photochemical emissions that produce low-altitude ozone and other smog-related contaminants cannot be reduced by conventional control techniques. They must be reduced by developing different chemicals and processes for such widely varying processes as dry cleaning and automotive engines.

HAZARDOUS POLLUTANTS

Hazardous air pollutants (HAPs) are emitted from a variety of sources, primarily industrial. They include inorganics such as lead used in gasolines and industrial processes. Organic HAPs include compounds such as benzene, contained in gasoline, and Perclene (tetrachloroethylene), used for degreasing and dry cleaning. Table 13.6 lists several examples of common sources of HAPs.

TABLE 13.6 Common Sources of Hazardous Air Pollutants

SOURCE	HAZARDOUS AIR POLLUTANTS
Dry cleaning	Tetrachloroethylene
Plastics production	Various volatile organics, including methylene chloride, phenol, and vinyl chloride
Electric motor manufacture	Organic solvents, organic vapors
Solvent degreasing (cleaning metal parts with organic solvents)	Various volatile organic compounds
Lead smelting	Particulate lead plus particulates from alloying metals such as antimony and arsenic, arsenic vapors
Major appliance manufacturing	Organic solvent vapors, inorganic vapors
Tire manufacturing	Organic vapors, solvent vapors

Hazardous Emission Control

Most hazardous emissions into the air originate from industrial processes. The best method of reducing these emissions is to use processes that do not produce or use dangerous compounds in the first place. This approach often requires a change in the manufacturing process. Alternative approaches include removal of the pollutants by scrubbers, filters, electrostatic precipitators, or carbon adsorption. The carbon adsorption process, first discussed in Chapter 4, Physical Chemistry, is essentially the same for air pollutants. One major difference is that hazardous air pollutants can often be removed from activated carbon by heating the carbon using steam [9]. The steam, water vapor, and hazardous air pollutant are then separated by condensation. The activated carbon can then be reused.

INDOOR AIR POLLUTION

Indoor air quality has become a more pronounced problem in recent years. The energy crisis in the early 1970s motivated most people to improve energy efficiency in their homes. This has led to tighter construction of residences and other structures. Indoor air pollution has a variety of sources, including off-gassing from building materials, carpets, and furniture, radon gas entering through floors and basements, fireplace and heater gases, home pesticides, and secondhand tobacco smoke. Some representative indoor air pollutants and their known or suspected health effects are shown in Table 13.7. These problems may not have come to light without technology providing an increased ability to monitor very low levels of airborne organic compounds and technical advances in the mathematics and statistics of epidemiology.

Modern construction materials often result in contamination of indoor air [10]. Urea formaldehyde insulation exudes formaldehyde vapors for a period of time after installation. Some plastics may emit vinyl chloride vapors after installation. New carpeting sends up organic vapors for a period of time after installation. The glues used in some particle board and plywood emit organic vapors. And there are other examples. The harmful effects of these vapors are compounded by the limited fresh air introduced into new homes by modern energy-saving, airtight construction.

Pesticides as well can be a problem in homes. Pesticides are used primarily for two reasons: to control termites and to prevent roaches, ants, and other insects from invading the home. Termiticides are applied to foundations and around basements during and/or after construction. Their purpose is to prevent termites from damaging the wood structures of the house. "Upstairs" pesticides are usually applied along baseboards and under and behind cabinets to prevent other insects such as roaches and ants from becoming a nuisance. Indoor air samples usually show residuals of such pesticides in the μg to ng/m^3 range in homes [11].

Radon gas is a natural radioactive pollutant present in many soils. It enters homes through foundations and floors that are not properly sealed. Radon was not recognized as a widespread problem until 1984 when Stanley Watras, an engineer working for Limerick Nuclear Power in eastern

TABLE 13.7   Common Sources of Indoor Air Pollution

SOURCE	CONTAMINANT	EFFECTS
Urethane foam building insulation	Formaldehyde	Carcinogenic
Poor foundation or basement seal	Radon gas	Carcinogenic, particularly lung cancer
Ventilation systems	Mold	Allergies
Old asbestos building insulation	Asbestos fibers	Carcinogenic
Permeation from soil through basement floors and walls	Radon gas	Carcinogenic
Heater and stove fumes	Products of incomplete combustion	Various
Household pesticides (for termites or interior insects)	Chlorinated or phosphorylated pesticides	Some chlorinated pesticides are known carcinogens
Secondhand tobacco smoke	Various, including nicotine, and nicotine decay products	Carcinogenic, increased bronchitis and pneumonia

Pennsylvania, began setting off nuclear monitors as he entered the plant [12]. These monitors are routinely used at nuclear facilities to detect contamination of workers from job-related activities. However, it was found that Mr. Watras was being exposed to extremely high levels of natural radon gas that had seeped into his basement and dispersed throughout his home. Since that time radon has been detected in a large percentage of homes throughout the nation. Radon contributes about half of all human exposure to radioactivity. Other natural sources contribute about one quarter, and all human-caused exposure, including medical and nuclear testing, the remainder.

Radon entry can be prevented by adding a good seal and integral vent layer to the basement or bottom floor of a house as it is constructed. The cost for such modification is minimal. However, the control of radon gas in existing homes is much more difficult. If often means adding some fraction of outside air, which increases heating and air conditioning expenses. Other measures include sealing the basement and crawl spaces of a home.

Modifications to existing homes are substantially more expensive than to new construction.

Exposure to secondhand tobacco smoke is a significant contributor to indoor air pollution where smoking is not prohibited. It has been estimated by the EPA that secondhand tobacco smoke causes an additional 3000 lung-cancer deaths each year. In addition, secondhand tobacco smoke causes 150,000 to 300,000 cases of bronchitis and pneumonia each year in children under 18 months [13].

Review Questions

1. Summarize the current air pollution regulations for new sources.

2. Make a drawing similar to Figure 13.1b for the Southern Hemisphere.

3. Sketch a low- and high-pressure region over the United States. Show the generalized circular movement for each pressure center.

4. Why does coal combustion add to acidic precipitation?

5. Describe the greenhouse effect. Where does it get its name?

6. Discuss the air quality in a nearby large metropolitan area. Sources for this information could include local newspapers, EPA publications, and newsmagazines.

7. Oxygen in the upper atmosphere exists almost exclusively in the dissociated form, O, not as combined O_2. The bond dissociation energy for O_2 is 495 kJ/mol. What range of light causes this dissociation?

8. Why is ozone both a protector of life and a pollutant? Where do these two differing actions occur?

9. Convert 250 ppm carbon dioxide [CO_2] to $\mu g/m^3$ at 20°C.

10. Convert 250 ppm of carbon dioxide [CO_2] to $\mu g/m^3$ at 25°C.

11. Estimate the amount of soda ash [$Na(OH)_2$] required per m^3 of exhaust gas to neutralize 1500 ppm of SO_2 at 20°C.

12. Estimate the amount of soda ash [$Na(OH)_2$] required per m^3 of exhaust gas to neutralize 20,000 $\mu g/m^3$ of SO_2 at 20°C.

13. An exhaust gas from a combustion source contains 20 ppm NO_2 and 300 ppm SO_2 at 20°C. Estimate the amount of sodium hydroxide [$NaOH$] per m^3 required to neutralize the exhaust gas stream. Hint: Calculate each gas separately and add the results.

References

1. M. Cresser and A. Edwards, *Acidification of Freshwaters* (New York: Cambridge University Press, 1987).

2. P. Winkler, "Trend Development of Precipitation—pH in Central Europe," *Proceedings of the Commission of the European Communities Workshop: Physico-Chemical Behaviour of Atmospheric Pollutants* (Berlin, Sept. 9, 1982), pp. 114–22.

3. E. Corcoran, "Cleaning Up Coal," *Scientific American* (May 1991), pp. 106–16.

4. K. Wark and C. F. Warner, *Air Pollution, Its Origin and Control*, 2nd ed. (San Francisco: Harper & Row, 1981).

5. National Research Council, Subcommittee on Airborne Particles, *Airborne Particles* (Baltimore: University Park Press, 1979).

6. National Academy of Sciences, *Policy Implications of Greenhouse Warming* (Washington, DC: National Academy Press, 1992).

7. R. A. Houghton, "The Global Effects of Tropical Deforestation," *Environmental Science and Technology*, vol. 24, no. 4 (1990), pp. 414–22.

8. M. Lippmann, "Health Effects of Tropospheric Ozone," *Environmental Science and Technology*, vol. 25, no. 12 (1991), pp. 1954–62.

9. *Handbook: Control Technologies for Hazardous Air Pollutants*, EPA/625/6-91/014 (Washington, DC: U.S. Environmental Protection Agency, June 1991).

10. C. Moseley, "Indoor Air Quality Problems," *Journal of Environmental Quality*, vol. 53, no. 3 (November/December 1990), pp. 19–22.

11. D. J. Anderson and R. A. Hites, "Indoor Air: Spatial Variations of Chlorinated Pesticides," *Atmospheric Environment*, vol. 23, no. 9 (1989), pp. 2063–66.

12. D. J. Hanson, "Radon Tagged as Cancer Hazard by Most Studies, Researchers," *Chemical and Engineering News* (Feb. 6, 1989), p. 7–13.

13. *Respiratory Health Effects of Passive Smoking: Lung Cancer and Other Disorders*, EPA 600/6-90/006f (Washington, DC: EPA, Dec., 1992).

14 🌐 Summing Up, the Global Picture

''EVENTUALLY WE'LL REALIZE THAT
IF WE DESTROY THE ECOSYSTEM,
WE DESTROY OURSELVES.''

JONAS SALK
Medical Researcher and Scientist
(Developed the Polio Vaccine)

Photograph courtesy of National Aeronautics and Space Administration (NASA).

INTRODUCTION

This final chapter provides a brief look at the broader aspects of environmental degradation, including global pollution, multimedia pollution, and pollutant interactions, as well as the challenges these conditions pose for environmental engineers and scientists. Finally, we provide a summary of accomplishments in the field and the challenges ahead for society in protecting the environment.

THE POPULATION DILEMMA

The population of Earth has increased at a tremendous rate over the last 200 to 300 years. The implications of this from an environmental standpoint are immense, for 100,000 years ago the human population of the world was likely less than that of any of the world's major cities today. Where the human population was only 2 million in 100,000 B.C., it exceeds 5 billion today and is continuing to increase at an exponential rate. Where one family existed 10,000 years ago, there will be a city of more than 100,000 people today. And while our species has continued to propagate unchecked, the standard of living for those of us fortunate enough to live in the developed world has grown at a phenomenal rate [1,2].

Prior to A.D 1500. the human population grew slowly. As humans became more adept at making tools and developed industrial and agricultural skills, however, the pace of population growth picked up and soon began to accelerate (see Figure 14.1) [1,3]. At the same time, humans began to consume energy and natural resources at an ever increasing rate. And in the modern era per person resource consumption and per person pollution have increased dramatically with improving technology. The combination of these trends—rapid population growth, the rise in individual consumption, and increases in both the extent and seriousness of pollution—has brought about notable declines in water, air, and land quality throughout the world.

Energy Consumption

Energy and natural resource consumption makes possible the higher standard of living enjoyed in the more developed countries. This high standard of living has occurred for only the past one or two centuries, increasing dramatically during the past 50 years. The human population of the world is using energy at an astounding rate. As noted earlier in this book, current estimates are that world energy reserves of petroleum will last for another 40 to 50 years, natural gas for possibly 60 years, and coal for another 225 to 250 years [4]. Thus, in the short span of a few centuries, humankind will have used up all of the fossil fuel energy stored on

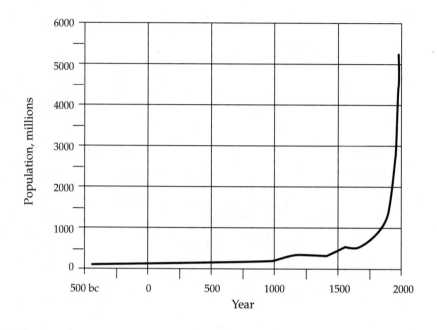

FIGURE 14.1 World population since 500 B.C.

earth — energy that took literally millions of years to produce. World and U.S. energy consumption is shown in Figure 14.2.

In 1930 per capita energy consumption in the U.S. was approximately 70 gigajoules per year. This steadily increased until around 1970. At that point, until the present, the consumption has stabilized at around 350 gigajoules per person per year [3]. In 1930 the U.S. population was 123 million. In 1990, the U.S. population was 249 million [5]. Thus, our nation's energy consumption has increased some 10 times in the past 40 years.

There are many ways to reduce the amount of energy we consume. As one example, some refrigerators currently available consume 80 to 90% less electricity than conventional models. The average new refrigerator available in 1990 cost $64 per year less to operate than comparable models produced in 1970 [6].

Efficient fluorescent lighting can reduce the lighting electricity demand by as much as 75 to 85%. Fluorescent lights cost more than incandescent lights, but they use significantly less power and last four to five times longer. They will often pay for themselves in 1 to 2 years of operation [6].

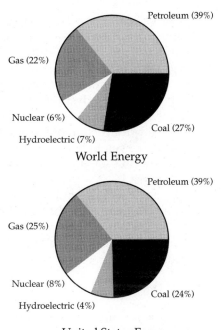

World Energy

United States Energy

FIGURE 14.2 Energy sources in 1990.

MULTIMEDIA POLLUTION

The many facets of the environment are not discrete elements but rather interconnected parts of a whole. Even the effort the control pollution contributes to pollution. Consider, for example, a wastewater treatment plant. The plant uses electricity to power pumps and blowers. The electricity production creates air pollution, solid wastes, and waste heat, and it disturbs the environment where the coal, oil, or natural gas is extracted. In addition, the aeration processes strip some volatile organic chemicals into the atmosphere, polluting the air. And the treatment processes create organic sludges. These are incinerated, landfilled, or land applied, so they create additional air, land, and/or water pollution.

Air pollution is probably the most far-reaching form of pollution, due to the transport and mixing that occurs in the atmosphere. In particular, CFCs have reduced the level of ozone in the upper atmosphere, and carbon dioxide and other greenhouse gases are warming the planet. Although there is real hope that we can signifiantly reduce the emission levels of CFCs, we have noted that those already in the atmosphere will remain for 50 to 400 years, continuing to destroy the protective ozone layer.

Carbon dioxide will probably continue to increase in the atmosphere for the foreseeable future—at least until humans have consumed the readily available carbon sources (coal, natural gas, and petroleum). This should occur sometime in the twenty-second century. Other greenhouse gases may come under control in the much nearer future. In fact, CFCs are doubly bad: They destroy the stratospheric ozone layer, and they trap reflected solar energy—that is, they are also greenhouse gases. The key to slowing the destruction of the environment is reducing population growth. It would be even better to reduce the actual population, but this does not appear possible in the foreseeable future because complex social issues work against it.

As noted earlier, our activities do not have single or simple consequences. We may see a pollutant source and think of it as affecting land, water, or air. But in most instances, the pollution encompasses several areas and has multiple impacts. To illustrate, we will consider two common human activities.

Book Printing and Binding

The textbook you are reading has altered the environment. Its manufacture consumed natural resources and produced pollutants that were emitted to the atmosphere, water, and land. A book is made of printed paper with stitching, glue, and covers made of paper, plastic, and/or cloth. The paper for this book was made from trees that were cut down for the pulp. The harvesting of trees increased soil erosion and degraded the land somewhere. It also increased suspended solids and turbidity of the nearby water and reduced the capacity of the environment there to consume carbon dioxide through photosynthesis. The paper-making process required strong chemicals to break the bonds of the wood fibers. These pulping processes involved strong solutions of sulfates, sulfites, or soda, often at temperatures of 300°C. The heating of pulp solutions used large amounts of energy from coal, hydroelectric, or nuclear power sources. The pulping process produced a wastewater high in organics and inorganic chemicals. This water had to be treated before discharge, consuming additional energy. Other strong chemicals were used to bleach the paper, making it white, rather than tan or light brown. The water from these processes contain chlorinated organics. The waste wood, such as bark, from the process had to be disposed of in some manner. If it was incinerated for energy, air pollution resulted from the combustion and land pollution from deposition of the ashes. If the leftovers were landfilled, additional land pollution resulted.

The actual paper-forming process produced a water with pulp or wood fines, wood fibers that pass through the screens used to form the paper. This wastewater had to be treated before discharge. A typical paper process requires from tens of thousands of gallons to more than one hundred thousand gallons of water per finished ton of paper. If we consider the manufacture of the chemicals used in the processes and the emissions from the power plants supplying the energy, the pollution process involved in making this book was even more complex.

Additionally, the printers used dyes and inks, materials that can be difficult to degrade. The binding process used glues and stitching to hold the paper in place. Glues often use toxic or carcinogenic solvents for carriers. Finally, the book's cover may be made of or coated with a plastic film. The plastic was manufactured from petrochemicals, requiring the extraction of crude oil and additional water, land, and air pollution at the plastics manufacturing plant.

Automotive Transportation

The everyday act of driving an automobile results in many different forms of pollution, including air, water, and soil pollution, and in the destruction of natural resources. Minerals must be extracted for the raw materials used to produce the car. These include iron ore, copper ore, and many other metals. Crude oil is the raw material used to produce the plastics for many parts of a car. The manufacture of glass for the windows uses silicates. Water pollution occurs from the cooling water used during the manufacture of the various components. Where plastic or metal parts are plated with metals, toxic compounds are usually added to the water. Vast amounts of coal, hydroelectric, or nuclear energy are required to supply the energy for smelting the ores to produce the metals. Coal-fired power plants produce significant amounts of air pollution and discharge vast amounts of heat to the environment, either by using cooling water from streams or lakes, or by dissipating heat into the atmosphere through cooling towers. Coal extraction greatly disturbs the natural environment. Hydroelectric power plants alter the natural environment of surrounding areas, creating lakes where free-flowing streams once ran. The water downstream from power plants is usually much colder than before, further altering the stream environment. And nuclear power plants produce small amounts of air pollution, discharge heat to the environment, and create nuclear wastes for which our nation has no disposal system. All of these effects, and more, come into play just in the manufacture of an automobile.

The actual act of driving the car requires petroleum (unless it is one of the few battery powered cars). The extraction of the crude oil used to make gasoline causes land pollution around the wells. Tankers taking crude oil across the seas produce small levels of air and water pollution, and if accidents occur, large areas of the sea and shore can be contaminated from petroleum spills. Oil refineries produce air, water, and land pollution. Gasoline stations may leak small (or large) amounts of fuel into the soil.

The act of driving produces additional air pollution as the fuel is burned. This air pollution, in the form of carbon dioxide and nitrous oxides, eventually falls on the soil as acid deposition (acid rain). Eventually, it will reach streams and lakes, depleting the water's alkalinity. And since automobiles must have their oil changed at regular intervals, waste oil is recycled, dumped, or landfilled. Even tire rubber causes pollution as it is worn off on roadways, finding its way into nearby streams.

The road system constructed in this country has changed a tremendous amount of land from its original state—mountains and plains, forests and grasslands, wetlands and coasts and other natural ecosystems. Concrete, steel, and asphalt require large amounts of natural resources for materials and energy for construction. The construction process puts silt and sediment in nearby waterways. Similar demands on the environment are required to maintain the highway system. And, in winter, deicing of roadways places salt and cinders in nearby waterways, increasing salinity and suspended solids.

Finally, the automobile must be disposed of after its useful life. Parts of some old automobiles are recycled, and other portions are landfilled. Yet some old cars seem to remain forever in "junk yards." Even in death, the automobile pollutes the environment.

Thus, the apparently simple act of driving an automobile is in reality a complex, multimedia form of pollution. Automobiles do not simply contaminate the air. They contribute to land and water pollution as well, and they consume valuable, irreplaceable, natural resources.

POLLUTANT INTERACTIONS

Pollutants have many interactions with other pollutants. In some cases these interactions can be reactions that occur in the environment. In other cases, the reactions may be indirect. One pollutant, or group of pollutants, may lower the resistance of an organism, enabling another pollutant to further weaken or kill it. Pollutants combine and react in the atmosphere, in water, and in the soil to form other pollutants, or in some cases to destroy desirable compounds. A classic example of this is the destruction of ozone.

Many chemicals are **synergistic**. That is, they work together. Their total effect when acting together is greater than the effect of the individual chemicals separately. Medicines are often used synergistically. Two or more drugs acting together may help the body overcome the effects of an illness better than the two drugs used separately. Unfortunately, pollutants often act on organisms, including humans, in a similar manner. Acid deposition causes a whole series of changes in the environment. It lessens the ability of trees and other vegetation to fight parasitic diseases. As soil is acidified, it increases the leaching of metals previously bound in it. These metals are then transported through the surface and subsurface channels into nearby streams and lakes. The pollutants entering water as a result of acid deposition have a synergistic effect upon the aquatic life. The reduction in alkalinity, lowering of the pH, and the increased levels of toxic metals all work together to harm and even kill fish, insects, crustaceans, shelled creatures, plants, and microorganisms that live in water. The totality of such negative effects is greater than the sum of individual effects—that is, if the aquatic organisms were exposed to these shocks separately.

THE PROGNOSIS

Much progress has been made in controlling some forms of pollution during the past 20 years. For the most part advancements have taken place in the United States, Canada, western Europe and a few other industrialized nations. In this country we have made great strides in protecting surfacewaters from municipal and industrial discharges, particularly from traditional pollutants such as BOD and suspended solids. Although the progress has to date been inadequate, we have taken significant steps to reduce air pollution. So indications are that the United States is making a notable contribution to pollution control and environmental protection. Yet much more remains to be done. We have been making progress in controlling the disposal of hazardous wastes for the past 15 years, but old disposal sites are not being cleaned up fast enough. Table 14.1 lists several areas in which the United States has achieved much, as well as several areas that must still be addressed.

And at the global level, many problems remain, and solutions are difficult. At present our only hope is to slow the degradation through better

TABLE 14.1   Environmental Progress and Problems in the United States

PROGRESS MADE	PROBLEMS REMAINING
Control of water pollution from municipal and industrial wastewater sources.	Control of water pollution from nonpoint sources.
Control of air pollutants, particularly the removal of lead from motor vehicle fuels.	Existing hazardous waste disposal sites need to be cleaned up and their groundwaters decontaminated.
Control of hazardous waste disposal	Reduction of forested areas needs to be slowed.
Control of municipal solid waste disposal	Wetlands need to be preserved.
Elimination of several hazardous chemicals in agriculture.	Carbon dioxide emissions must be reduced.

engineered systems, recycling of materials, protection of our remaining valuable natural resources, and strenuous efforts to reduce the population growth rate. Few of these goals can be accomplished without cooperation among nations. There are many good things to be said about the changes in eastern Europe during the past few years, particularly regarding the reduced threat of global nuclear war. But new problems, many with environmental impacts, are arising out of the political changes. Eastern European countries and the former Soviet Union are trying to develop an industrial base to raise their people's standard of living. In that endeavor it is tempting to exploit their natural resources further. Meanwhile, many people in Africa can barely survive at best, with many dying of starvation on a daily basis.

These are complex and sometimes daunting social, economic, and environmental issues. Yet the stratospheric ozone problem is one shining example of what can be done when nations agree to cooperate in saving the environment. Although some ozone-depleting chemicals will be in the atmosphere for decades, it is encouraging to know that through the Montreal Protocol substantial reductions in the production of these chemicals are already occurring, and most nations will stop production of the worst ones within the next few years. The world needs many additional agreements and similar international cooperation in solving the multitude of problems—including greenhouse gas emissions, deforestation, and others—that remain.

Review Questions

1.   As a means of reducing severe air pollution in highly populated areas such as Los Angeles and New York City, federal and state officials are encouraging the development of electric automobiles. Describe the different forms of air, water, and land pollution created by the use of these electric vehicles. Include raw material consumption, manufacturing, use, power generation, and disposal. List materials consumed, chemicals emitted or disposed of, and the repositories involved.

2.   Describe the different forms of air, water, and land pollution created by your college or university. Include raw material consumption, manufacturing, use, and disposal.

3.   What are some ways to reduce the amount of pollution created in producing textbooks?

4.   What are some less-polluting alternatives to the automobile? What modifications could be made in automobiles to reduce the amount of pollution they represent?

5. What are some reasonable alternatives to reduce the quantity of water consumed in the United States?

6. Investigate another nation (one you choose or one assigned by your instructor). Describe current environmental protection measures, the status of the country's natural resources, its population, the stability of its government, and any changes in the works that might improve the welfare of the people and the natural environment.

7. Table 14.1 lists several accomplishments in our efforts to protect the environment. Use your library's resources to identify others.

References

1. C. McEvedy and R. Jones, *Atlas of World Population History,* (Harmondsworth, Middlesex, England: Penguin Books, Ltd., 1978).

2. N. Keyfitz, "The Growing Human Population," *Scientific American,* vol. 261 (September 1989), pp. 118–26.

3. K. Arms, *Environmental Science* (Philadelphia: Saunders College Publishing, 1990).

4. *BP Statistical Review of World Energy* (London: The British Petroleum Company, 1991).

5. *Statistical Abstract of the United States,* 111th ed., The National Data Book (Washington, DC: U.S. Department of Commerce, Bureau of the Census, 1991).

6. A. P. Fickett, C. W. Gellings, and A. B. Lovins, "Efficient Use of Electricity," *Scientific American* (September 1990)

Appendix

Saturation Dissolved Oxygen in Distilled Water.

TEMPERATURE, °C	DO, mg/L	TEMPERATURE, °C	DO, mg/L
0	14.6	16	9.9
1	14.2	17	9.7
2	13.9	18	9.5
3	13.5	19	9.3
4	13.1	20	9.1
5	12.8	21	8.9
6	12.5	22	8.7
7	12.1	23	8.6
8	11.8	24	8.4
9	11.6	25	8.3
10	11.3	26	8.1
11	11.0	27	8.0
12	10.8	28	7.8
13	10.5	29	7.7
14	10.3	30	7.6
15	10.1		

Properties of Water

TEMPERATURE, °C	DENSITY, kg/m³	VISCOSITY × 10³, N-s/m²	KINEMATIC VISCOSITY × 10⁶, m²/s
0	999.8	1.781	1.785
5	1000.0	1.518	1.519
10	999.7	1.307	1.306
15	999.1	1.139	1.139
20	998.2	1.002	1.003
25	997.0	1.002	1.003
30	995.7	0.798	0.800
35	994.0	0.725	0.729
40	992.2	0.653	0.658

Glossary

Activated sludge An active population of microorganisms used to treat wastewater, or the process in which the organisms are employed.

Adsorption The concentration or collection of a soluble material (the solute) at the surface of another substance (the adsorbent). The solute can be a gas, a liquid, or a dissolved substance.

Advanced wastewater treatment The removal of any dissolved or suspended contaminants beyond secondary treatment. Often this is the removal of the nutrients nitrogen and/or phosphorus.

Aeration Intimate contact of the atmosphere and water to add air (oxygen) to the water. The term is also applied to gas stripping, where an undesirable gas is removed from the water. (See gas stripping.)

Aerobes Organisms that require molecular oxygen as an electron acceptor for energy production. See anaerobes.

Aerobic process A process requiring molecular oxygen.

Alcohol An organic compound with one or more hydroxyl "-OH" groups.

Aldehyde An organic compound with a carbonyl at one end of a hydrocarbon group.

Alkalinity The capacity of a water to neutralize acids.

Amine A functional group consisting of "-NH₂."

Amino acid   A functional group that consists of a carbon combined with a carboxylic acid, "-COOH," and an amine, "-NH$_2$." Amino acids are the building blocks for proteins.

Anabolism   Biosynthesis, the use of energy to produce new cellular materials from organic or inorganic chemicals.

Anaerobes   Organisms that do not require molecular oxygen to sustain life. Like all known life forms, these organisms require oxygen, but they obtain it from inorganic ions such as nitrate or sulfate or from protein.

Anaerobic process   A process that occurs only in the absence of molecular oxygen.

Anoxic process   A process that occurs only at very low levels of molecular oxygen or in its absence.

Anthropogenic   Indicative of, made, or caused by human activity or actions.

Aromatic   A form of bonding in which ring compounds share electrons with more than two atoms. Such "delocalized" electrons give rings unusual stability.

Attached growth reactor   A reactor in which the microorganisms are attached to engineered surfaces. Examples are the trickling filter and the rotating biological contactor. (See suspended growth reactor.)

Autotrophic   Organisms that utilize inorganic carbon for synthesis of protoplasm. Ecologists narrow the definition further by requiring that autotrophs obtain their energy from the sun. In microbiologist parlance, this would be a photoautotroph. See photoautotrophic and chemoautotrophic.

Autotrophs   Organisms including plants and algae that can obtain carbon for synthesis from inorganic carbon sources such as carbon dioxide and its dissolved species (the carbonates).

Bacteria   One-celled microorganisms that do not have nuclear membranes.

Baghouse filter   A fabric device used to remove particulate air pollutants from industrial smokestack emissions.

Biochemical oxygen demand (BOD)   The amount of oxygen required to oxidize any organic matter present in a water during a specified period of time, usually five days. BOD is an indirect measure of the amount of organic matter present in a water.

Biofilm A coating of microorganisms attached to a surface, such as that on a trickling filter, rotating biological contactor, or rocks in streams.

Biogeochemical cycle The cycle of elements through the biotic and abiotic environment.

Biosynthesis Catabolism, the production of new cellular materials from other organic or inorganic chemicals.

Carbonaceous biochemical oxygen demand (CBOD) The amount of oxygen required to oxidize any carbon-containing matter present in a water.

Carbonyl A functional group with an oxygen atom double bonded to a carbon atom.

Catabolism The production of energy by the degradation of cellular materials.

Cell A unit of varying dimensions in a landfill, isolated from the environment by 6 to 12 inches of soil cover. A cell is one day's waste or less, covered with soil at the end of each day.

CFCs The acronym for chlorofluorocarbons, chemicals that deplete the ozone layer in the upper atmosphere.

Chemical fixation (or stabilization/solidification) A term for several different methods of chemically immobilizing hazardous materials into a cement, plastic, or other solid matrix.

Chemical oxygen demand (COD) The equivalent amount of oxygen required to oxidize any organic matter in a water sample using chromic acid, a strong chemical oxidizing agent.

Chemoautotrophic Organisms that utilize inorganic carbon (carbon dioxide or carbonates) for synthesis and inorganic chemicals for energy. See autotrophic and photoautotrophic.

Chemotroph An organism that obtains energy from the metabolism of chemicals, either organic or inorganic.

Chlorofluorocarbons Synthetic organic compounds used for refrigerants, aerosol propellants (prohibited in the United States), and blowing agents in the manufacture of plastic foams. CFCs rise into the upper atmosphere, destroying ozone and increasing global warming. Typical atmospheric residence times for them are 50 to 200 years.

Clarifier (sedimentation basin) A tank in which quiescent settling occurs, allowing solid particles suspended in the water to agglomerate and sink to the bottom, the resulting solids to be removed as a sludge.

Climatology The study of the climate—weather processes and patterns in the earth's atmosphere over long periods of time.

Closure The act of preparing a landfill for long-term inactivity, including placement of a cover over it to prevent infiltration of surface water.

Coagulation Particle destabilization to enhance agglomeration.

Colloids Small particles with negligible settling velocities in a fluid. Sizes of typical colloidal particles typically range from $10^{-3}\,\mu$m to $1\,\mu$m.

Complexation The ionic bonding of one or more central ions or molecules by one or more surrounding ions or molecules.

Component A part of a mixture or solution.

Composting The controlled aerobic degradation of organic wastes into a material that can be used for landscaping, landfill cover, or soil conditioning.

Compound A substance composed of two or more elements.

Compression settling Sedimentation that occurs in the lower reaches of clarifiers where particle concentrations are highest and particles can settle only by compressing the mass of particles below.

Consumers Organisms that consume protoplasm produced from photosynthesis or consume organisms from higher levels that indirectly consume protoplasm from photosynthesis.

Conversion The fraction of a species converted to product once it has entered a system.

Corrosive waste One that is outside the pH range of 2 to 12.5, or one that corrodes steel at a rate greater than 6.35 mm (0.25 in.) per year. This is one of EPA's four hazardous waste characteristics.

Covalent bond   A bond in which electrons are shared approximately equally by two atoms.

Cybernetic   Systems that change in response to feedback.

Decomposers   Organisms that utilize energy from wastes or dead organisms. Decomposers complete the cycle by returning nutrients to the soil or water and carbon dioxide to the air or water.

Denitrification   The anoxic biological conversion of nitrate to nitrogen gas. The process occurs naturally in surface waters low in oxygen, and it can be engineered in wastewater treatment systems.

Deoxygenation   The consumption of oxygen by aquatic organisms as they oxidize materials in the aquatic environment.

Discrete settling   Sedimentation in which individual particles sink independently, neither agglomerating or interfering with the settling of the other particles. This occurs in waters with a low concentration of particles.

Disease   Any impairment of normal functions in an organism.

Disinfection   The destruction or inactivation of pathogenic microorganisms. (See sterilization.)

Dispersion   A stable mixture of particles suspended in a fluid medium.

Dissolved oxygen (DO)   The amount of molecular oxygen dissolved in water.

Dump   An illegal and uncontrolled area where wastes have been placed on or in the ground. (See landfill.)

Ecology   The study of living organisms and their environment or habitat.

Ecosystem   An organism or group of organisms and their surroundings. The boundary of an ecosystem may be arbitrarily chosen to suit the area of interest or study.

Effluent   The fluid exiting a system, process, or tank. The outflow from one process can be an inflow, or influent, to another process. (See influent.)

Electronegativity   The potential of an atom to attract electrons when bonded in a compound. The scale of measurement for the force of attraction involved is 0 to 4, with 0 being the most electropositive (low attraction) and 4 being the most electronegative (high attraction).

Electrostatic precipitator An air pollution control device that uses an electric field to trap particulate pollutants.

Elementary reaction A reaction in which the rate expression corresponds to the stoichiometric equation.

Epilimnion The top layer of a lake.

Equivalent mass The mass of the compound that will produce 1 mol of available reacting substance. Thus, for an acid, it would be the mass of acid that will produce 1 mol of H^+. For a base it would be the mass needed to produce 1 mol of OH^-.

Ether An organic compound that has two hydrocarbon groups bound by an interior oxygen atom. The general formula is R'-O-R''.

Eucaryotic organisms Those that have a nuclear membrane. This includes all known organisms except viruses and bacteria.

Evaporation The conversion of liquid water to water vapor. It occurs on the surface of water bodies such as lakes and rivers and immediately after precipitation events in small depressions and other storage areas.

Evapotranspiration The sum of evaporation and transpiration. Since it is difficult to measure the two terms independently, they are often grouped as one value.

Facultative organisms Those that prefer or preferentially use molecular oxygen when it is available but will use other pathways for energy and synthesis if it is not available.

Fermentation Energy production without the benefit of oxygen as a terminal electron acceptor—oxidation in which the net effect is one organic compound oxidizing another. See respiration.

Fixed solids (FS) Those that do not volatilize at 550°C.

Fixed suspended solids (FSS) The matter remaining in suspended solids analysis that will not burn at 550°C. This represents the nonfilterable inorganic residue in a sample.

Flash point   The lowest temperature at which enough vapor is produced to cause combustion if an ignition source is present.

Flocculant settling   Sedimentation in which particle concentrations are sufficiently high that agglomeration occurs. This reduces the number of particles and increases the average particle mass. As particles stick together, higher settling velocities result.

Fluidization   The suspension of particles by upward velocity of a fluid. Buoyancy and fluid friction overcome gravity operating on the particles.

Flux   The movement of a mass past a surface, plane, or boundary. The units are mass per unit area per unit time or $[kg/m^2 \cdot hr]$.

Gas stripping   Transfer of an undesirable gas from a water stream to the atmosphere.

Global warming   The long-term warming of the planet caused by increases in "greenhouse gases," which trap reflected heat and prevent it from exiting to space.

Greenhouse gases   Atmospheric gases that trap solar radiation. Of the solar energy entering the earth's atmosphere and striking the earth, a portion is reflected back and a portion penetrates the earth's surface. The reflected portion travels at a wavelength different from its wavelength in space. Carbon dioxide and other gases absorb this reflected radiation, containing its heat and increasing the earth's temperature. The effect is much like a greenhouse, hence the name.

Groundwater   Water contained in geologic strata.

HAPs   An acronym for hazardous air pollutants

Hardness   The concentration of multivalent cations measured as calcium carbonate $[CaCO_3]$ and expressed as meq/L or mg $CaCO_3$/L. It is important because hard waters require increased amounts of soap for bathing or washing clothes and because they form scale on piping, cooking vessels, boilers, heat exchangers, and other equipment.

Heterotrophic organisms   Those that obtain carbon for synthesis from other organic matter or proteins.

Hindered (zone) settling   Sedimentation in which particle concentrations are sufficient that they interfere with the settling of other particles. Particles thus sink together as a body and the fluid is required to move through remaining spaces between particles and clumps of particles.

Hydrocarbon Any organic compound composed entirely of carbon and hydrogen. Two examples are methane gas and octane.

Hypolimnion The lower layer of a lake.

Ignitable waste One that may cause a fire during storage, transportation, or disposal. This is one of EPA's four hazardous waste characteristics.

Infectious disease One caused by pathogenic organisms.

Infiltration The movement of water from the surface of the land through the unsaturated zone and into the groundwater. This occurs during and immediately after precipitation events. It can also occur at the bottom of lakes and rivers.

Influent The fluid entering a system, process, or tank. The effluent from one process can be an influent to another process. (See effluent.)

Infrastructure The network of facilities society has created to maintain or improve the standard of living, including highways, buildings, railroads, bridges, water treatment plants, power plants, air pollution control facilities, and wastewater treatment plants.

In situ **treatment** Treatment of a waste in place, as opposed to pumping or digging it up and then treating it.

Ion exchange An adsorption process in which one ion is exchanged for another ion of like charge. Thus, there is an equivalence of exchanged charge.

Irreversible reaction One in which the reactant(s) proceed to product(s), but the products do not react at an appreciable rate to reform reactant(s).

Isomers Two or more different compounds with the same chemical formula but different structure and characteristics.

Kerogen A fossilized organic material present in oil shale and some other sedimentary rocks.

Ketones Organic compounds with two hydrocarbon groups bonded to a carbonyl group.

Landfilling The placement of wastes into the ground under controlled conditions to minimize their migration or effect on the surrounding environment. (See dump.)

Leachate   The contaminated liquid that migrates through or from a soil or waste, transporting the soluble contaminants with it as it moves. It is the result of water seeping into and through the wastes and dissolving part of the organic and inorganic matter in the landfill.

Leaching   Dissolving the soluble portion of a solid mixture by a solvent. An example is the dissolving of inorganic or organic contaminants from refuse in a landfill by the infiltration of rainwater.

Ligand   The ion or molecule that surrounds or complexes with a central atom or ion.

Limnology   The study of freshwater ecosystems.

Mass balance   An organized accounting of all inputs and outputs in an arbitrary but defined system. Stated in other terms, the rate of mass accumulation within a system is equal to the rate of mass input less the rate of mass output plus the rate of mass generation within the system.

Maximum contaminant level (MCL)   The maximum allowable concentration of a given constituent in potable water.

Mercaptans   See Thiols.

Metabolism   Processes that sustain an organism, including energy production synthesis of proteins for cell repair and replication.

Metalimnion   The middle layer of a lake.

Meteorology   The study of the atmosphere and weather below 100 km.

Mixed liquor suspended solids (MLSS)   The total suspended solids concentration in an activated sludge tank.

Mixed liquor volatile suspended solids (MLVSS)   The volatile suspended solids concentration in an activated sludge tank.

Nitrification   The biological oxidation of ammonia and ammonium sequentially to nitrite and then nitrate. The process occurs naturally in surface water and can be engineered in wastewater treatment systems. The purpose of nitrification in wastewater treatment systems is to reduce the oxygen demand resulting from ammonia in the system.

Nitrogen fixation   The conversion of atmospheric (or dissolved) nitrogen gas into nitrate by microorganisms

Nitrogeneous oxygen demand (NOD)   The amount of oxygen required to oxidize any ammonia present in a water.

Nonpoint source pollution (NPSP)   Any pollution that cannot be attributed to a particular discharge point—from agricultural crops, city streets, construction sites, for example.

NPDES   The National Pollutant Discharge Elimination System. The discharge criteria and permitting system established by the EPA in response to the Clean Water Act and its amendments, or the permit required by each discharger as a result of the Clean Water Act.

Organic compound   Any compound containing carbon except for the carbonates (carbon dioxide, the carbonates and bicarbonates), the cyanides, and cyanates.

Organic nitrogen   Nitrogen contained as amines in organic compounds such as amino acids and proteins.

Oxidative phosphorylation   The synthesis of the energy storage compound adenosine triphosphate (ATP) from adenosine diphosphate (ADP) using a chemical substrate and molecular oxygen.

Packed tower   See trickling filter.

Pathogenic organism   An organism capable of causing infection.

Phenol   An aromatic benzene ring with a hydroxyl substituted for one hydrogen.

Phenyl-   A benzene ring named as a constituent group, C_6H_5-.

Phosphorylation   The synthesis of the energy storage compound adenosine triphosphate (ATP) from adenosine diphosphate (ADP).

Photoautotrophic   Organisms that utilize inorganic carbon dioxide for protoplasm synthesis and light for an energy source. See autotrophic and chemoautotrophic.

Photochemical pollutants   Chemicals that react photochemically (in the presence of sunlight) to destroy ozone in the stratosphere.

Photophosphorylation The synthesis of the energy storage compound adenosine triphosphate (ATP) from adenosine diphosphate (ADP) using solar energy.

Phototrophs Organisms that obtain energy from light using photooxidation.

Pollution Any condition or substance, resulting from human activity, that adversely affects the quality of the environment.

Potable water Water that does not contain harmful or objectionable impurities and that is safe and aesthetically pleasing to drink.

POTW (publicly owned treatment works) Any municipally owned wastewater treatment facility.

Precipitation The falling to earth of condensed water vapor in the form of rain, snow, sleet, or hail.

Primary ambient air quality standards Air quality criteria designed to protect human health.

Primary particulate pollutant A particulate pollutant prior to aggregating with other particulates following combustion.

Primary standards Drinking water quality criteria related directly to human health and enforceable by the U.S. Environmental Protection Agency. (See secondary standards.)

Primary treatment Treatment that includes all operations prior to and including primary treatment—bar screening, grit removal, comminution, and primary sedimentation.

Procaryotic organisms Organisms that do not have a cellular membrane.

Producers Autotrophic organisms that produce protoplasm using inorganic carbon and energy from the sun.

Publicly owned treatment works (POTW) A municipally owned wastewater treatment facility.

Reaeration The dissolving of molecular oxygen from the atmosphere into a water.

Reactive waste One that reacts violently with water, forms potentially explosive mixtures with water, is normally unstable, contains cyanide or sulfide in sufficient quantity to evolve toxic fumes at high or low pH, can explode if heated under pressure, or is an explosive compound listed in Department of Transportation (DoT) regulations. This is one of EPA's four hazardous waste characteristics.

Recycling The reuse of a waste product, either by directly reusing or reclaiming it, that would otherwise be thrown away.

Refuse All forms of solid waste.

Refuse derived fuel A fuel derived from the combustible portion of municipal solid waste—often processed into small briquettes, similar in size to charcoal.

Respiration Energy production in which oxygen is the terminal electron acceptor—oxidation to produce energy where oxygen is the oxidizing agent. See fermentation.

Reversible reaction One in which the reactant(s) proceed to product(s), but the products react at an appreciable rate to reform reactant(s).

Runoff The water that flows overland to lakes or streams during and shortly after a precipitation event.

Saltwater intrusion The gradual replacement of freshwater by saltwater in coastal areas where excessive pumping of groundwater occurs.

Secondary ambient air quality standards Air quality criteria designed to protect infrastructure and the environment.

Secondary particulate pollutant An aggregated mass of particulate pollutants, produced in combustion. Secondary particulate pollutants have a higher mass and lower number than primary particulate pollutants.

Secondary standards EPA-recommended drinking water quality criteria relating to aesthetics and/or human health. (See primary standards.)

Secondary treatment The conversion of the suspended, colloidal, and dissolved organics remaining after primary treatment into a microbial mass that is removed in a second stage. Secondary treatment includes both biological processes and associated sedimentation.

Secured landfill A landfill that has containment devices such as liners and a leachate collection system so materials placed in it will not migrate into the surrounding soil, air, and water.

Sedimentation   The gravity settling, and thus removal, of materials more dense than a suspending fluid.

Sedimentation basin   See clarifier.

Shock load   Wastewater with an unusually high organic content and/or high flow rate as it enters a treatment plant.

Site remediation   Cleaning up a hazardous waste disposal site that has been abandoned or whose owners (or the responsible parties) either refuse to or are financially unable to clean up.

Siting   Obtaining government (federal, state, and local) permission to construct an environmental processing, treatment, or disposal facility at a given site.

Sludge   A liquid to semisolid stream of waste material and water. Sludge results from the concentration of contaminants in water and wastewater treatment processes. Typical wastewater sludge contains from 0.5 to 10% solid matter. Typical water treatment sludge contains 8 to 10% solids.

Softening   The removal of divalent cations from water by precipitation or ion exchange.

Solidification   See chemical fixation.

Solids flux   See flux.

Source reduction   The elimination or reduction of a waste at its source by modifying the process that produces it. (See waste minimalization.)

Species   An ion or molecule in solution.

Stabilization   See chemical fixation.

Sterilization   The destruction or inactivation of all microorganisms. (See disinfection.)

Storage   The short-term retention of water after a precipitation event.

Stratosphere   The atmosphere from approximately 12 km to 70 km. Air temperature increases in this region.

Strong acid   One that, for practical purposes, ionizes completely under the conditions of interest. Common strong acids are hydrochloric, nitric, and sulfuric.

Substrate level phosphorylation The synthesis of the energy storage compound adenosine triphosphate (ATP) from adenosine diphosphate (ADP) using organic substrates without molecular oxygen.

Surface water Water in lakes, rivers, and oceans.

Suspended growth reactor A reactor in which the microorganisms are suspended in the wastewater. Examples of suspended growth reactors are activated sludge reactors and anaerobic digesters. (See attached growth reactor.)

Synergism The act of working together. When the effect of chemicals working in concert is greater than the sum of their individual effects, they are said to be synergistic. The combined effect can be either positive or negative.

System An arbitrarily defined area or volume surrounded by a boundary and possessing specific inputs, outputs, and reactions.

Theoretical oxygen demand (ThOD) The amount of oxygen, calculated using stoichiometry, required to convert the material to end products of carbon dioxide and water.

Thermocline The depth at which an inflection point occurs in a lake temperature profile.

Thiols Organic compounds containing the "-SH" functional group. Thiols are also called mercaptans.

Total dissolved solids (TDS) The amount of dissolved matter in a water.

Total solids (TS) The amount of organic and inorganic matter contained in a water.

Toxic waste One that exhibits the hazardous waste characteristic of toxicity as defined by the toxicity characteristic leaching procedure (TCLP). In the procedure a waste is extracted for 24 hours with an acetic acid solution. The extract is then analyzed for the presence of specific contaminants. This is one of the EPA's four hazardous waste characteristics.

Trace contaminants Polluting or poisonous substances found in very low levels in water.

Transpiration The loss of water from plants through leaves and other parts. This loss can be a significant amount of water during very dry periods.

Trickling filter   An attached-growth biological process in which the microbiol film adheres to stationary rock or plastic media.

Trophic level   A level in the food chain. The first trophic level consists of the primary producers, autotrophs. The second trophic level is vegetarians, which consume autotrophic organisms.

Troposphere   The lower atmosphere, from the earth's surface to approximately 12 km. This portion of the earth's atmosphere contains about 95% of the atmospheric gases. Temperature gradually declines through this region.

Ultimate biochemical oxygen demand (BOD_u)   The total amount of oxygen required to oxidize any organic matter present in a water—that is, after an extended period such as 20 or 30 days.

Ultimate disposal   The process of returning residuals to the environment in a form that will have minimal or at least reduced negative environmental impacts.

Virion   A virus particle. Viral DNA or RNA enclosed in an organic capsule, in other words.

Virus   A submicroscopic genetic constituent that can alternate between two distinct phases. As a virus particle, or virion, it is DNA or RNA enveloped in an organic capsule. As an intracellular virus, it is viral DNA or RNA inserted into the host organism's DNA or RNA.

Volatile   Easily vaporized.

Volatile solids (VS)   The amount of matter that volatilizes (or burns) when a water sample is heated to 550°C.

Volatile suspended solids (VSS)   The nonfilterable residue from the firing of total suspended solids at 550°C. See total suspended solids and fixed suspended solids.

Waste minimization   Changes in industrial processes to reduce or eliminate waste. (See source reduction.)

Wastewater   Consumed or used water that contains dissolved and/or suspended matter.

Weak acid   One that does not ionize completely under the conditions of interest. Examples include acetic acid, carbonic acid, and hypochlorous acid.

Wetland Semiaquatic land that is either inundated or saturated by water for varying periods of time during each year and that supports aquatic vegetation which is specifically adapted for saturated soil conditions.

Zone settling See hindered settling.

Index